Literature of Analytical Chemistry: A Scientometric Evaluation

Authors

Tibor Braun
Ernő Bujdosó
András Schubert

CRC Press is an imprint of the
Taylor & Francis Group, an **informa** business

CRC Press
Taylor & Francis Group
6000 Broken Sound Parkway NW, Suite 300
Boca Raton, FL 33487-2742

Reissued 2019 by CRC Press

A Library of Congress record exists under LC control number:

ISBN 13: 978-0-367-22306-9 (hbk)
ISBN 13: 978-0-367-22309-0 (pbk)
ISBN 13: 978-0-429-27437-4 (ebk)

Visit the Taylor & Francis Web site at http://www.taylorandfrancis.com and the CRC Press Web site at http://www.crcpress.com

THE AUTHORS

Tibor Braun, Ph.D., D.Sc., is a Professor in the Institute of Inorganic and Analytical Chemistry of the L. Eötvös University, Budapest, Hungary. He is a Deputy Director of the Library of the Hungarian Academy of Sciences and Head of the Information Science and Scientometrics Research Unit of that Library.

Dr. Braun received his higher education at the V. Babes University, Cluj. His thesis research was completed at the Hungarian Academy of Sciences and was dedicated to nuclear analytical methods and radioanalytical separations.

Upon graduation he entered the Institute of Atomic Physics of the Romanian Academy of Sciences, Bucharest and his research there concerned nuclear radiation effects on semiconductor oxide catalysts, and uses of radioisotopes in analytical chemistry. Several publications in journals such as *Nature, Journal of Physics and Chemistry of Solids, Talanta, and Microchimica Acta* came from his research.

In 1963 he joined the Institute of Inorganic and Analytical Chemistry in charge with teaching and research in the field of analytical and nuclear analytical chemistry.

In 1968 Dr. Braun founded the international *Journal of Radioanalytical Chemistry* and in 1969 the international journal entitled *Radiochemical and Radioanalytical Letters* serving as editor for both journals. (In 1984 the two journals changed title and became the *Journal of Radioanalytical and Nuclear Chemistry, Articles* and the *Journal of Radioanalytical and Nuclear Chemistry, Letters.* Dr. Braun has been appointed Editor-in-Chief of both journals.) In 1975 Dr. Braun founded the international journal entitled *Scientometrics* and was appointed its managing editor.

Over 100 publications in journals, books, and conference proceedings have resulted from the work at the Institute. Major monographs on "Radiometric Titrations", "Isotope Dilution Analysis", "Extraction Chromatography", and "Polyurethane Foam Sorbents in Separation Science" have been published in collaboration with other investigators.

In 1978 Dr. Braun joined, as a part time job, the Library of the Hungarian Academy of Sciences. A major portion of his effort was devoted to build up an information science and scientometrics research group and program. A series of publications in journals and many monographs came from this research.

Dr. Braun's research, in collaboration with colleagues at the Institute and Library is primarily in radiochemistry, radioanalytical chemistry, analytical chemistry, and quantitative studies of science. Dr. Braun has published or presented over 110 papers related to the above-mentioned topics.

Dr. Braun has been a visiting professor in Peru (1969), Jamaica (1975), and Japan (1985) in charge of research and teaching activities at the Junta de Control de Energia Atomica (Lima), University of the West Indies (Kingston), and Tohoku University (Sendai).

Ernő Bujdosó, Ph.D., C.Sc.Phys., is the Head of the Department for Information Science, Library of the Hungarian Academy of Sciences.

Dr. Bujdosó graduated in 1955 from L. Kossuth University, Debrecen, Hungary. He received his Ph.D in 1958 after which he joined the Nuclear Research Institute of the Hungarian Academy of Sciences where his research concerned the investigation of low energy nuclear reactions.

In 1967 he entered the Research Institute for Non-Ferrous Metals as the Head of the Radioisotope Laboratory. He worked on the development and application of nuclear methods in the research and production of the Hungarian Aluminum Industry. He has also been active in the organizational and teaching tasks of health physics bearing posts in the Committee of International Radiation Protection Association (IRPA).

In 1979 he joined the Library of the Hungarian Academy of Sciences to establish a new computerized science information system. Between 1982 and 1984 he was visiting professor in the Department of Chemistry, University of Montana, Missoula, engaged in the study of oscillating chemical reactions.

Dr. Bujdosó is the Editor of the *Journal of Radioanalytical and Nuclear Chemistry*, and is on the Editorial Boards of the international journal of *Scientometrics* and of other Hungarian journals of library and information science. He lectures on information science and scientometrics in the Universities of L. Eötvös, Budapest and L. Kossuth, Debrecen.

Dr. Bujdosó has published over 100 scientific papers relating to nuclear physics, application of nuclear methods, and scientometrics, and has edited or coauthored 12 books.

András Schubert, Ph.D., is the Head of the Department for Scientometrics of the Information Science and Scientometrics Research Unit (ISSRU) at the Library of the Hungarian Academy of Sciences.

He studied Chemical Engineering at the Budapest Technical University and received his M.D. and Ph.D in Physical Chemistry (in 1970 and 1978, respectively).

In 1970 he entered the Agricultural University at Gödöllő (Hungary), where he became a lecturer at the Department of Physics. In his research, he dealt mainly with the thermodynamics and kinetics of chemical reactions and membrane phenomena. He authored several research papers and a book entitled *Kinetics of Homogeneous Reactions* in this topic. He is a member of the Presidential Board of the Hungarian Biophysical Society.

Dr. Schubert joined the ISSRU in 1979. His main concern is the theory and use of scientometric indicators in the analysis and assessment of scientific research. He is a member of the Editorial Board of the journal *Scientometrics*, in which much of his research is published. He is a coauthor of several books including *Scientometric Indicators: A 32-Country Comparative Evaluation of Publishing Performance and Citation Impact* published by World Scientific Publ. Co. Pte. Ltd., Singapore and Philadelphia, 1985.

TABLE OF CONTENTS

Chapter 1

INTRODUCTION

In the foreword of his book *Little Science — Big Science,* Derek De Solla Price[1] says,

> My goal is not a discussion of the content of science or even a humanistic analysis of its relations. Rather, I want to clarify these more usual approaches by treating separately all the scientific analyses that may be made of science. Why should we not turn the tools of science on science itself? Why not measure and generalize, make hypotheses and derive conclusions? My approach will be to deal statistically, in a not very mathematical fashion, with general problems of the shape and size of science and the ground rules governing growth and behavior of science-in-the-large. That is to say, I shall not discuss any part of the detail of scientific discoveries, their use and interrelations. I shall not even discuss specific scientists. Rather, treating science as a measurable entity, I shall attempt to develop a calculus of scientific manpower, literature, talent and expenditure on a national and on an international scale.

Price is cited at this length because we think his ideas were the first to express explicitly the endeavor toward the quantification of science, which has led to scientometrics as it is now known and interpreted.

"Quantification of science? A peculiar notion," the reader will say. Have the natural sciences not been quantified a long time ago? Indeed they have, to varying degrees, and so the subject of scientometrics is not the quantification of the natural sciences themselves but the quantification of the science of those sciences, which is so young that it has not had time to quantify itself. The science of science in fact is a recent hybrid (some readers will like to call it interdisciplinary) field involving the study of science-making from philosophical, methodological, historical, economic, managerial, administrative, policy-making, and sociological angles. It began to show the first signs of an identity perhaps 2 decades ago. By no means is all of the field devoted to the quantification of science. There has been, nevertheless, a significant movement to try to approach the science-making process with the help of some quantitative tools.

What is it about science that we want to quantify? It is helpful in many respects to look at the process of science-making as an input-output phenomenon (see Figure 1), and accordingly, we can try to quantify the input and the output. The primary input consists of money and manpower. The application of these two items leads to secondary input items which can also be quantified: the number of pencils used by theorists, liters of liquid nitrogen bought, computer hours clocked, laboratory buildings needed, etc. Finally, there is the scientific output which consists of knowledge.[2]

In Figure 1 knowledge is represented under the heading of "recorded knowledge", and this is in fact nothing else than the scientific subject literature. The literature — the body of writings on a subject — is the prime means of communication in any subject; it is a representation and record of the knowledge and activities in the subject. So, the importance of the quantitative analysis of the literature of a subject field lies in the fact that it contributes to the understanding of and insight into that field. The notion that scientific subject literatures are phenomena susceptible to systematic investigation in much the same way as are physical or biological phenomena is at least 60 years old. One of the earliest papers of this type was that of Cole and Eales[3] in 1917 in which they described and interpreted a count of the literature of comparative anatomy from the years 1543 through 1860. Cole and Eales were interested in measuring the relative contributions and performance of the participating countries over 3 centuries. Their study had a clearly defined objective: to determine which groups of animals and which aspects of anatomy engaged the attention of workers and to trace the influence

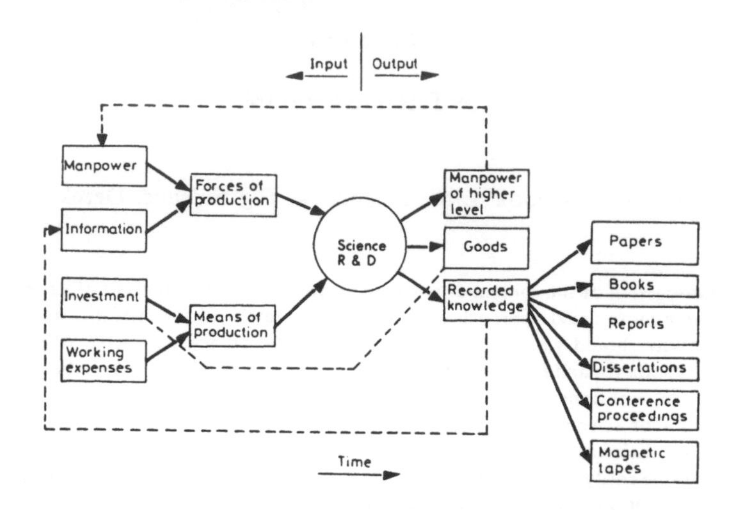

FIGURE 1. Flow chart of the simplified working "mechanism" of scientific research.

of contemporary events on the history of anatomical thought. They attempted to detach and plot separately the performance of each European country. They summed these goals by stating that " . . . it seemed possible to reduce to geometrical form the activities of the corporate body of anatomical research, and the relative importance from time to time of each country and division of the subject."[3]

Cole and Eales[3] were acutely aware of the limitations of their techniques: "A chart represents numerical values only, and may by itself be seriously misleading. The author of 50 small ephemeral papers, judged by figures, is of greater importance than William Harvey, represented by two entries, both of great significance. It is hence necessary that any conclusions drawn from the charts should be checked by an examination of the literature dealt with."

The paper of Cole and Eales was followed by a whole series of studies dealing with similar investigations sometimes extending to the study of various scientific fields and/or specialties. The bibliography of Hjerppe[4] on this topic already lists several hundred papers. If this bibliography is scanned, one can see that in the titles of papers the word "bibliometrics" occurs quite frequently. Pritchard,[5] who is generally given credit for coining the term "bibliometrics", defines it as follows: "The definition and purpose of bibliometrics is to shed light on the process of written communication and of the nature and course of a discipline (in so far as this is displayed through written communication) by means of counting and analysing the various facets of written communication."

It is our opinion that although the methods of bibliometrics and scientometrics are very similar, sometimes even perfectly identical, it is advisable to distinguish them according to the subject and the purpose of their topics. Bibliometrics considers books, periodicals, etc. as formal and tangible documents, its major goal being the quantitative analysis of library collections and services in order to improve scientific documentation, information, and communication activities. Scientometrics analyzes the quantitative aspects of the generation, propagation, and utilization of scientific information in order to contribute to a better understanding of the mechanism of scientific research activities.

The terms "bibliometrics" and/or "scientometrics" are also sometimes used to denote an investigation of a field of science or of scientific activity based on the analysis of scientific publications through the references from and the citations to them. As such, a reference is defined as the acknowledgment one scientific article gives to another, and a citation is the acknowledgment one article receives from another. Thus, for instance, the fifth reference noted in this book is considered to be a reference from this article to the article of Pritchard as well as a citation to the article of Pritchard from this book.

On the other hand, it was Nalimov[6] who first coined the term "scientometrics" and outlined its subject field. His views agree with those of Price in that science is a process developing in time, and as such it can obviously be subjected to quantitative investigations just as the time-dependent processes of biology, chemistry, or physics. Phenomenologically, as said by Nalimov,[6] science is a process of searching for basically new pieces of information. This process has a cumulative and collective character: any scientific result is built, to a certain extent, on a set of previously declared principles. Novel scientific results are born as a development or reevaluation of previous ones. Science is a self-organizing system, the development of which is governed by its information flows. According to Nalimov,[6] scientometrics stands for those quantitative methods which deal with the investigation of science viewed as an information process. This approach is cybernetic in character. As is known, complicated systems may be studied by observing the information flows governing the system.

The primary purpose of this monograph is collecting and presenting in an organized manner the most pertinent sciento- and bibliometric information dealing with the statistical evaluation of the literature of analytical chemistry. Some aspects of the literature of analytical chemistry were investigated statistically soon after World War II. This book does not follow the chronological order of such investigations; our intention was rather to classify and analyze the set of knowledge gathered so far according to various biblio- or scientometric techniques and laws.

In addition to the above-mentioned purpose, Chapter 10 presents the results of an original study representing the first attempt toward the building of scientometric indicators based on reliable statistical methodologies. These indicators are used for the comparative evaluation of the analytical chemistry productivity and impact of different countries.

At the end of this introduction, there is an important issue to be clarified. We consider it probable that Crane,[7] the late director of *Chemical Abstracts,* has launched the type of investigation which later has become the source of certain misinterpretations. Crane, quite correctly and logically from the viewpoints of the head of Chemical Abstracts Service and the managing of that abstract journal, endeavored to get an insight into the development of world chemical literature by grouping the papers processed by *Chemical Abstracts* according to various statistical points of view. These statistical studies and some later investigations of Crane[8] inspired the work of Strong[9] published in 1947 under the title "Trends in Quantitative Analysis. A Survey of Papers for the Year 1946". In his investigations Strong attempted to follow the progress of quantitative analytical chemistry through a statistical analysis of the papers abstracted in *Chemical Abstracts,* and this (i.e., that the trends are followed exclusively from the survey of the literature) was duly stressed in the title of the work.

This paper of Strong inspired most probably the work of Fisher et al.[10] published in 1956 entitled "Trends in Analytical Chemistry — 1955". This was followed by a further study by Fisher[11] entitled "Trends in Analytical Chemistry — 1965" and a later study in 1975 by Brooks and Smythe[12] entitled "The Progress of Analytical Chemistry 1910—1970".

As can be seen in the titles of these papers dealing with the analysis of analytical subject literature, the words "trends" and "progress" stand without the mention of "of the literature", and this could, and in fact did, create the impression that the progress or the trends of the scientific field are concerned. This is, however, not the case. Therefore, commenting on the paper of Brooks and Smythe,[12] Braun[13] emphasizes that

> ... The authors have tried to approach a very much up-to-date problem with methods and means applied only in very few cases to analytical chemistry. To follow the progress of analytical chemistry between 1910 and 1970 they consider the number of publications, its time-dependent growth, and the distribution of publications with respect to various factors (country, language, subfield, elements) in order to evaluate some of the long-term trends of this century. Now, the statistical distribution and growth of scientific publications, citations, and manpower has been dealt with in some basic works[1,6,14,15] and it has been shown in these that the statistical data on the number of scientific publications can be correlated to the data on new scientific achievements, although scientific achievement and scientific information are different concepts. However, one should avoid the misjudgment that the number of publications is the only and fundamental measure (indicator) of the "size of science" or its "development", "trend" or "progress". Thus as a title, "Progress (or Trends) of the Literature of Analytical Chemistry 1910—1970" would have been more relevant.

Chapter 2

VOLUME OF THE LITERATURE ON ANALYTICAL CHEMISTRY

In order to investigate the subject literature of analytical chemistry, it would be useful, obviously, to know the complete volume of the literature. In our opinion, however, accurate measurements in this respect are almost impossible. The problems which arise in attempting to assess the total volume of the analytical literature are so great and diverse as to cast doubts on the value of any such estimate. In all the measurements known up to the moment, the volume of the analytical literature was derived largely from study of review journals, i.e., *Chemical Abstracts, Referativnyj Zhurnal Khimiya,* and/or *Analytical Abstracts.*

Since 1961 Baker, the director of Chemical Abstracts Service, has been publishing data (five reports annually) on the size and growth of the worldwide chemical literature,[16-19] reflecting in fact the development of *Chemical Abstracts* itself. These papers can be regarded as a continuation of the already mentioned investigations of Crane[7,8] but on a better methodological basis.

The counting of the items representing the total chemical literature processed in *Chemical Abstracts* is not too complicated a task. Determining the share of analytical chemistry in the total literature and its accurate measurement is, however, a much more complex problem. For a measurement of this type it is hardly conceivable that someone would be able to count all the abstracts connected with analytical chemistry in all sections of *Chemical Abstracts.* It was found by Fisher[11] that the "Analytical Chemistry" section of the 1965 volumes of *Chemical Abstracts* contained only 48% of the abstracts concerned with analytical chemistry.

In possession of this fact, Brooks and Smythe[12] mention that the volume of the literature on analytical chemistry can be calculated by multiplying the number of items in the "Analytical Chemistry" section of *Chemical Abstracts* by 2.1. For this to be true, it must be assumed that the 48% share has not changed since 1965. Instead of this method, Brooks and Smythe[12] counted all entries dealing with analytical aspects for 25 common elements for the entire years 1955, 1960, 1965, and 1970 in both *Chemical Abstracts* and *Analytical Abstracts.* The measurements showed that *Analytical Abstracts* contained only 54% of the total analytical items for these elements which appeared in *Chemical Abstracts.* Entries for *Analytical Abstracts* were therefore multiplied by 1.86 to give values which in turn were found to be 2.49 times the total number of entries in the "Analytical Chemistry" section of *Chemical Abstracts.* To obtain an estimate for the total number of analytical papers for any particular year, therefore, the total number of entries for the "Analytical Chemistry" section of *Chemical Abstracts* was counted and multiplied by 2.5.

According to the above authors, who are in our opinion more or less correct, the percentage of analytical literature was reasonably constant over the 1910 to 1970 period, and apart from 1920 it never fell outside the limits of 5.6 to 7.5. Their final, correct conclusion: "Overall it appears that analytical chemistry is at least maintaining itself in relation to chemistry as a whole."[12] It was pointed out by Braun et al.[20] that in all investigations carried out on the basis of abstracting journals, the recorded data cover the literature from a certain period only, i.e., from the beginning of the abstracting service involved. Failure to include earlier literature, i.e., that published before abstracting began, may cause serious errors in assessing total cumulative values. In order to bridge the gap, a correction procedure was applied by Braun et al.[20] which will be discussed in slightly more detail in Chapter 3. Perhaps it is worth mentioning that the total estimated number of chemical papers from around the world was, according to the calculations, about 198×10^3 items in 1910, and that of the analytical literature in 1915, about 16×10^3 items. In turn, worldwide chemical literature in 1970 had a volume of about 5224×10^3, and that of analytical chemistry, about 312×10^3; this means a share of 6% for analytical chemistry in the worldwide body of chemical literature, in agreement with the value suggested by Brooks and Smythe.[12]

Chapter 3

GROWTH OF THE LITERATURE ON ANALYTICAL CHEMISTRY

In the natural sciences, if a mathematical model is to be constructed to describe the phenomenon under study the following is generally done: one or more possible hypotheses are set up first on the basis of preliminary, logically interpreted information arising from previous experiments. These hypotheses are translated into the language of mathematics and compared to the experimental data. The investigation is concluded by selecting the appropriate hypothesis which is corrected according to the new experimental data. It may happen that in setting up the hypothesis the analogy of known and already studied natural phenomena is also utilized.

The same procedure may be followed in studying the increase of information flow in analytical chemistry. On the basis of Price's[1] work it is stated by Nalimov and Mulchenko[6] that in the absence of limiting factors the growth rate of publications regarded as carriers of scientific information, i.e., the increase of literature, is determined by the actual level of science. All sound and new scientific theories should give rise to a certain amount of new scientific work in which these theories are developed, supported, or refuted. Consequently, the growth rate of scientific literature reflected in the number of papers p(t) can be described by the following differential equation:[6]

$$\frac{dp}{dt} = k \cdot p \tag{1}$$

This equation shows that the growth rate, dp/dt, is proportional to the actual level of p, i.e., the relative growth rate $^1/_p \cdot {}^{dp}/_{dt}$ is constant. The solution of the above differential equation with the initial condition that $p = p_0$ at time $t = 0$ is an exponential function.

$$p(t) = p_0e^{kt}, \text{ with } k > 0 \tag{2}$$

The exponential curve is well characterized by time during which the value of p is doubled (doubling time).

In general, the growth may be exponential only until the external conditions under which the literature of the given field develops do not change substantially. Changes, e.g., those caused by war, inevitably disturb exponential growth, but the growth rate is later regenerated. The growth process described by the above exponential relationship cannot continue forever; limiting factors, e.g., shortage of money or manpower, must and do gradually manifest their effect. In this case the mechanism of growth will be better described by the Pearl-Reed logistic function or the Gompertz differential equations.

The Pearl-Reed logistic function has the form of the differential equation

$$dp/dt = kp(b - p) \qquad 0 < p < b \quad \text{and} \quad k > 0 \tag{3}$$

Here the growth is limited, since b is considered to be the limit of p.

The relative growth rate

$$\frac{1}{p} \cdot \frac{dp}{dt} = k(b - p) \tag{4}$$

is already not constant, being a linear function of p. This means that the higher the level of p, the lower the growth rate. By solving the above differential equations with the former initial conditions, one obtains the logistic equation

$$p(t) = \frac{b}{1 + a\exp(-kbt)} \quad k > 0 \tag{5}$$

where

$$a = \frac{b}{p} - 1 \tag{6}$$

In the initial stage, when $p \ll b$, the logistic curve practically coincides with the exponential one. The $p = b$ and $p = 0$ lines are asymptotes of the logistic curve. The curve has an inflection at $p = b/2$, where the sign of acceleration changes.

The Gompertz function has the following form:

$$p(t) = b\left(\frac{p_0}{b}\right)^{e^{-kt}} \tag{7}$$

which can be obtained by integration of the differential equation

$$\frac{dp}{dt} = kp\ln\frac{b}{p} \tag{8}$$

where the symbols used are the same as above. Whereas the logistic function has a symmetric shape, the Gompertz function has the inflexion point shifted towards 0; the S-curve is shortened and more curved at the first bend than at the second one.

Another model of growth was suggested by Moravcsik,[21] who considered science a spherical body of knowledge growing at its epidermis, i.e., along the research fronts based on knowledge acquired in the most recent research in an n-dimensional space, where the dimensions can be identified with the various directions in which science can grow independently of the other directions

$$\frac{dQ}{dt} = C_n \cdot S(t) \tag{9}$$

Q(t) being the amount of scientific knowledge; S(t), its surface area at time t; and C, a constant.

The solution of this equation is

$$(Q)t = \beta t^n \tag{10}$$

where β is a proportionality constant. This equation gives an exact exponential growth if n, i.e., the number of freedom in the system of scientific research, goes to infinity.

Let us now see how the above mathematical models agree with the investigations on the growth of analytical chemistry subject literature. This question was studied almost simultaneously by Orient[22] and Brooks and Smythe.[12] According to the latter authors, in the analytical literature there is "an exponential increase for 1920—30. The exponential increase is not maintained for 1930—35, probably because of the influence of the Great Depression. An expected decrease for the war years is followed by an exponential increase from 1950 onwards."[12]

In our opinion these statements are incorrect, mostly for methodological reasons, since the authors drew their conclusions on the basis of a growth rate curve obtained by plotting the yearly number of analytical papers against the years. However, exponential growth and the corresponding doubling times may be determined reliably only on the basis of cumulative growth curves.

Orient[22] counted the abstracts in the "Analytical Chemistry" section of *Referativnyj Zhurnal Khimiya*. The results between 1953 (the first publication year of the journal) and 1972 indicate, too, that the growth was exponential with approximate doubling times of 8 years between 1964 and 1972 and 6 years between 1953 and 1964. Orient also studied the growth rate of U.S.S.R. analytical chemical literature (in relation to the growth rate of world analytical literature) and the growth rate of literature dealing with organic and inorganic compounds and with the theory of analytical chemistry. Credit can be given to these investigations for proving for the first time that the growth of analytical chemical literature indeed follows an exponential model. The numerical results on doubling times are, however, incorrect. The origin of the errors, also committed by other authors investigating the exponential growth of other subject literatures, is a methodological one, pointed out first by May[23] and also mentioned briefly at the beginning of this chapter. Namely, in all data on the basis of which Orient calculated the results, the literature before 1953, i.e., before the starting point of investigations, was neglected.

Failure to include earlier literature can cause serious errors in determination of growth rates. Braun et al.[20] have used the data of Brooks and Smythe[12] and Baker[16-19] to compute the corrected dynamic growth rates of the output of chemical and analytical literature both worldwide and from various countries. The procedure used was to multiply the yearly number of abstracts of analytical papers in *Chemical Abstracts* by the interpolated share values of the countries given in Table 1. The logarithms of the cumulated number of papers were then plotted vs. time. Figures 2 and 4 show the results. The upper ends of the curves indicate a constant rate exponential growth with the doubling times shown. But these apparent doubling times are incorrect as a result of the same reason mentioned above for Orient's data. Braun et al.[20] have assumed that the growth rate is constant over the whole historical period and have applied to the data a multistep correction procedure. The total estimated correction (number of analytical papers published before 1915) was found to be 11.053. When this value is added to each point of the total analytical data curves and interpolated national fractions obtained are added to each national point, the semilog data plots shown in Figures 3 and 5 are obtained. Note that the data of Figures 3 and 5 now have a straight line fit over the entire time period, not just at the upper end. Note also — and it is very significant — that the apparent doubling time has increased. The same procedure was carried out with the chemical literature. The raw and corrected data are shown in Figures 6 and 8, and 7 and 9, respectively. Tables 2 and 3 show the corrected data for total worldwide publication and for a number of different countries. Note that the doubling times of the worldwide body of chemical and analytical chemical literature are 14.5 and 13.9 years, respectively, and compare these values with Orient's data of about 7 years on the average. If we suppose some relationship or correlation between the growth rate of the analytical literature and that of the analytical knowledge (or progress), we can only speculate about the consequences such a 100% difference can have on our planning, forecasting, and evaluation of any educational, research, or development activities. Note also that the U.S.S.R. and U.S. were essentially equal in total analytical publications in 1970, but the U.S. was still far ahead in the amount of chemistry publications even in 1975. With a doubling time in chemical publications of 6.7 years vs. 15 for the U.S., the U.S.S.R. can be expected to pass the U.S. in 1986. This has obviously already occurred in analytical chemistry. From the data one can

Table 1
PERCENTAGE OF PUBLISHED ANALYTICAL CHEMISTRY RESEARCH IN VARIOUS COUNTRIES

Country	Year 1965	Year 1970
U.S.S.R.	25.4	28.4
U.S.	15.8	17.7
Japan	11.0	7.7
Germany[a]	6.4	6.1
U.K.	4.3	5.9
Czechoslovakia	5.3	5.6
France	3.5	2.6
India	3.5	2.6
Scandinavia	0.7	2.1
Romania	3.5	2.0
Poland	4.1	1.8
Spain	1.8	1.5
The Netherlands	0.8	1.3
Italy	1.7	1.0
China	3.1	—
Rest of the world	9.1	11.1

[a] Both East and West Germany.

From Brooks, R. R. and Smythe, I. E., *Talanta*, 22, 495, 1975. With permission.

also calculate the relationship between the growth rates of all chemical and analytical chemical literature; this is given in Table 4. It is also enlightening to compare doubling times in analytical chemistry and chemistry with those of other disciplines. Holt and Schrank[24] studied growth rates of journal literature in various disciplines. Table 5 shows their data along with those for analytical chemistry and chemistry.[20]

In the above-discussed cases the growth laws of the total analytical literature of the world or various countries were reviewed. As could be seen, in measuring growth rates with sufficient accuracy, problems are caused by the lack of numerical data on the literature not processed by abstract journals, i.e., literature before 1910. This problem could be solved only by the corrections described above, and it necessarily follows from this procedure that even the corrected data can be regarded only as informative. More reliable data can be expected, however, from the growth studies of the literature of younger subfields of analytical chemistry for which the publication times of the first papers can be determined accurately. The increase in the number of publications is easier to follow from *Chemical Abstracts,* other abstract journals, or a carefully prepared bibliography. Some of these results follows.

Orient and Markusova[25] studied the growth of world literature of electroanalytical chemistry and related it to the growth of the same field in the U.S.S.R. As data base, the "Electrochemistry" section of *Chemical Abstracts* was used. The observed functions describing the cumulative growth of electroanalytical literature between 1950 and 1968 can be seen in Figures 10 and 11. The curves follow the exponential growth model already mentioned and appear to indicate that the electroanalytical literature of the U.S.S.R. grew somewhat faster than that of the world. The authors, using the method of the least squares, approximated the curves by the $p = ae^{kt}$ equation, with $a = 1771.2$,

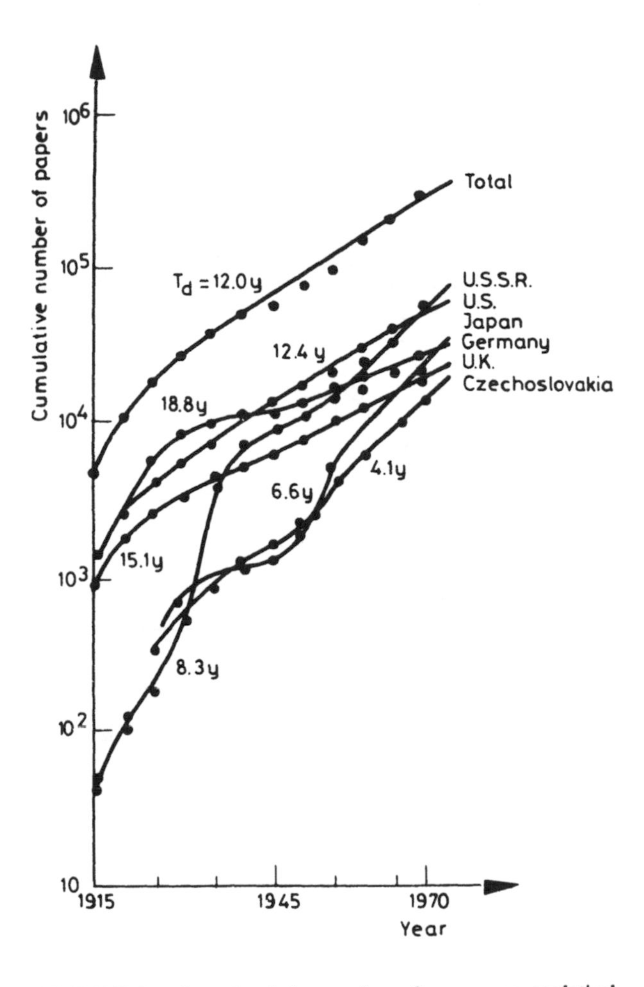

FIGURE 2. Growth of the number of papers on analytical chemistry in some leading countries (uncorrected data). T_d = doubling time, years. (From Braun, T., Bujdosó, E., and Lyon, W. S., *Anal. Chem.*, 52(6), 617A, 1980. With permission. Copyright 1980, American Chemical Society.)

k = 0.186, and σ = 17% for the curve of Figure 10, and a = 196.4, k = 0.249, and σ = 21% for the curve of Figure 11 (σ is the mean square error). It is characteristic for both curves that between 1950 and 1965 they are well described by the $p = ae^{kt}$ equation. The authors also show that from 1965 the theoretical curve rises more steeply than the observed one. This (i.e., some recession in electrochemical literature) also appears to be supported by the results of Orient and Pats[26] on the growth of the literature on polarography between 1950 and 1974, drawn from *Referativnyj Zhurnal Khimiya* as a data base (Figure 12). As seen in the figure, the slope of the exponential curve changes in 1965, i.e., the growth decelerates.

The growth of radioanalytical literature was investigated in particular detail by Braun et al.[20,27,28] Some of the results are shown in Figures 13 to 17. In connection with these growth curves perhaps two points are worth noting. The first concerns the extremely fast growth rate of the literature of some subfields of radioanalytical chemistry activation analysis, e.g., the use of semiconductor detectors in radioanalytical chemistry. For example, it can be seen that the literature of activation analysis doubled every 2.2 years, and this growth rate remained constant for more than 10 doubling periods. This seems to correlate with the extremely dynamic progress of the field.

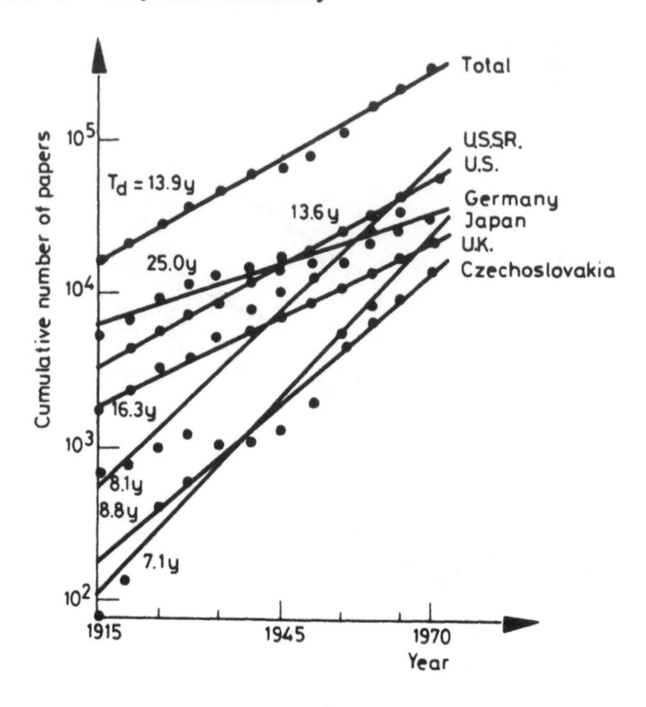

FIGURE 3. Growth of the number of papers on analytical chemistry in some leading countries (corrected data). T_d = doubling time, years. (From Braun, T., Bujdosó, E., and Lyon, W. S., *Anal. Chem.*, 52(6), 617A, 1980. With permission. Copyright 1980, American Chemical Society.)

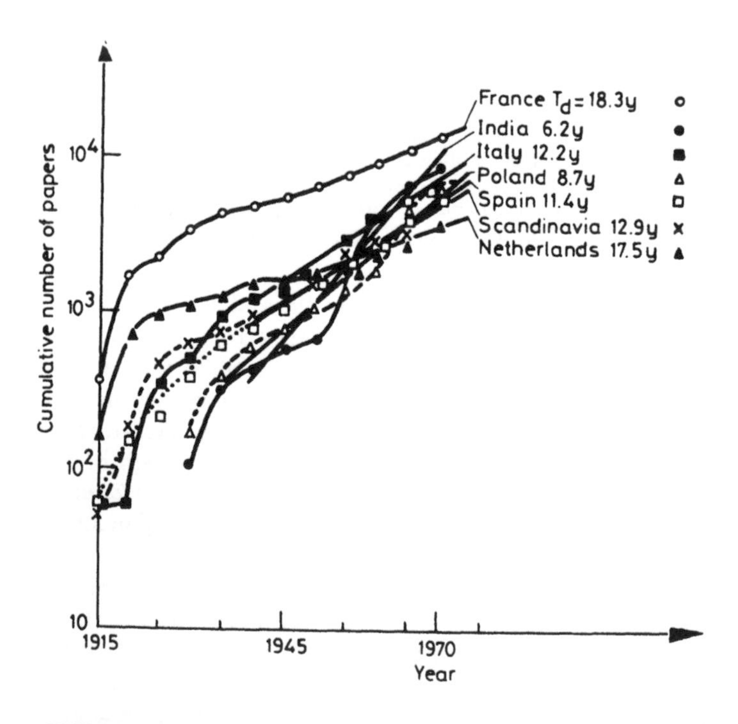

FIGURE 4. Growth of the number of papers on analytical chemistry in some leading countries (uncorrected data). T_d = doubling time, years. (From Braun, T., Bujdosó, E., and Lyon, W. S., *Anal. Chem.*, 52(6), 617A, 1980. With permission. Copyright 1980, American Chemical Society.)

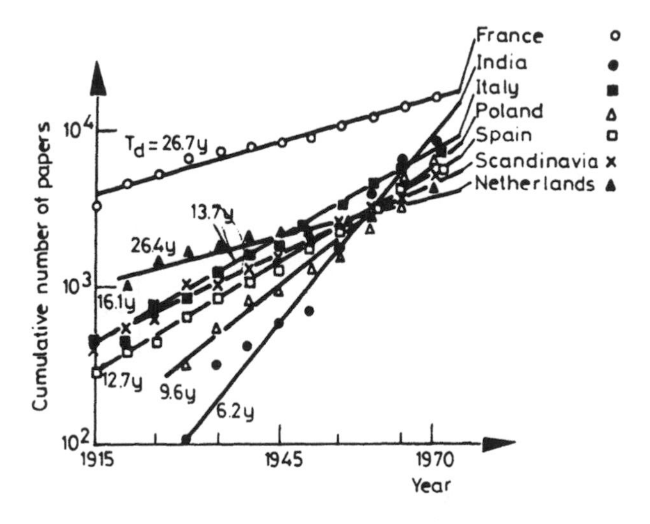

FIGURE 5. Growth of the number of papers on analytical chemistry in some leading countries (corrected data). T_d = doubling time, years.

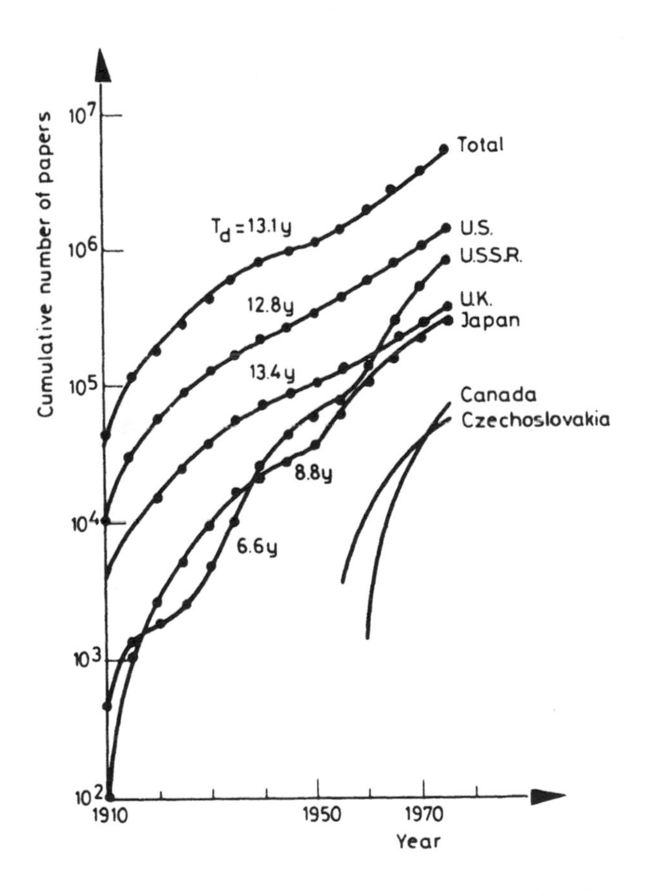

FIGURE 6. Growth of the number of papers on all branches of chemistry in some leading countries (uncorrected data). T_d = doubling time, years. (From Braun, T., Bujdosó, E., and Lyon, W. S., *Anal. Chem.*, 52(6), 617A, 1980. With permission. Copyright 1980, American Chemical Society.)

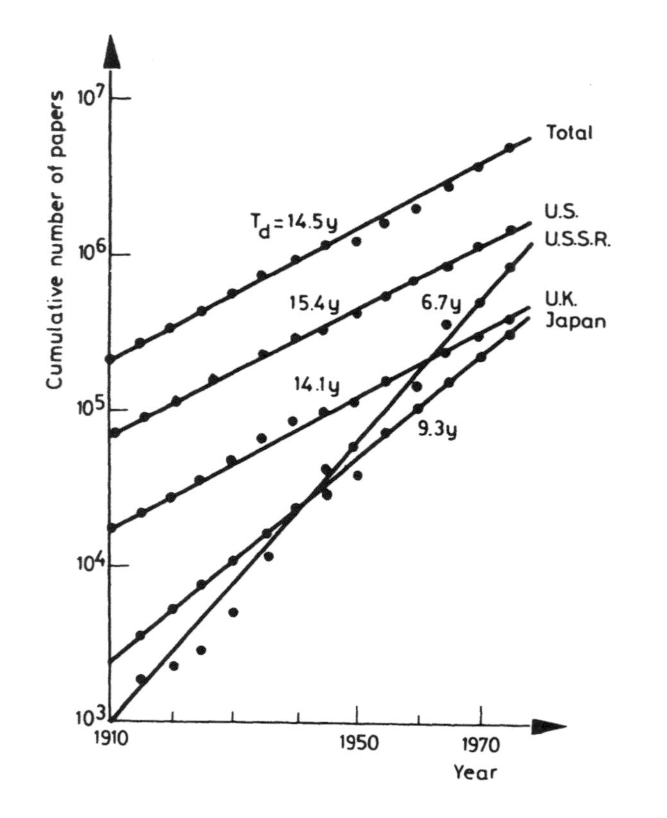

FIGURE 7. Growth of the number of papers on all branches of chemistry in some leading countries (corrected data). T_d = doubling time, years. (From Braun, T., Bujdosó, E., and Lyon, W. S., *Anal. Chem.*, 52(6), 617A, 1980. With permission. Copyright 1980, American Chemical Society.)

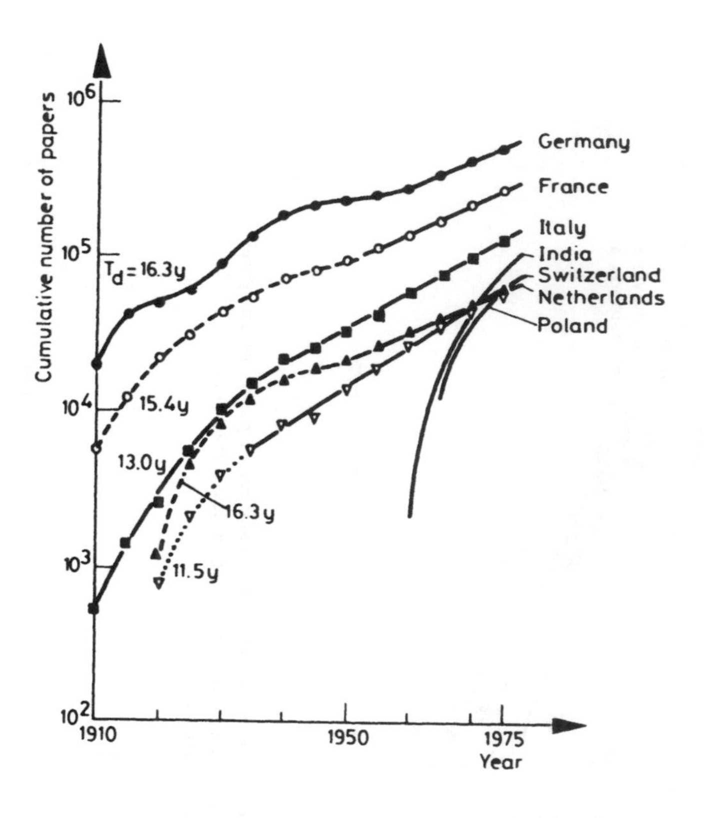

FIGURE 8. Growth of the number of papers on all branches of chemistry in some leading countries (uncorrected data). T_d = doubling time, years. (From Braun, T., Bujdosó, E., and Lyon, W. S., *Anal. Chem.*, 52(6), 617A, 1980. With permission. Copyright 1980, American Chemical Society.)

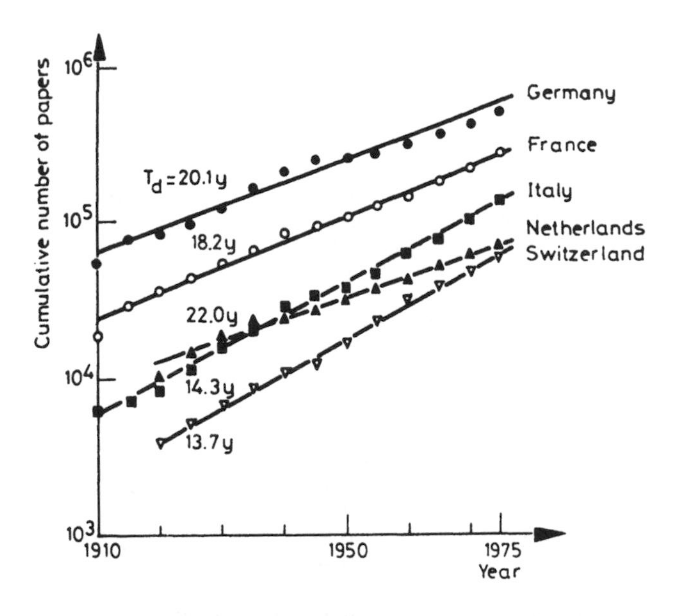

FIGURE 9. Growth of the number of papers on all branches of chemistry in some leading countries (corrected data). T_d = doubling time, years.

Table 2
DATA FOR ANALYTICAL LITERATURE CORRECTED FOR "COUNTING LOSS"

Country	Number of publications in 1915 × 10³	Number of publications in 1970 × 10³	Share of the total analytical literature (%)	Growth rate (doubling time in years)
U.S.S.R.	0.70	58.61	18.8	8.1
U.S.	3.18	59.43	19.0	13.6
Japan	0.08	23.17	7.4	7.1
Germany[a]	5.58	31.77	10.2	25.0
U.K.	1.79	31.83	10.2	16.3
Czechoslovakia	0.18	14.59	4.7	8.8
France	3.32	16.71	3.4	26.7
India	—	8.65	2.8	6.2
Scandinavia	0.42	5.46	1.7	16.1
Poland	—	6.67	2.1	9.6
Spain	0.28	5.70	1.8	12.7
The Netherlands	0.67	4.54	1.4	26.4
Italy	0.46	6.53	—	—
World literature	16.03	311.89	—	13.9

[a] Both East and West Germany.

From Braun, T., Bujdosó, E., and Lyon, W. S., *Anal. Chem.*, 52(6), 617A, 1980. With permission.

Table 3
DATA FOR CHEMICAL LITERATURE CORRECTED FOR "COUNTING LOSS"

Country	Total number of publications in 1910 × 10^3	Total number of publications in 1975 × 10^3	Share of the total chemical literature (%)	Growth rate (doubling time in years)
U.S.	74.64	1,520.55	29.1	15.4
U.S.S.R.	0.97	862.98	16.5	6.7
U.K.[a]	16.75	394.46	7.5	14.1
Japan	2.57	339.63	6.5	9.3
Germany[b]	55.08	367.54	7.0	20.1
France	19.32	269.15	5.1	18.2
Italy	6.37	141.25	2.7	14.3
The Netherlands	3.97[c]	75.16	1.4	22.0
Switzerland	3.38[c]	63.97	1.2	13.7
World literature	197.70	5,223.96	—	14.5

[a] Data before 1951 calculated as U.K. = 0.65 × British Commonwealth.
[b] Both East and West Germany.
[c] Extrapolated from 1920 data.

From Braun, T., Bujdosó, E., and Lyon, W. S., *Anal. Chem.*, 52(6), 617A, 1980. With permission. Copyright 1980, American Chemical Society.

Table 4
RELATIONSHIP BETWEEN CHEMICAL AND ANALYTICAL CHEMICAL LITERATURE

Country	Analytical literature as percent of total chemistry in 1970	Relative growth rate of the development of analytical literature (T_d analytical/T_d chemical)
U.S.S.R.	11.2	1.2
U.S.	5.1	0.9
Japan	9.8	0.8
Germany[a]	6.7	1.2
U.K.	10.3	1.1
France	7.1	1.5
The Netherlands	7.4	1.2
Italy	5.9	0.9
World literature	8.2	1.0

[a] Both East and West Germany.

From Braun, T., Bujdosó, E., and Lyon, W. S., *Anal. Chem.*, 52(6), 617A, 1980. With permission. Copyright 1980, American Chemical Society.

The other characteristic growth behavior can be seen in the case of 14-MeV neutron generator activation analysis (Figures 16 and 17), which reflects the logistic model mentioned at the beginning of this chapter.

A very rapid doubling time of about 2 years also can be noticed in the growth rate of the literature on flow injection analysis (FIA) at least in the first 6 to 7 years of its existence[29] (Figure 18). In later years the growth curve shows traits similar to those mentioned in the case of growth in ion chromatography journal literature (Figure 19).

Table 5
LONGTERM GROWTH RATES OF JOURNAL LITERATURE IN VARIOUS DISCIPLINES

Field	Range of data	Growth rate (%/year)
Biology	1927—1964	4.39
Economics	1886—1959	5.50
Electrical engineering	1903—1962	3.50
Physics	1903—1964	3.73
Psychology	1927—1964	2.90

From Braun, T., Bujdosó, E., and Lyon, W. S., *Anal. Chem.*, 52(6), 617A, 1980. With permission. Copyright 1980, American Chemical Society.

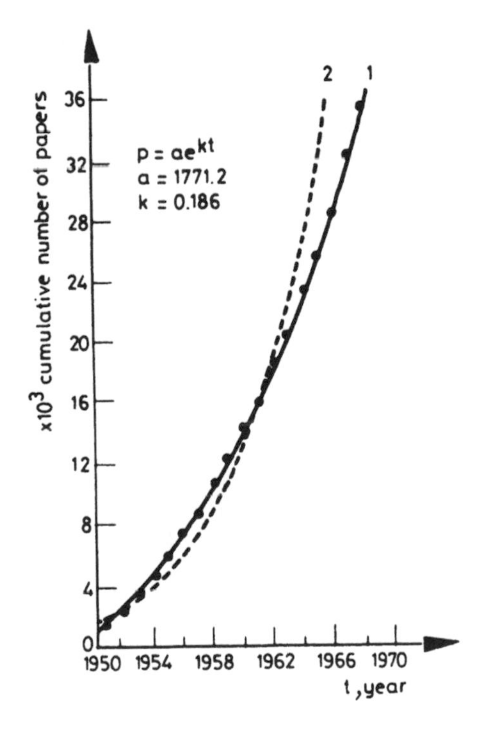

FIGURE 10. Growth of world literature on electroanalytical chemistry between 1950 and 1968. Curve 1, experimental; curve 2, theoretical. (From Orient, I. M. and Markusova, V. A., *Electrochemical Methods of Analysis* (in Russian), Sbornik, Izd. Metallurgiya, Moscow, 1972. With permission.)

Among those literature topics with extremely rapid growth rates, perhaps atomic absorption must also be mentioned. For this field a doubling time of 2.5 years was found by Orient et al.[30] in the U.S.S.R.

In a recent paper Braun[31] studied some factors governing publication growth in ion chromatography. In Figure 19, curve A shows the growth of journal papers dealing

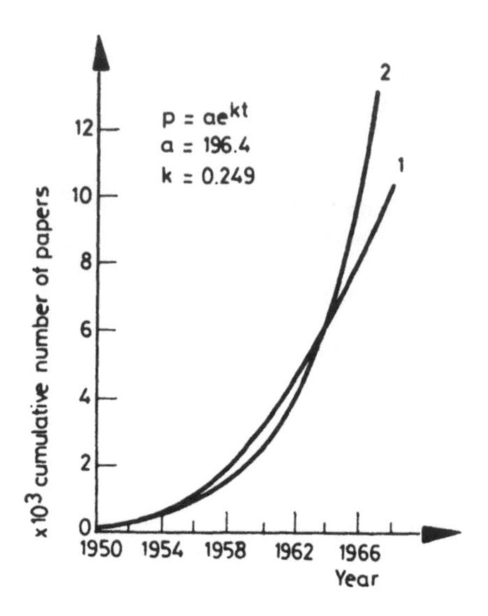

FIGURE 11. Growth of Soviet literature on electroanalytical chemistry between 1950 and 1968. Curve 1, experimental; curve 2, theoretical. (From Orient, I. M. and Markusova, V. A., *Electrochemical Methods of Analysis* (in Russian), Sbornik, Izd. Metallurgiya, Moscow, 1972. With permission.)

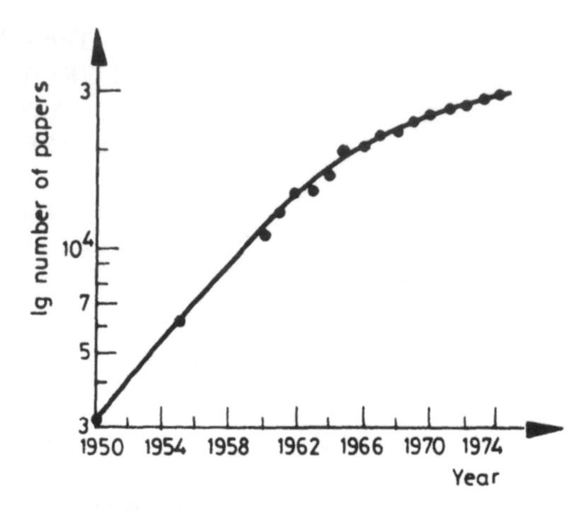

FIGURE 12. Growth of world literature on polarography between 1950 and 1974. (From Orient, I. M. and Pats, R. G., *Polarography, Problems and Trends* (in Russian), Strabynya, Ya. P. and Majranovskiy, S. G., Eds., Sbornik, Izd. Zinatne, Riga, 1977, 388. With permission.)

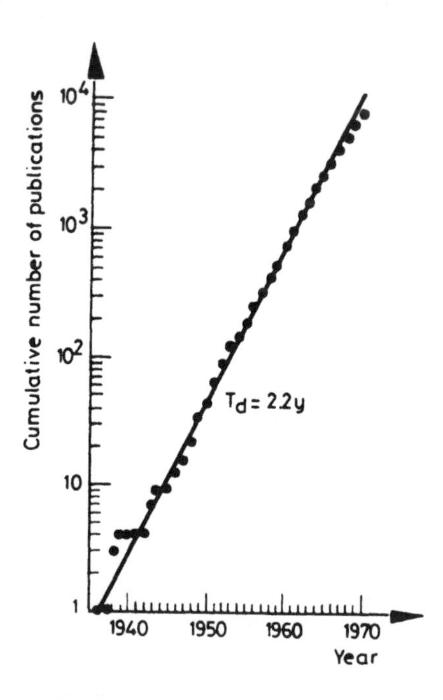

FIGURE 13. Growth of publications on activation analysis. (From Braun, T., Lyon, W. S., and Bujdosó, E., *Anal. Chem.*, 49, 682A, 1977. With permission. Copyright 1977, American Chemical Society.)

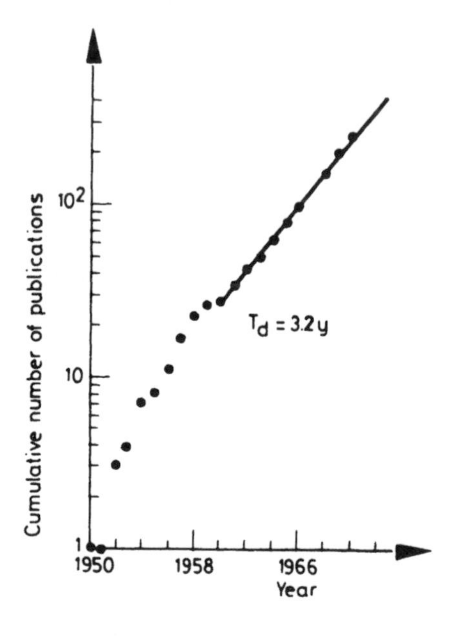

FIGURE 14. Growth of publications on charged particle activation analysis. (From Braun, T., Lyon, W. S., and Bujdosó, E., *Anal. Chem.*, 49, 682A, 1977. With permission. Copyright 1977, American Chemical Society.)

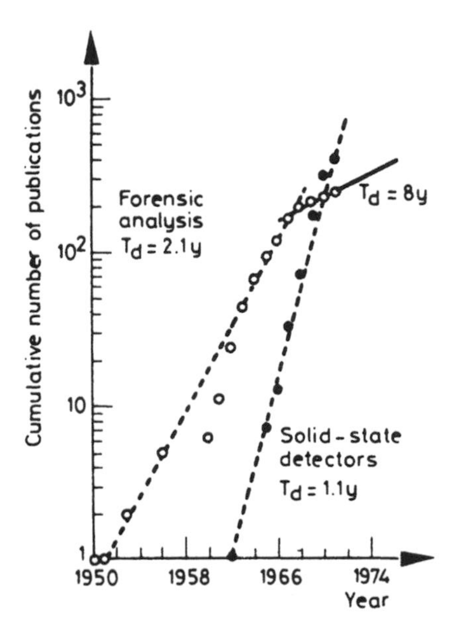

FIGURE 15. Growth of publications on forensic analysis and solid-state detectors. (From Braun, T., Lyon, W. S., and Bujdosó, E., *Anal. Chem.*, 49, 682A, 1977. With permission. Copyright 1977, American Chemical Society.)

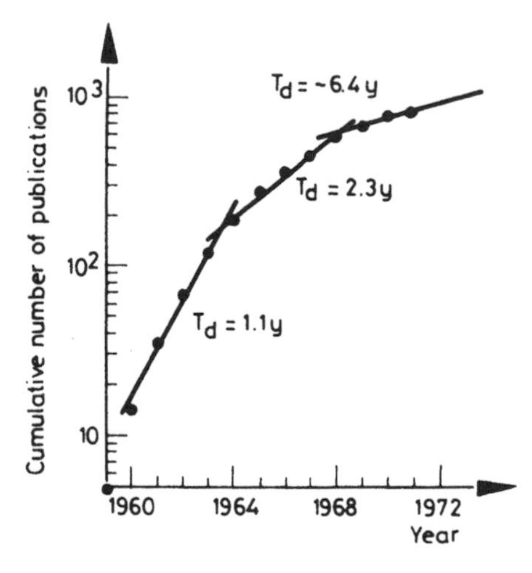

FIGURE 16. Growth of publications on activation analysis by 14-MeV neutron generators. (From Braun, T., Lyon, W. S., and Bujdosó, E., *Anal. Chem.*, 49, 682A, 1977. With permission. Copyright 1977, American Chemical Society.)

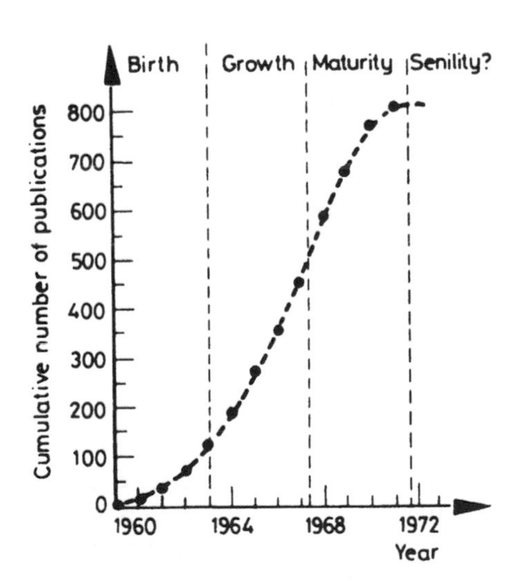

FIGURE 17. Four ages of neutron generators. (From Braun, T., Lyon, W. S., and Bujdosó, E., *Anal. Chem.*, 49, 682A, 1977. With permission. Copyright 1977, American Chemical Society.)

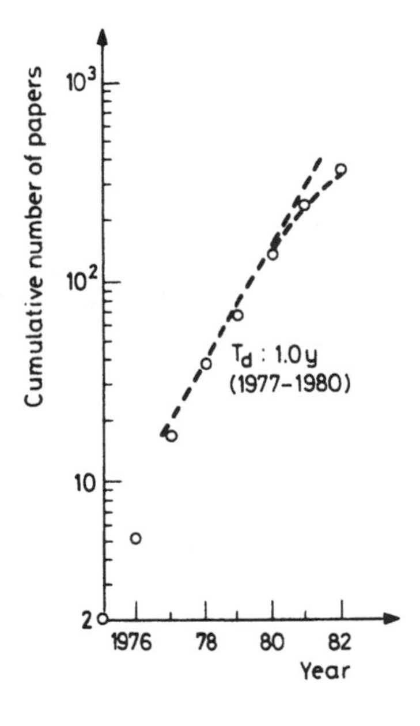

FIGURE 18. Growth of publications on FIA. (From Braun, T. and Lyon, W. S., unpublished results, 1983. With permission.)

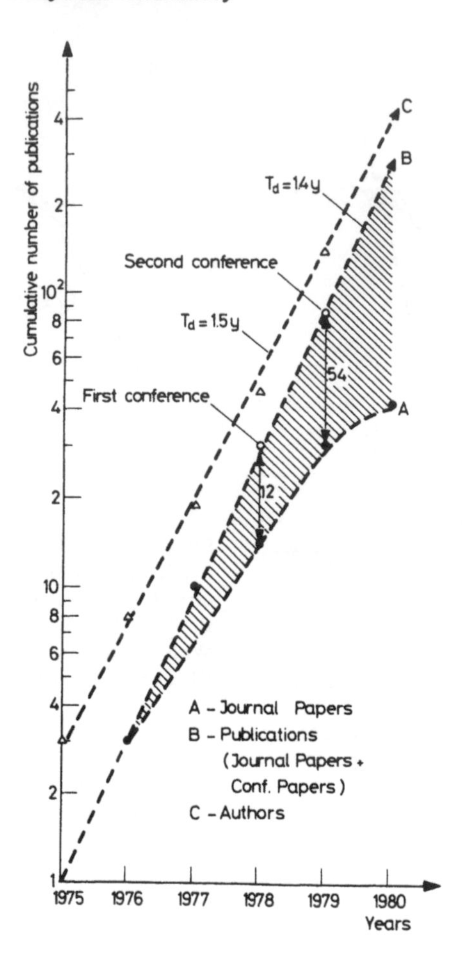

FIGURE 19. Growth of publications on ion chromatography. Curve A shows papers in journals; curve B, papers in journals plus conference papers; and curve C, numbers of authors. T_d = doubling time. (From Braun, T., *Anal. Proc.*, 19, 352, 1982. With permission.)

with ion chromatography between 1975 and 1980. Although the curve indicates a significant development, it is evident that the growth rate is not a constant-rate exponential. Curve B, which is already of constant-rate exponential character, depicts the increase in the total publishing activity (journal papers and conference reports in print). The doubling time of the growth is 1.4 years, a "skyrocketing" rate. From a comparison between the two curves it is also evident that the extra publications that are necessary for sustaining the very rapid exponential growth rate originate from conferences. Had the "stochastic" publishing activity in ion chromatography not been "helped" and "orientated" by the organization of conferences and by the publishing of papers given at these conferences, the development of publishing activity in this field would probably not have shown the exponential skyrocketing effect. In Figure 19 it can also be seen that the "helping" and "orientating" injections that are necessary for the sustenance of the exponential growth of publishing activity in this topic had to increase (in 1978, 12 papers; in 1979, 54 papers). For the sake of interest, it could perhaps be mentioned that the number of new researchers (authors) entering the field also increased exponentially (T_d = 1.5 years) between 1975 and 1980, as shown in curve

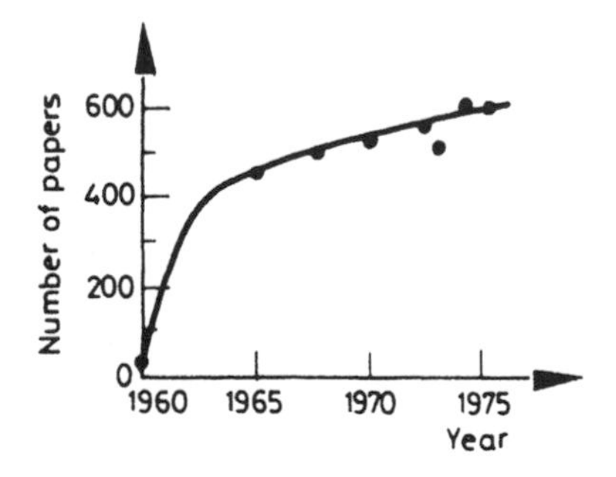

FIGURE 20. Growth of the literature on organic reagents between 1960 and 1975. (From Orient, I. M., *Zh. Anal. Khim.*, 32, 502, 1977. With permission.)

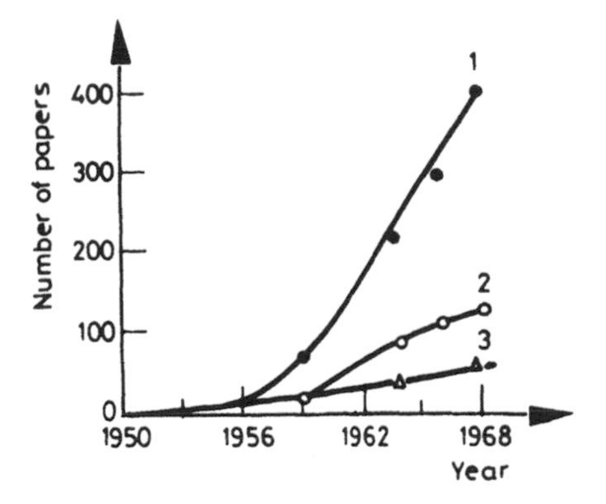

FIGURE 21. Growth of the literature on separation of traces by zone melting (2) and directed crystallization (3) for purification. Curve 1 is the sum of curves 2 and 3. (From Melikhov, I. V. and Berdonosova, D. G., *Zh. Anal. Chem.*, 31, 809, 1976. With permission.)

C of Figure 19, and that this doubling time is almost equal to the doubling time of the increase in the number of publications.

In showing the growth characteristics of analytical subject literature, a third model is worth mentioning in which the growth is linear. The growth of literature on organic reagents was investigated by Orient[32] (see Figure 20) and was found to be linear between 1964 and 1975. A linear growth was also observed by Melikhov and Berdonosova[33] for the literature on the separation of mixtures by zone melting as well (Figure 21).

As we have seen, the growth of the total body of analytical chemical literature follows the exponential or logistic model, and so does the literature of some analytical subfields. At the same time, however, the linear growth model also occurs. Magyar,[34] to reconcile the linear and exponential growth models in another field, postulated that it is the creation of some new subfields added to the linear growth of already existing fields that might lead to overall exponential growth. This may hold for analytical chemistry, too.

As a direct outcome of the results outlined in this chapter, we considered it worthy to mention the views of Menard[15] who states that if we can measure different growth rates in scientific literature and suppose them to be related in some way to the growth of science or knowledge, we are in a position to evaluate how they affect the careers and lives of scientists working in those fields (see also Chapter 7, Section I).

He noted that the literature of different scientific subfields is growing at highly variable rates and for illustration considered three models in which doubling times were on the average 15 years and at the extreme, 5 and 45 years. The professional life of a scientist may be taken as beginning when he becomes a graduate student and ending 45 years later when he retires at 65. Thus, in a slow subfield in which the literature doubles in 45 years, the number of scientists is constant because new appointments merely balance retirements. Typically, such a subfield might have 200 specialists, if we suppose that manpower grows parallel to the growth of the literature and follows the same mathematical model. It never varies, nor can it long continue. Either the doubling time increases, or the scientists double their output each 45 years. In a field doubling in 15 years, retirements amount to only one eighth of new recruitments during a 45-year career and thus are almost negligible. The number of new people in the field increases rapidly. If the initial number is 100, it is 200 in only 15 years, and after 45 years it is 700, allowing for retirement. Such a rapidly doubling subfield, starting smaller than a slowly doubling one, would quickly outgrow the slower subfield.

The growth of a subfield in 5 years or less is truly spectacular. If only 10 men transfer into it at the beginning, there are more than 1000 in 35 years. These growth rates have major effects on the age distribution among the specialists in a subfield, and thus on the queuing position of a new person after a given time. The median age in the slow subfield is halfway from beginning to end, which is 42 years. In the average and fast subfields it is the initial age plus one doubling period, 35 and 25 years, respectively. This means that in the fast field a student is at the median age when he receives his doctorate after 5 years of graduate study. So rapidly does the field expand that he is in the senior one eighth of his profession by the time he is 35, whereas in both the average and slow subfields he would have to wait until he is about 60.

We can estimate the living literature existing in a subfield as approximately that produced in the three previous doubling periods by the average population of specialists during the period. In the fast model there are 10 people and thus a total literature of 190 papers. In the average subfield there are 100 people and almost 6000 papers. In the slow model we have to assume that an equilibrium population of 200 has just been established, and in the past the population and literature have expanded at the same rate; the literature consists of 27,000 papers. We may now consider the activities of a student entering into graduate study in one of these subfields. A diligent speed reader, he plunges into the literature of the fast subfield and emerges 38 days later, tired but "au courant". In the average subfield a student would have to read 5 papers a day for 3 years to catch up. What then of the student in the old and slow subfield who is confronted with 27,000 papers? They hang over him for his entire professional life. Not only students but also skilled senior investigators are kept busy compiling ever more massive bibliographies.

We can also consider the growth of the literature during the 5 years a student is engaged in graduate studies. Another 190 papers come out in the fast field, and the student absorbs them gradually. Thus he has seen one half of the literature develop around him. In the average subfield another 1200 papers have appeared, or about 1 every 2nd day, which is simply too much to read. The situation is even worse in the slow subfield in which the literature increases by almost 3000 papers. From the very beginning of their careers, therefore, specialists in average and slow subfields are losing ground relative to the growing volume of unread literature. The situation changes with continuing expansion. Within 20 years all the subfields have a backlog of literature too large, and each is increasing by 3000 papers in 5 years. After 10 years more the fast field would acquire 12,000 papers during 5 years of graduate study. Long before, it would have broken into a group of narrower and diverging sub-subfields. It appears that the time for a student or young scientist to get into a fast subfield is somewhere between the third doubling period, when the subfield is identifiable, and the sixth, when the amount of literature begins to be unwieldy.

Let us consider now the effort necessary to become established in the subfield. At three-papers-per-year productivity and the age distributions we have calculated, a newcomer in a fast subfield need only write six papers before arriving at the median age. In an average subfield 36 papers represent the effort of the middle man in the queue, but in the slow subfield it is 57 papers by age 42. Thus the sustained effort to become established as a sound, experienced specialist varies enormously. After only 26 papers, a man in a fast subfield is a patriarch in the oldest one eighth of the population. In the average and slow subfields, 110 papers are required. Even with single-minded dedication, many cannot sustain that level of production of scientific papers for a lifetime. At some point the scientist drops out of the race to enter a different one or to become a spectator. Few people write 50 papers. Consequently, most people in average and slow subfields drop below the normal level of scientific productivity before reaching the midpoints of their careers. Hardly any of the senior group are still active in research. In contrast, the senior persons in a fast subfield are much younger, and most engage in research even though they are also administrators and committeemen. The administration and evaluation of research in the fast subfield consequently is guided by people engaged in it, which presumably is as effective as the system can be.

The growth of modern analytical chemistry as reflected in the statistical evaluation of its subject literature has been reviewed. It has been shown that in general the analytical literature follows an exponential growth law with an overall doubling time of about 13 years. In other words the published analytical information grows so quickly that in the next 13 years we will see an accumulation of an amount equaling that accumulated during the whole history of the field. One of the main questions of concern regarding this rather impressive growth rate asks what in fact this growth reflects. Does it reflect the growth of analytical knowledge?

There is no easy answer to this question. Nevertheless the growth rate of the analytical literature would serve as a reflection of knowledge produced by analytical authors, provided that one could make two assumptions: first, that all the knowledge obtained by these authors is to be found in the literature; and second, that every one of the papers in that literature contains either an equal or a known proportion of the knowledge. Neither of these assumptions will stand examination. One conclusion is clear; growth of analytical knowledge must be distinguished from growth of analytical literature. The former is a more abstract concept and hence not so directly assessed. Moravcsik[35] distinguishes three relevant quantities of interest when studying growth of science: scientific activity, productivity, and progress. He illustrates the meaning of these quantities by using an analogy. Let us imagine that we want to get from one spot in a dense forest to a preassigned second spot 5 mi away. *Activity* would then correspond to the amount of work we do, e.g., thrashing about in the undergrowth, cutting trails among the trees, running off in exploratory directions to reconnoiter, etc. *Productivity* would correspond to the amount by which, as a result of all these activities, we have come closer to our final target point. Finally, *progress* would correspond to the ratio of this productivity to the total task, i.e., how much closer we have come to the target divided by the total 5 mi. In other words, the first relevant quantity of interest, scientific activity, is concerned with the consumption of input resources and is, therefore, related to such factors as the number of scientists involved, the expenditure on their research, and percentage of their time spent on research, and the number of supporting staff. The second, research productivity, refers to the extent to which this consumption of resources creates a body of scientific results (see Figure 1, Chapter 1). These results are usually manifested in the form of research communications, i.e., the subject literature, although scientists do communicate through other informal channels, i.e., letters, seminars, and personal communication. The third, scientific progress, refers to the extent to which the scientific activity actually results in substantive contributions to scientific knowledge.

Seemingly, one can consider the literature of analytical chemistry reflecting the *productivity* of the analytical subject field. The question arises, therefore, whether the analytical literature has been evolving in the direction of the dinosaur, becoming ever larger in size but smaller in brain. It seems clear that the total size has been increasing exponentially, but the relative change in quality remains unknown.

Rescher[36] defines the λ-quality level, $(0 < \lambda < 1)$ of a publication or result as follows: if there are $p(t)$ publications in all at time t, then there will be $[p(t)]^{\lambda}$ publications at the λ-level. He characterizes specific values as follows:

$\lambda = 1$ at least routine,

$\lambda = 0.75$ at least significant,

$\lambda = 0.50$ at least important,

$\lambda = 0.25$ at least very important.

Nevertheless, Price[1] seems to be the first to point out that the number of important contributions is the square root of the total number of contributions.

If now the total analytical literature (assuming, not without a certain trace of optimism, that anything published is at least routine) is growing exponentially as shown with a doubling time T_d, then the literature of λ-quality, for $\lambda > 0$, is also growing exponentially with the doubling time of T_d/λ. This means clearly that as the quality scale is ascended, exponential growth slows down.

Thus if we take about 300,000 publications as the size of the analytical literature in 1970, in term of Rescher's λ-levels there would be about:[37] 300,000 at least routine, 12,819 at least significant, 548 at least important, and 23 at least very important publications.

Taking now the doubling time of the world analytical literature as 13 years, the corresponding doubling times for each λ-level group of analytical publications would be about 17.33 years for at least significant, 26.00 years for at least important, and 52.00 years for very important publications.

These seem to be excessively critical figures. On the other hand, the exponential increase in the amount of the literature has to be regarded not as useless verbiage but as the useful and necessary inputs needed for genuine advances.

Empirical investigations on the quality of analytical literature (and we can add that this holds for the quality of the literature of any subject field) are practically nonexistent. This is quite surprising, if we consider their importance to quantitative approaches to the growth of knowledge. Of course, such analyses are very time consuming and require expert knowledge. A criticism can be made that the assignment to categories is very subjective. Also, such a categorization fails to recognize that some duplication is necessary to ensure that new results reach a variety of audiences.

The main conclusion from the above-mentioned is that publications containing significant new analytical ideas, results, applications, and systematizations make up a tiny fraction of the total literature, and their growth rate is rather slow. The great bulk consists of duplications, trivia, and low-quality texts and expositions that make no or only a very modest contribution to the subject.

The end conclusion poses a serious question. Do duplication, trivia, and low-quality publications serve a useful educational task? They surely represent activity, but much of it is misdirected in the sense that it spreads as much misinformation as knowledge and clogs communication channels.

Better compilations of statistics on the literature of analytical chemistry are obviously needed for informed, critical assessments and refinements of the present scientometric techniques so that valid evaluations of analytical knowledge growth may be obtained.

At the end of this chapter it is perhaps also worth mentioning two investigations, in instrumental analytical subfields, of the correlation between the number of papers published in the subfield and the number of instruments extant in the subfield. Brooks and Smythe[12] carried out such investigations in atomic absorption analysis, and Braun et al.[27] in the subfield of activation analysis. The curves reflecting the correlations are shown in Figures 22 and 23. The reasons governing the mechanism are not clear yet, but the S-shaped curve obtained for atomic absorption is interpreted by Brooks and

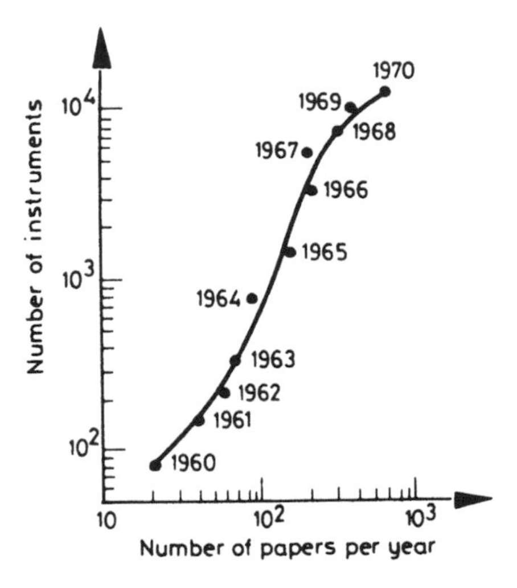

FIGURE 22. Relationship between the total number of atomic-absorption spectrophotometers in existence and the total number of atomic-absorption papers published. (From Brooks, R. R. and Smythe, I. E., *Talanta*, 22, 495, 1975. With permission.)

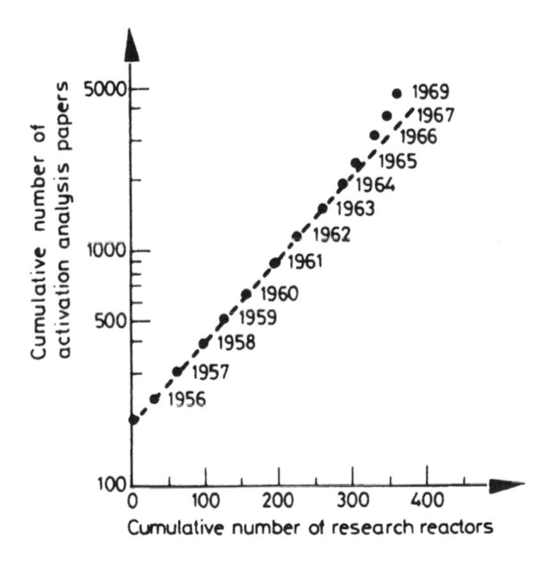

FIGURE 23. Relationship between worldwide output of activation analysis publications and the number of research reactors. (From Braun, T. and Bujdosó, E., *J. Radioanal. Chem.*, 50, 9, 1979. With permission.)

Smythe[12] to mean that at first there is a rush for scientists to publish in a new field, i.e., atomic absorption. Brooks and Smythe refer to this as the "bandwagon effect". Later in time when the method becomes well established, the ever-increasing number of instruments are used more and more for routine analysis rather than for producing "publishable" research topics.

Chapter 4

OBSOLESCENCE AND REFERENCE HALF-LIVES OF THE LITERATURE ON ANALYTICAL CHEMISTRY

If the number of references taken from the end of a certain number of analytical papers (e.g., from all the papers in a yearly volume of a certain analytical journal) older than time t are plotted against time, an exponentially decaying curve is obtained:[38]

$$R(t) = N \exp - \frac{0.693}{T_{1/2}} t \qquad (11)$$

with R(t) the number of references older than t years, N the total number of references, and $T_{1/2}$ the half-life of the literature.

This equation is very similar in character to that of radioactive decay. It is valid for papers published in periodicals only, as these are the references which can be cited continuously in time. The analogy to radioactive decay is only partial, since a paper does not disappear or disintegrate after being cited as a nucleus does by decay, and a paper can potentially be cited at any point in time. The half-life of a subject literature indicates the rate with which the frequency of citation (or use of that literature) decreases, i.e., the literature is used less frequently or becomes obsolete.

The model so far outlined assumes that the decline in frequency of citation (i.e., referencing) can be ascribed to "obsolescence", i.e., lessening of interest (reflected in referencing) in papers (or scientific results) as they grow older. Line[39] pointed out that this is too simple a model. As we have seen in Chapter 3, the volume of the literature is itself growing, apparently exponentially, so the number of papers available for use (and referencing) declines with age. In principle, the whole "obsolescence" of the literature might be ascribable to literature growth. Figure 24 gives a schematic view of a subject literature that is growing exponentially; the area within the curves represents the volume of publication with the number of items published per year doubling in 5 years. The squares represent recorded uses (citations) during the current year. The dots represent hypothetical uses — those that would occur if every published item had an equal chance of being used. The half-life of actual use, within which the squares lie, is m. The median age of the published literature, within which half the dots lie, is p. Now $m < p$ if the more recent literature is more likely to be useful and the older items are less likely to be used. It is the difference between m and p that truly represents "obsolescence".

Orient[22] and Brown[40] were concerned with the obsolescence of the total literature of analytical chemistry. Orient[22] investigated the distribution in time of the references in the papers published in the 1974 volumes of *Zhurnal Analiticheskoi Khimii* (Figure 25).

As can be seen there is a short, rising curve (Orient calls its maximum the "modus") followed by a descending curve, concave upwards and flattening as age increases (its half-life is called the "median"). The whole curve can be interpreted as an increase of use as the knowledge of papers diffuses followed by a steady decline in use as papers grow older. The second part of the curve can be plotted semilogarithmically as one of exponential decay, and a half-life can be calculated as mentioned at the beginning of this section.

The decay curves obtained by Orient indicated that in the periodical literature of analytical chemistry, the median, i.e., the half-life of the citations to foreign authors,

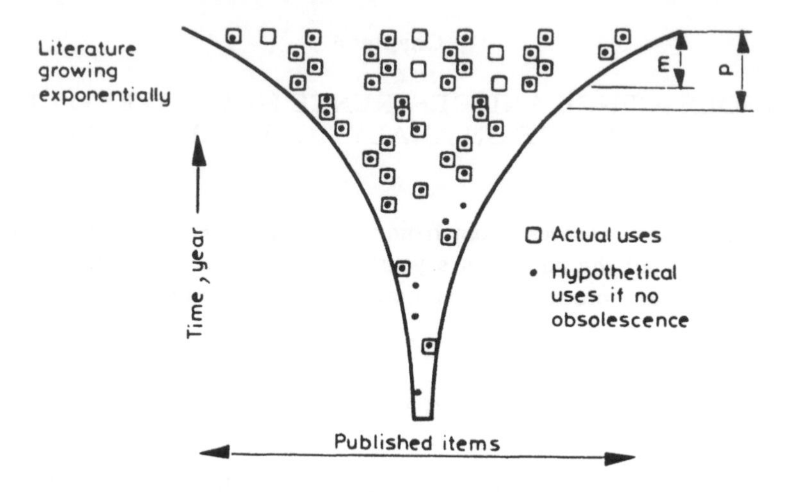

FIGURE 24. Literature growth and use. (From Line, M. B., *J. Doc.*, 26, 46, 1970. With permission.)

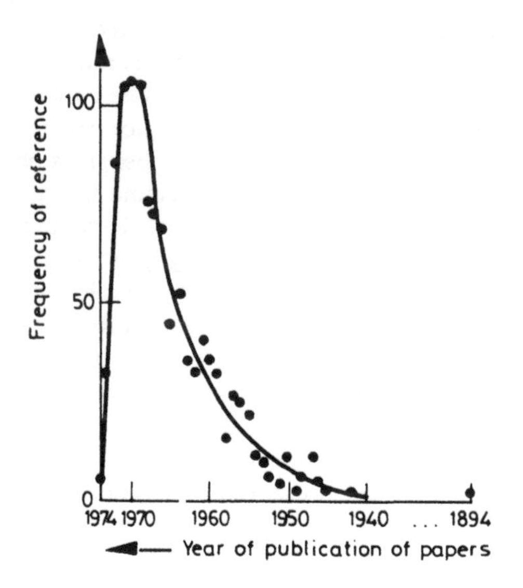

FIGURE 25. The distribution of references of papers published in *Zhurnal Analiticheskoi Khimii*, 1974. (From Orient, I. M., (in Russian), *Zavodsk. Lab.*, 41(9), 1071, 1975; (English transl.), *Ind. Lab. USSR*, 41, 1327, 1975. With permission.)

is 7.6 ± 0.6 years and the modus, 3.5 ± 0.5 years. In the case of U.S.S.R. authors these values are 6.3 ± 0.6 and 3.2 ± 0.4 years, respectively.

Brown[40] calculated a 9.3-year half-life for the chemical literature from the Poisson distribution of the average citation per article of 11 chemical journals.

Braun et al.[28] were dealing with the decay of radioanalytical literature. Their results can be seen in Figures 26 and 27. We may note on Figure 26 that from 1972 to 1974 in the *Journal of Radioanalytical Chemistry*, 85% of the references were <10 years old.

Burton and Kebler[41] suggest the half-lives collected in Table 6. Although their data for individual fields appear to be accurate, the comparison between various fields shows a probable range of figures rather than absolute ones. However, the half-lives

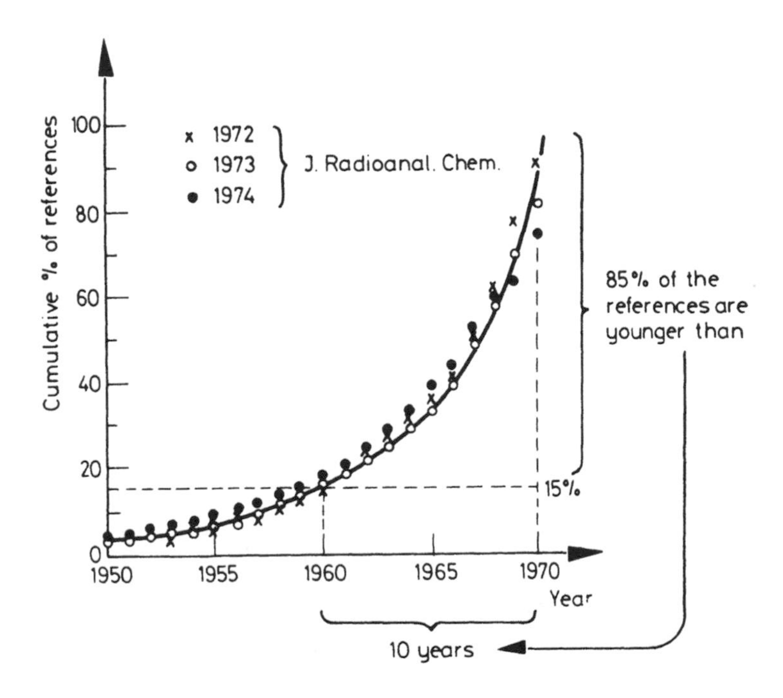

FIGURE 26. Decay of radioanalytical references. (From Braun, T., Lyon, W. S., and Bujdosó, E., *Anal. Chem.*, 49, 682A, 1977. With permission. Copyright 1977, American Chemical Society.)

in Table 6 can give reasonable indications of the sort of time scale in different subject areas.

Price[42] has called the over-citation of recent literature the "immediacy effect" in scientific literature. He proposed that the immediacy effect reflects the existence of two types of literature — the ephemeral and the classical. The former type is only cited for a limited time after which its results become in some way superseded, and it sinks into oblivion. The classical paper, on the other hand, retains its value, and therefore its citability, over a much longer period. The immediacy effect reflects the extent to which the ephemeral literature is being cited. Price also divides the literature into "research front" and "archival" papers. The former are simply the papers published during the period covered by the immediacy effect, while the latter were published prior to this time. Under this definition, the research front includes both ephemeral and classical papers. This division implies that separate probabilities of being cited apply to research front and archival papers. Research front papers are obviously cited relatively more heavily than archival. The probability of an archival paper being cited in a given year is somewhat over one half, whereas the probability for research front papers is 5:3.

"Research front" papers form a tighter citation linkage, characteristic of new analytical material in the process of absorption into analytical knowledge (and distinct from the older archival material that has already been assimilated). Price has suggested that after about every 50 original contributions, it becomes necessary to write a review paper so as to package the accumulated amount of research into a digestible form. Correspondingly, the research front material is at this point absorbed from the "skin" of the analytical subject and incorporated into the main body.[43]

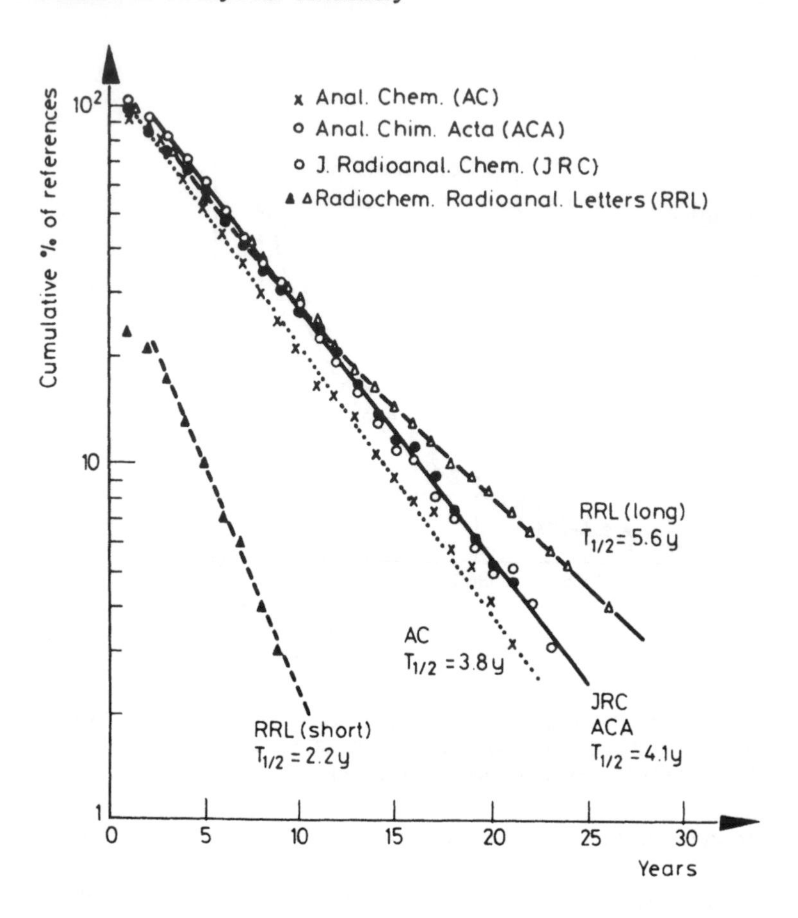

FIGURE 27. Component-resolved decay curve of radioanalytical references. (From Braun, T., Lyon, W. S., and Bujdosó, E., *Anal. Chem.*, 49, 682A, 1977. With permission. Copyright 1977, American Chemical Society.)

Table 6
THE HALF-LIVES OF SOME MAJOR SUBJECT DISCIPLINES

Subject	Half-life (years)
Metallurgical engineering	3.9
Physics	4.6
Chemical engineering	4.8
Mechanical engineering	5.2
Physiology	7.2
Chemistry	8.1
Botany	10.0
Mathematics	10.5
Geology	11.8

Reprinted, by permission of John Wiley & Sons, Inc., from Burton, R. E. and Kebler, R. W., *Am. Doc.*, 11, 18, 1960. Copyright © 1960 by the American Society for Information Science.

Chapter 5

DISTRIBUTION OF THE LITERATURE ON ANALYTICAL CHEMISTRY

I. INTRODUCTION

In Chapter 3 the publications dealing with the statistical analysis of the growth of analytical subject literature were reviewed. In this chapter the distribution of literature according to various aspects will be subjected to similar formal statistical analysis with respect to countries, language, subfields, topics, and techniques.

II. WITH RESPECT TO COUNTRIES

The distribution of the worldwide body of literature on analytical chemistry throughout various countries has been analyzed by Boig and Howerton,[44] Fisher,[10,11] and Brooks and Smythe.[12] Their main results are collected in Tables 7 to 9. The tables show sometimes significant differences among the data of different authors. Nevertheless, from the measurements a general view can be obtained on the publication efforts in analytical chemical research in various countries. It should be stressed that all the data on literature distribution by countries are comparative rather than absolute. Thus, a decline in German analytical chemistry means a decline in the fraction of worldwide analytical chemical literature that is German and not an absolute decline in the amount of German analytical chemical research or publication. The tables show that at the end of the last century, one half or more of all analytical chemistry publications were German. Second to Germany were France and the U.K. As one moves through the 20th century, the positions change significantly. Immediately following World War II, the prime country was the U.S., while Germany showed the massive effect of the war. The data also show a steady postwar rise in Soviet publications and a decline in the U.S. position. The relative German contribution to analytical chemical publication today is a pale shadow of its level 80 years ago. It is also striking that the position of the U.S.S.R. superseded that of the U.S. The French contribution by 1970 was remarkably low as compared to its position prior to World War I. The above data pertain to the total subject literature of analytical chemistry. It would not be irrelevant to have similar results on the literature of various analytical subfields. Unfortunately, very few such results are available.

Futekov et al.[45] investigated the contributions of various countries to the literature on the analysis of selenium and tellurium. It is interesting to observe here that the ranking of countries follows with almost complete fidelity the 1970 ranking concerning the total analytical literature, i.e., the U.S.S.R. is in first place followed by the U.S., Japan, Germany (East and West), England, and India. The ranks of the various countries in the literature of instrumental and noninstrumental methods used for the analysis of selenium and tellurium were also set up by the above authors. They have shown that in the application of X-ray fluorescence and spectographic, radioanalytical, electroanalytical, and fluorimetric methods, the U.S.S.R. is in first place, and the U.S. occupies this place only in the literature on the application of atomic absorption.

A more recent study by the same major contributors shows that in the analytical investigation of selenium and tellurium, the picture did not change too much during the 1970 to 1981 period.[46]

Brooks and Smythe[47] investigated the distribution throughout various countries of the literature on atomic absorption in 1962 and 1970 (Table 10). As can be seen from

Table 7
PERCENTAGE OF ANALYTICAL WORK CARRIED OUT IN VARIOUS COUNTRIES

Country	1877		1887		1897		1907		1917		1927		1937		1947		1948		1949		1950	
	Rank	%	Rank	%	Rank	%	Rank	%	Rank	%	Rank	%	Rank	%	Rank	%	Rank	%	Rank	%	Rank	%
U.S.	6	1.66	5	4.56	4	8.82	3	10.74	1	41.25	2	15.89	3	17.42	1	26.43	1	25.22	1	23.07	1	25.56
U.S.S.R.	—	0.03	8	1.91	7	2.33	10	0.50	—	0.63	7	3.01	1	25.00	2	11.76	4	9.22	2	22.69	2	20.18
England	3	12.96	2	12.06	3	10.25	4	8.08	3	13.44	4	10.82	4	6.23	3	10.92	3	11.48	3	9.62	3	10.18
France	2	17.28	3	10.44	2	18.94	2	17.99	4	8.75	3	11.37	5	6.16	4	10.61	2	13.11	6	7.58	4	6.61
Germany	1	58.14	1	57.80	1	48.00	1	49.01	2	21.88	1	36.84	2	19.05	6	5.31	5	7.50	4	7.71	5	6.43
The Netherlands	—	—	—	0.29	9	1.04	6	3.08	5	5.31	5	3.56	—	0.99	8	2.39	6	5.88	5	7.65	6	5.50
Czechoslovakia	—	—	7	1.91	—	0.39	9	0.50	—	—	12	0.96	9	1.70	—	0.31	12	1.63	13	1.27	7	3.51
Austria	4	5.32	6	3.82	6	4.02	7	1.92	—	0.94	9	2.33	6	5.24	12	0.21	9	3.44	9	2.04	8	2.98
Japan	—	—	—	—	—	—	—	0.08	7	18.75	11	1.64	7	2.83	5	7.60	8	4.34	12	1.47	9	2.92
Spain	—	—	—	—	—	—	—	—	—	0.31	17	0.55	—	0.71	7	4.68	7	5.06	7	3.63	10	2.69
Belgium	—	0.33	—	0.29	8	1.04	8	1.25	—	—	8	2.74	—	1.13	15	1.52	16	0.90	—	0.76	11	2.11
India	—	—	—	—	—	—	—	—	6	1.88	—	—	—	1.06	16	1.25	11	1.99	11	1.47	12	1.81
Italy	5	1.99	4	4.71	5	4.41	5	5.41	—	—	6	3.01	8	2.05	9	2.39	13	1.45	8	3.00	13	1.34
Argentina	—	—	—	—	—	—	—	—	—	—	13	0.82	—	0.64	14	1.46	10	2.17	10	1.91	14	1.16
Denmark	—	—	—	—	—	—	—	—	—	—	—	—	—	0.07	—	0.52	—	0.36	—	0.06	15	1.05
Canada	—	—	—	—	—	—	—	0.08	—	—	—	0.27	—	0.28	—	1.14	17	0.72	—	0.44	16	0.99
Sweden	—	0.33	—	0.15	—	—	—	—	—	0.63	15	0.68	—	0.64	10	2.19	14	1.27	15	0.88	17	0.70
Switzerland	—	0.33	9	1.03	—	0.39	11	0.42	—	0.63	10	1.92	—	1.06	11	2.19	15	1.27	14	1.21	18	0.58

From Boig, F. S. and Howerton, P. W., *Science*, 115, 555, 1952. With permission.

Table 8
PERCENTAGE OF ANALYTICAL WORK CARRIED OUT IN VARIOUS COUNTRIES

	Percentage of total				Percentage of total		
Country	1946	1955[a]	1965[b]	Country	1946	1955[a]	1965[b]
U.S.S.R.	12.3	9.1	21.7	Israel	—	0.4	0.4
U.S.	41.6	23.8	0.2	Holland	—	0.1	0.3
Germany	0.2	10.2	0.0	Chile	0.2	0.6	0.2
Japan	0.2	12.3	6.8	Greece	0.2	0.1	0.2
Poland	—	0.6	5.1	Scotland	—	0.3	0.2
				Portugal	—	0.1	0.2
Great Britain	14.6	7.4	4.3				
France	8.0	4.7	4.2	Brazil	0.9	0.2	0.1
Czechoslovakia	0.2	3.7	3.6	Egypt	0.1	0.4	0.1
Italy	2.5	5.2	3.0	Korea	—	—	0.1
Hungary	0.1	1.1	2.6				
India	1.7	3.9	2.5	Norway	0.2	0.4	0.1
China	0.1	0.1	1.3	Peru	0.1	0.2	0.1
Austria	—	1.7	1.2	Puerto Rico	0.1	—	0.1
Canada	1.1	2.2	1.2	South Africa	0.2	0.3	0.1
Spain	2.2	1.7	1.2	Turkey	—	—	0.1
Australia	1.4	0.6	1.1	Wales	—	—	0.1
				Argentina	0.9	0.9	0.04
Bulgaria	—	—	1.1				
Romania	0.2	—	1.1	Ceylon	—	—	0.04
Switzerland	2.4	1.2	1.1				
Belgium	1.0	1.3	0.9				
				Lebanon	—	0.1	0.04
Sweden	3.3	1.7	0.9				
The Netherlands	1.8	1.0	0.6	Lichtenstein	—	—	0.04
Yugoslavia	0.2	0.5	0.5	Mexico	0.1	0.1	0.04
Denmark	0.8	0.6	0.4	United Arab Republic	—	—	0.04
Finland	0.3	0.6	0.4	Venezuela	—	—	0.04

[a] Eight additional countries were represented in the 1955 survey, each at 0.1% of the total.
[b] Based on 2261 entries.

From Fischer, R. B., *Anal. Chem.*, 37(13), 27A, 1965. With permission. Copyright 1965, American Chemical Society.

the table, the leading role is held by the U.S., and its advantage has been increasing. The progress of England, Czechoslovakia, and Canada can also be considered significant as well as the decline of Australia, Germany, and particularly South Africa.

III. WITH RESPECT TO LANGUAGE

The distribution of the world literature of analytical chemistry with respect to languages has been investigated in both many older and more recent papers. It was noted by Strong[9] in 1947 that "there is much discussion in educational circles today on the relative importance to scientists of the various languages, especially Russian." His results, pertaining to 1946 and calculated on the basis of *Chemical Abstracts*, can be seen in Table 11. The contention that knowledge of the Russian language is and will be important to analytical chemists is born out in the table. Russian papers are next to English in frequency, though one sixth their number; French papers are a close third.

In 1952 Boig and Hoverton[44] gave a much deeper analysis of the language of analyt-

Table 9
PERCENTAGE OF ANALYTICAL WORK CARRIED OUT IN VARIOUS COUNTRIES

Country	1910	1915	1920	1925	1930	1935	1940	1945	1950	1955	1960	1965	1970
U.S.S.R.	1.0	—	—	0.8	5.7	29.4	30.8	18.2	17.8	13.0	22.9	25.4	28.4
U.S.	28.9	30.4	25.3	13.4	18.8	14.6	25.0	48.3	19.9	18.0	20.7	15.8	17.7
Japan	1.0	1.0	—	3.2	2.6	3.1	2.9	—	5.0	12.3	7.7	11.0	7.7
Germany[a]	31.9	30.4	19.9	39.7	26.3	16.4	10.5	2.5	6.7	12.1	4.8	6.4	6.1
U.K.	17.6	20.2	12.3	11.9	10.5	6.4	7.1	12.4	12.0	8.2	6.0	4.3	5.9
Czechoslovakia	—	—	—	4.8	3.5	2.8	1.7	—	6.0	8.1	3.8	5.3	5.6
France	10.3	4.5	21.0	7.1	14.5	7.6	3.8	2.5	9.2	4.7	3.1	3.5	2.6
India	—	—	—	—	1.3	1.8	0.8	2.1	0.6	4.3	5.0	3.5	2.6
Scandinavia	—	1.1	1.8	3.9	1.8	0.8	2.1	2.6	3.3	2.4	1.0	0.7	2.1
Romania	—	2.3	3.5	4.8	—	0.8	0.4	—	—	0.8	2.0	3.5	2.0
Poland	—	—	—	—	2.2	2.0	—	—	—	1.6	1.5	4.1	1.8
Spain	2.1	1.1	1.8	0.8	2.2	2.1	2.5	1.6	4.2	1.8	1.7	1.8	1.5
The Netherlands	—	3.4	8.8	3.2	1.3	1.5	2.9	0.5	0.8	0.8	0.8	0.8	1.3
Italy	—	1.1	—	4.0	1.8	4.1	2.5	—	2.3	4.2	2.5	1.7	1.0
China	—	—	—	—	—	2.6	—	—	0.8	—	5.6	3.1	—
Rest of the world	7.2	4.4	5.1	2.4	7.7	4.0	7.0	9.3	11.4	7.7	10.9	9.1	11.1

[a] Both East and West Germany.

From Brooks, R. R. and Smythe, I. E., *Talanta*, 22, 495, 1975. With permission.

Table 10
COUNTRIES IN WHICH ATOMIC ABSORPTION RESEARCH WAS CARRIED OUT

Country	Percentage of atomic absorption papers	
	1962	1970
U.S.	35.5	43.8
U.K.	3.4	14.0
U.S.S.R.	7.0	5.6
Czechoslovakia	1.7	5.5
Canada	1.7	5.1
Australia	15.8	4.9
Japan	7.0	4.5
France	1.7	4.3
Germany[a]	10.5	3.0
South Africa	14.0	0.3
Others	1.7	9.0

[a] Both East and West Germany.

From Brooks, R. R. and Smythe, L. E., *Anal. Chim. Acta*, 74, 35, 1975. With permission.

ical papers. Their results can be found in Table 12. The table shows that German was the most important language until World War I, at which time it lost the lead to English. However, German led once more by 1927, only to lose first place again to English by 1937. English has been the most important language ever since. Russian, second in 1952, has become the leading foreign language. French has usually been second or

Table 11
LANGUAGE IN WHICH ANALYTICAL PAPERS WERE PUBLISHED IN 1946

Language	Number of papers	Percent of papers
English	870	66.2
Russian	153	11.6
French	137	10.4
Spanish	57	4.3
Italian	31	2.4
German	23	1.7
Swedish	19	1.4
Dutch (Flemish)	16	1.2
Danish	3	0.2
Japanese	2	0.2
Norwegian	1	0.1
Czech	1	0.1
Yugoslavian	1	0.1
Total	1,314	99.9

From Strong, F. C., *Anal. Chem.*, 19, 968, 1947. With permission. Copyright 1947, American Chemical Society.

third, except in 1937 and 1949, when it was in fourth place. Spanish was fifth in 1952 followed in order by Japanese, Czech, Italian (fourth in 1877), Dutch, and Portuguese.

Not in complete numerical agreement but similar in conclusion are the results reported by Fisher[10,11] and Brooks and Smythe.[12] The results of the latter authors are shown in Table 13. The order of importance of the various major languages used in analytical chemistry in 1970 was English, Russian, German, Japanese, Czech, French, and Spanish. This is almost the same as the order pertaining to chemistry literature as a whole, i.e., English, Russian, German, French, Japanese, Polish, and Italian. In 1975, 95.7% of the worldwide body of chemical literature was published in these 7 languages.[19]

IV. WITH RESPECT TO SUBFIELDS, TOPICS, AND TECHNIQUES

A convenient, objective indication of the future practical importance of the various analytical subfields, topics, and techniques is considered obtainable through a statistical analysis of the research literature.

In order to prove the validity of this statement, we should know the distribution of analytical literature among the various forms of publication. The investigations of Earle and Vickery[48] indicated, as can be seen in Table 14, that about 80% of the literature in the whole field of science is represented by papers in periodicals, the relative ratio decreasing proportionally towards technology and the so-called soft sciences. In our knowledge, no such investigation was carried out on the whole field of analytical chemistry, but Braun and Bujdosó[49] have shown that about 75% of the literature of radioanalytical chemistry is represented by papers published in periodicals. On the other hand, Orient[22] disclosed some data indicating that the share of nonperiodical publications in analytical literature shows an increasing tendency between 1946 and 1972 (Figure 28). We are inclined to state that the distribution of the whole of analytical literature is reflected better by the figure given by Braun and Bujdosó, because the data of Orient is distorted by the effect of a peculiar situation. Namely, in the

Table 12
LANGUAGES IN WHICH ANALYTICAL PAPERS WERE PUBLISHED, 1877 TO 1950

Language	1877		1887		1897		1907		1917		1927		1937		1947		1948		1949		1950	
	Rank	%	Rank	%	Rank	%	Rank	%	Rank	%	Rank	%	Rank	%	Rank	%	Rank	%	Rank	%	Rank	%
English	3	14.95	2	16.62	3	19.07	2	19.27	1	55.9	2	29.32	1	28.82	1	43.70	1	44.39	1	39.77	1	43.9
Russian	—	—	6	0.74	5	2.33	6	0.5	8	0.06	6	2.33	2	24.36	3	11.34	4	9.22	2	22.69	2	20.18
French	2	17.61	3	11.47	2	19.97	3	19.23	3	9.06	3	14.11	4	8.92	2	14.67	2	15.82	4	11.15	3	10.70
German	1	64.12	1	64.70	1	53.05	1	51.87	2	23.43	1	41.37	3	24.22	4	0.83	3	12.38	3	11.34	4	10.64
Spanish	—	—	—	—	—	—	—	—	6	1.88	7	2.19	6	0.23	6	6.45	5	7.50	5	5.74	5	4.44
Japanese	—	—	—	—	—	—	—	—	7	0.94	9	0.82	5	2.41	5	6.97	6	4.25	7	1.27	6	2.80
Czech	—	—	5	1.62	—	—	—	—	—	—	8	0.96	8	1.20	—	0.21	10	0.81	12	0.57	7	2.75
Italian	4	—	4	4.71	4	4.41	4	5.41	5	1.88	5	3.01	7	2.05	7	2.39	7	1.45	6	3.00	8	1.40
Dutch	—	1.99	—	—	6	1.04	5	3.08	4	5.31	4	3.56	10	0.71	10	0.83	8	1.36	8	0.89	9	1.17
Portuguese	—	—	—	—	—	—	—	—	10	0.31	14	0.03	14	0.42	8	2.08	11	0.63	13	0.57	10	0.64
Yugoslavian (Croatian)	—	—	—	—	—	—	—	—	—	—	—	—	—	—	—	—	—	—	—	—	11	0.29
Swedish	—	—	—	—	—	—	—	—	9	0.06	10	0.07	12	0.5	9	1.66	9	1.18	9	0.83	12	0.29
Danish	—	—	—	—	—	—	—	—	—	—	—	—	—	0.07	—	0.21	12	0.27	—	—	13	0.18
Hungarian	—	—	—	—	—	—	—	—	—	—	—	—	9	1.06	11	0.62	13	0.27	11	0.70	14	0.18
Bulgarian	—	—	—	—	—	—	—	—	—	—	—	—	17	0.21	—	—	—	—	—	—	15	0.12
Polish	—	—	—	—	—	—	—	—	—	—	13	0.03	11	0.50	—	—	14	0.27	10	0.76	16	—

From Boig, F. S. and Howerton, P. W., *Science*, 115, 555, 1952. With permission.

Table 13
LANGUAGES IN WHICH ANALYTICAL PAPERS WERE PUBLISHED, 1910 TO 1970

Language	1910	1915	1920	1925	1930	1935	1940	1945	1950	1955	1960	1965	1970
English	50.6	51.7	40.4	26.1	32.8	23.8	34.1	64.9	35.4	32.3	34.2	24.8	30.3
Russian	1.0	—	—	0.8	5.7	29.4	30.8	18.2	17.8	13.0	22.9	25.4	28.4
German	32.9	31.5	21.7	39.7	27.6	16.7	12.6	3.0	9.2	14.0	6.7	9.8	8.1
Japanese	1.0	1.1	—	3.2	2.6	3.1	2.9	—	5.0	12.3	7.7	11.0	7.7
Czech	—	—	—	4.8	3.5	2.8	1.7	—	6.0	8.1	3.8	5.3	5.6
French	12.4	5.6	21.0	7.9	16.7	8.4	4.2	2.5	11.7	5.2	3.5	4.2	3.6
Spanish	2.1	2.2	2.3	1.6	4.0	2.9	5.4	7.8	5.9	3.1	3.7	1.8	2.6
Scandinavian	—	1.1	1.8	3.9	1.8	0.8	2.1	2.6	3.3	2.4	1.0	0.7	2.1
Romanian	—	2.3	3.5	4.8	—	0.8	0.4	—	—	0.8	2.0	3.5	2.0
Polish	—	—	—	—	2.2	2.0	—	—	—	1.6	1.5	4.1	1.8
Hungarian	—	—	—	—	—	0.8	—	—	0.8	0.2	1.9	1.0	1.8
Dutch	—	3.4	8.8	3.2	1.3	1.5	2.9	0.5	0.8	0.8	0.8	0.8	1.3
Italian	—	1.1	0.5	4.0	1.8	4.1	2.5	—	2.3	4.2	2.5	1.7	1.0
Chinese	—	—	—	—	—	2.6	—	—	0.8	—	5.6	3.1	—
Other	—	—	—	—	—	0.3	0.4	0.5	1.0	2.0	2.2	2.8	3.7

From Brooks, R. R. and Smythe, I. E., *Talanta*, 22, 495, 1975. With permission.

Table 14
DISTRIBUTION OF CITATIONS TO DIFFERENT TYPES OF PUBLICATIONS IN SCIENCE, TECHNOLOGY, AND SOCIAL SCIENCES

Type of publication	Science (%)	Technology (%)	Social sciences (%)
Books	12	14	46
Periodicals	82	70	29
Other (including reports)	6	16	25

From Earle, P. and Vickery, B. C., *ASLIB Proc.*, 21, 237, 1969. With permission.

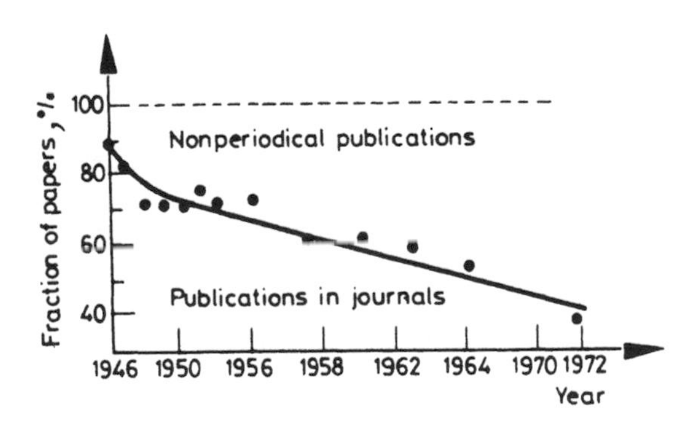

FIGURE 28. Distribution of analytical publications between periodicals and nonperiodicals. (From Orient, I. M., (in Russian), *Zavodsk. Lab.*, 41(9), 1071, 1975; (English transl.), *Ind. Lab. USSR*, 41, 1327, 1975. With permission.)

U.S.S.R., particularly in the field of analytical chemistry, a sudden recent increase can be witnessed in the forms of publication such as "Obzors" and "Sborniks". According to the data of Orient,[51] 62% of the papers published in the U.S.S.R. in 1972 appeared in such publication forms. It must be taken into account, as noticed by Orient, that in the international literature very few references are made to these papers, since they represent only a channel of information within the country. Incidentally, the paper in which Orient discusses this problem was published also in a "Sbornik".[51]

The specific weight of various subfields within the literature of analytical chemistry has been investigated for 1946, 1955, and 1965 by Fisher[11] using *Chemical Abstracts* as a data base (Table 15).

The same was investigated by Berezkin and Chernysheva[52,53] for the years 1965, 1970, 1975, and 1980 (Table 16). The difference between the two sets of data arises probably from the fact that the latter authors carried out their measurements on the basis of *Chemical Abstracts, Analytical Abstracts,* and *Journal of Chromatography.*

On the basis of a third data base, the 1970 volumes of *Referativnyj Zhurnal Khimiya,* calculations were done by Orient.[54] Her results are given in Table 17. Berezkin and Chernysheva[52,53] dealt also with the distribution of literature pertaining to the analysis of organic compounds according to various points of view. Their results are shown in Tables 18 to 21. It is interesting to compare these data with the results of Beyermann[55] calculated on the basis of the 1976 volumes of *Analytical Abstracts* (Tables 22 to 24). Without any detailed evaluation of the data, it can be stated safely that the most striking relative progress is still shown by the literature of chromatography. This is also proved by Petruzzi,[56] who states that " . . . if the health of a scientific specialty can be measured by the number of journals and meetings on the subject, then we can be sure that chromatography is robust. The growth of chromatographic separation techniques is also attested by other measures such as current instrument sales and projections based on studies of the growth areas of analytical instrumentation."

Beyermann,[55] on the basis of the 1976 volumes of *Analytical Abstracts,* also investigated the distribution of analytical literature among the analysis of organic and inorganic compounds. His results show that the distribution of publications on analytical chemistry in 1976 was as follows: organic analysis, 57%; inorganic analysis, 31%; and analytical techniques, 12%. From his data the author points out that although there are about 10^3 times more known organic compounds than inorganic ones, the literature dealing with the analysis of organic compounds is slightly less than double that of inorganic compounds.

Brooks and Smythe[12] were concerned with the distribution of analytical literature among the analysis of various matrices, as was their investigation of the 1955 to 1970 volumes of *Analytical Abstracts.* Their results are shown in Figure 29. As the authors state, " . . . there is an appreciable increase in the relative proportion of the literature of biochemical analysis and technical apparatus, and a smaller increase in analysis of water, air and effluent. These trends appear to be at the expense of inorganic and organic analysis which both show significant downward trends except for the last three years of inorganic analysis. These trends were not unexpected, because of the phenomenal rise of biochemistry during the past 15 years."

Kabanova and Kurilina[57] investigated the distribution of papers on the electroanalytical methods of inorganic compounds in the 1955 to 1973 period on the basis of *Referativnyj Zhurnal Khimiya.* Part of their results are shown in Figure 30. The number of papers belonging to total electroanalytical chemistry, coulometry, and conductometry increases, monotonously disregarding the maximum in the number of conductometric papers in 1971. Between 1960 and 1965, papers on amperometry and potentiometry dominated. From 1970 to 1971 the number of papers on potentiometry shows a maximum, due to the first avalanche of papers concerning ion-selective electrodes. The sharp rise of voltammetric papers starts in 1965, and reaches a maximum

Table 15
DISTRIBUTION OF ANALYTICAL CHEMISTRY LITERATURE ACCORDING TO SUBFIELDS

Method	Percent of total		
	1946	1955	1965[a]
Optical methods	36.7	43.2	40.5
Colorimetry	23.0	20.2	15.2
Spectrophotometry	5.7	15.5	13.3
Visible	[b]	5.9	7.2
UV	[b]	5.8	3.6
IR	[b]	2.8	2.4
Raman	[b]	0.6	0.1
Other	[b]	0.3	—
Emission	5.3	4.2	6.7
Arc and spark	[b]	2.9	4.3
Flame	[b]	1.3	2.4
Fluorescence	1.6	1.3	3.4
Light scattering	1.1	2.0	0.9
Refractometry	[c]	[c]	0.6
Optical activity	[c]	[c]	0.4
Titrimetry	25.6	22.0	12.6
Visual	[b]	17.2	8.1
Potentiometric	[b]	3.2	2.7
Amperometric	[b]	0.9	1.0
Conductometric	[b]	0.3	0.3
Other titrimetry	[b]	0.4	0.5
Gas analysis	1.4	2.3	12.0
Gas chromatography	—	[c]	9.5
Other gas analysis	1.4	2.3	2.5
Electrical methods	4.4	6.2	10.2
Polarography	4.0	4.8	7.3
Coulometry	[c]	[c]	1.1
Potentiometry	[c]	[c]	1.1
Electrical conductivity	0.4	1.4	0.5
Chronopotentiometry	—	[c]	0.2
Radioactivity	1.1	2.0	7.4
Gravimetric	8.5	6.5	3.6
X-ray methods	0.6	3.3	3.4
Fluorescence	[b]	0.6	2.2
Diffraction	[b]	2.3	0.8
Absorption	[b]	0.3	0.4
Biological assay	3.6	2.8	2.7
Thermal analysis	[c]	[c]	1.9
Mass spectrometry	[c]	[c]	1.4
Kinetic methods	[c]	[c]	1.1
Electron probe	—	—	0.8
Microscopy	0.5	1.6	0.6
NMR and ESR	—	—	0.4
Data handling	[c]	[c]	0.4
All other methods	17.8	9.5	1.0

[a] Based on 2052 entries.
[b] Not available as separate data, but included in appropriate group totals.
[c] Not available as separate data, but probably very small and included in "All other methods" category.

From Fischer, R. B., *Anal. Chem.*, 37(13), 27A, 1965. With permission.

Table 16
DISTRIBUTION OF PAPERS AMONG VARIOUS SUBFIELDS OF ANALYTICAL CHEMISTRY

Subfield	1965 (%)	1970 (%)	1975 (%)	1980 (%)
Chromatography	24	30	27	36
Gas chromatography	8	11	9	13
Spectroscopy (ESR, NMR, and photometry)	36	37	36	40
Electroanalytical chemistry	10	13	20	12
Gravimetry, titrimetry, etc.	30	20	17	12

From Berezkin, V. G. and Chernysheva, T. Yu., *J. Chromatogr.*, 141, 241, 1977. With permission.

Table 17
DISTRIBUTION OF PAPERS AMONG VARIOUS SUBFIELDS OF ANALYTICAL CHEMISTRY IN 1970

Subfield	%
Photometry	15.7
Fluorescence and luminescence	4.0
Spectral	8.7
Radioactivation	7.5
Electroanalysis	12.0
Polarography	6.6
Potentiometry	2.4
Coulometry	1.0
Conductometry	0.2
Others	1.8

From Orient, I. M., *Trends in the Logics of Development and Scientometrics in Chemistry* (in Russian), Kabanov, V. A., Ed., Moscow State University Press, Moscow, 1976, 90. With permission.

Table 18
DISTRIBUTION OF PAPERS ON ANALYTICAL CHEMISTRY OF ORGANIC COMPOUNDS ACCORDING TO METHOD USED

Subfield	1965 (%)	1970 (%)	1975 (%)	1980 (%)
Chromatography	39	48	44	50
Gas chromatography	14	17	15	15
Spectroscopy (ESR, NMR, and photometry)	29	29	31	32
Electroanalytical chemistry	17	16	18	11
Gravimetry, titrimetry	15	8	7	

From Berezkin, V. G. and Chernysheva, T. Yu., *J. Chromatogr.*, 141, 241, 1977. With permission.

Table 19
DISTRIBUTION OF PAPERS ON CHROMATOGRAPHIC ANALYSIS OF ORGANIC COMPOUNDS

Chromatography subfield	1970 (%)	1975 (%)	1980 (%)
Gas	29	25	20
Paper	13	7	3
Thin-layer	32	28	25
Liquid column	26	39	52

From Berezkin, V. G. and Chernysheva, T. Yu., *J. Chromatogr.*, 141, 241, 1977. With permission.

Table 20
DISTRIBUTION OF PUBLICATIONS (%) AMONG INDIVIDUAL ASPECTS OF EXPERIMENTAL TECHNIQUES IN THE CHROMATOGRAPHIC ANALYSIS OF ORGANIC COMPOUNDS

	Gas chromatography			Liquid (column) chromatography		Thin-layer chromatography	
Technique	1970	1975	1980	1975	1980	1975	1980
Detectors	21	19	21	16	18	4	5
Column efficiency and preparation	17	22	31	23	30	—	16
Apparatus and materials	22	24	24	14	11	23	26
Determination of physicochemical characteristics	13	16	7	8	6	—	2
Other experimental aspects	17	10	7	15	10	53	32
Automation		9	10	24	25	20	19

From Berezkin, V. G. and Chernysheva, T. Yu., *J. Chromatogr.*, 141, 241, 1977; Berezkin, V. G., Chernysheva, T. Yu., and Bolotov, S. L., *J. Chromatogr.*, 251, 227, 1982. With permission.

in the 1968 to 1971 period. In constructing this curve, the authors also took into account papers on mercury electrode inversion voltammetry, but disregarded papers concerning dropping mercury electrodes. In 1968 the number of the latter was higher than that of potentiometric papers (about 40%). Figure 31 shows the share of individual methods within the literature of electroanalytical chemistry. As can be seen, maximum share is taken by potentiometry (30% in 1955 and 40% between 1970 and 1973). About the same is the share of amperometry between 1955 and 1965 (20 to 25%), but this decreases to 10% of all papers in the 1967 to 1972 period. The contributions of conductometry, coulometry, and electrogravimetry were 10 to 15% between 1955 and 1965, but this has decreased to 3 to 10% in the past years. The number of papers on theoretically new methods was very small in the period studied (1 to 4%). Such methods were, e.g., the analytical techniques based on the measurement of impedance or dielectric permeability. More recent data on electroanalytical literature are presented in Table 25.

Two papers were concerned with the scientometric analysis of the literature on atomic absorption methods in the past few years.[30,47] Brooks and Smythe,[47] on the basis of the *Atomic Absorption Newsletter* as data base, investigated the distribution of literature according to the matrix analyzed (Table 26) and according to the various atomic absorption techniques (Table 27).

Table 21
DISTRIBUTION OF PAPERS (%) ON CHROMATOGRAPHIC ANALYSIS OF ORGANIC COMPOUNDS

	Gas chromatography			Liquid (column) chromatography			Paper chromatography			Thin-layer chromatography		
Compound	1970	1975	1980	1970	1975	1980	1970	1975	1980	1970	1975	1980
Aliphatic and aromatic hydrocarbons, polymer synthesis products	19	13	18	5	2	9	6	2	7	10	3	11
Oxygenated compounds (phenols, carbohydrates, organic acids, alcohol, etc.)	28	19	11	14	13	12	24	24	29	28	28	26
Biologically active compounds and drugs	8	25	26	4	5	18	14	13	13	16	20	20
Nitrogenous compounds	9	13	11	67	64	48	36	34	24	22	21	27
Insecticides and pesticides	12	6	5	1	2	2	2	3	—	6	12	7
Other	24	24	29	9	14	11	19	2	25	18	16	9

From Berezkin, V. G. and Chernysheva, T. Yu., *J. Chromatogr.*, 141, 241, 1977; Berezkin, V. G., Chernysheva, T. Yu., and Bolotov, S. L., *J. Chromatogr.*, 251, 227, 1982. With permission.

Table 22
DISTRIBUTION OF PAPERS ON ANALYSIS OF ORGANIC COMPOUNDS

Type of matrix	1976 (%)	Publications dealing with traces (%)
Biochemical	20	4
Industrial organic chemicals	13	1
Air and water	8	1
Pharmaceuticals	7	1
Food	6	1
Agricultural	3	1

From Beyermann, K., *Pure Appl. Chem.*, 50, 87, 1978. With permission.

Table 23
DISTRIBUTION OF PAPERS ON SEPARATION IN ORGANIC TRACE ANALYSIS

Method	1976 (%)
Liquid-liquid extraction	33
Gas chromatography	29
Thin-layer chromatography	18
Chromatography	16
HPLC	6
Ion-exchange chromatography	3
Other	4

From Beyermann, K., *Pure Appl. Chem.*, 50, 87, 1978. With permission.

Table 24
DISTRIBUTION OF PAPERS ON DETERMINATION OF TRACES OF ORGANIC COMPOUNDS

Method	1976 (%)	Sensitivity
Gas chromatography	30	pg
UV spectrometry	20	μg
Fluorescence spectrometry	19	μg
Gas chromatography and mass spectrometry	14	μg
Thin-layer scanning methods	9	μg
Radioimmunoassay	5	pg
Electroanalytical	3	μg

From Beyermann, K., *Pure Appl. Chem.*, 50, 87, 1978. With permission.

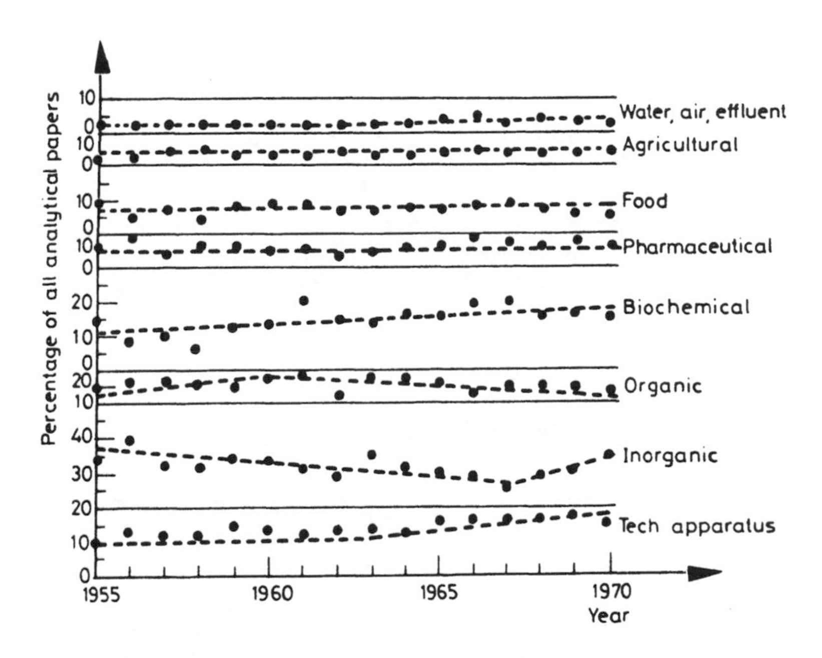

FIGURE 29. Broad trends of the literature of analytical chemistry for eight subfields for the period 1955 to 1970. (From Brooks, R. R. and Smythe, I. E., *Talanta*, 22, 495, 1975. With permission.)

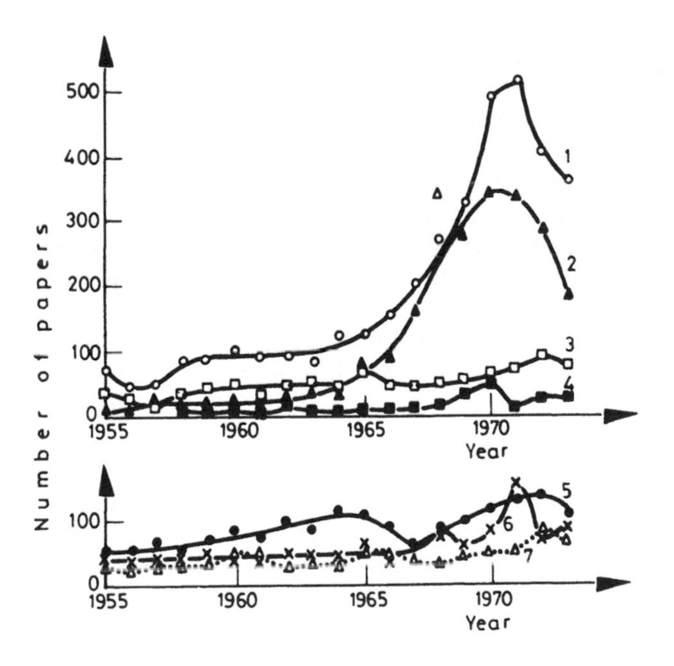

FIGURE 30. Yearly output of electroanalytical papers for the period 1955 to 1976. Curves: 1, potentiometry; 2, voltammetry; 3, coulometry; 4, new electroanalytical methods; 5, amperometry; 6, conductometry; 7, electroanalysis. (From Kabanova, O. L. and Kurilina, N. A., (in Russian), *Zh. Anal. Khim.*, 30, 2432, 1975. With permission.)

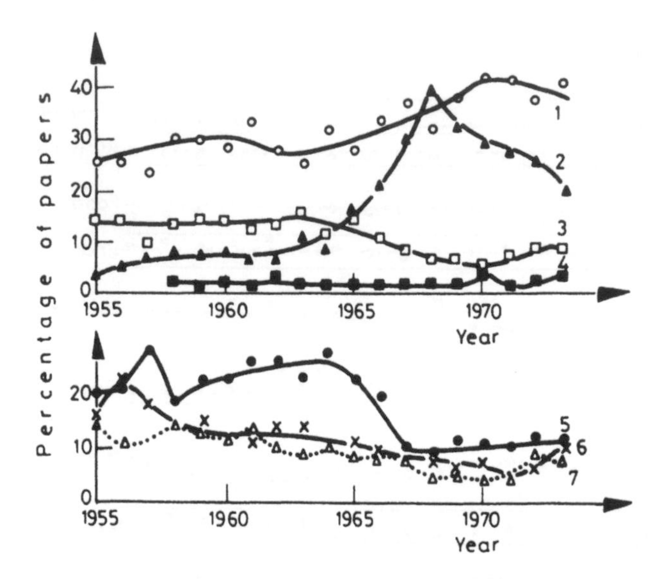

FIGURE 31. Distribution of electroanalytical papers for the period 1955 to 1973. Curves: 1, potentiometry; 2, voltammetry; 3, coulometry; 4, new electroanalytical methods; 5, amperometry; 6, conductometry; 7, electroanalysis. (From Kabanova, O. L. and Kurilina, N. A., (in Russian), *Zh. Anal. Khim.*, 30, 2432, 1975. With permission.)

It can be seen that in the period 1961 to 1971, nearly one half of the research effort was devoted to instrumental developments. All other categories except "Geochemical" and "Reviews" have maintained a relatively constant proportion during the period 1965 to 1971. The sharp increase in the "Geochemical" category after 1963 is almost certainly due to the worldwide mineral boom, when atomic absorption spectrometry afforded a ready, speedy, accurate, and inexpensive method for analyzing very large numbers of geochemical samples, i.e., soils, stream sediments, rocks, vegetation, and water. It is noteworthy that the number of reviews has fallen sharply and consistently since the euphoria of the early days. Altogether, a total of nearly 100 reviews have appeared in the period 1960 to 1971, one half of which appeared during the first 4 years. From Table 27 it is evident that the proportion of research into hollow-cathode lamps has dropped steadily from 17.8% in 1965 to 3.0% in 1971. This is probably because of diversion of effort to other sources, i.e., electrodeless discharge tubes, and because hollow cathodes have reached a degree of refinement which does not warrant such extensive work. Work on other "Sources" has also dropped from 17.8% in 1965 to 0.7% in 1971, perhaps because most other possible sources have been investigated and found to be less suitable than hollow cathodes or electrodeless discharge tubes. Developments in nebulization seem to have slackened, as the proportion of work in this field has dropped steadily since 1963. Perhaps the most interesting finding to emerge from the survey is the enormous decrease in work on new instruments — from 50.0% in 1961 to 2.6% in 1971. This trend confirms the coming of age of atomic absorption spectrometry, when basic instrument design has become more or less standardized and where further work is devoted to refinement rather than innovation. Table 27 also shows the increasing interest in nonflame excitation (carbon rod atomizer, etc.) and in flameless atomic absorption analysis of mercury. The large increase in "Other techniques" is almost entirely due to atomic fluorescence research (11.0% in 1969 and 13.2% in 1971).

Table 25
SUBJECT THEMES OF ELECTROANALYTICAL PAPERS ABSTRACTED IN *ANALYTICAL ABSTRACTS*

Theme	1969	1970	1979	1980
Potentiometry				
Ion-selective and membrane electrodes	34	52	95	120
Ion-selective field effect transistors	—	—	2	4
pH electrodes	13	14	16	12
Enzyme-based sensors	—	4	10	16
Bacteria-based sensors	—	—	2	5
Gas sensors (membrane and air-gap)	—	—	—	—
General aspects of potentiometry	21	28	19	16
Subtotals	68	98	149	195
Polarography and voltammetry				
Polarography (AC, DC, etc.)	38	39	24	12
Square-wave, pulse, and differential-pulse polarography	1	4	8	10
Voltammetry (general)	10	9	12	10
Cyclic voltammetry	1	2	4	2
Amperometry	10	10	9	10
Electrode systems for voltammetry (not in above)	10	8	16	6
Subtotals	70	72	73	50
Electrometric stripping analysis				
Cathodic and anodic stripping	6	7	15	16
Potentiometric stripping	—	—	4	2
Subtotals	6	7	19	18
Other electroanalytical methods				
Conductometry and oscillometry	14	7	3	7
Chronopotentiometry	7	3	4	5
Coulometry	20	20	15	20
Electrolysis	2	6	2	5
Gas and vapor sensors (nonpotentiometric)	1	1	2	8
Spectroelectrochemistry	4	—	2	8
Miscellaneous (Karl Fischer titrations, potentiostats, amplifiers, etc.)	6	7	7	13
Subtotals	54	43	36	62
General aspects[a]				
Reference electrodes	4	4	2	1
Titrimeters	8	2	5	—
Computerization and use of microprocessors	2	2	9	11
Automation	8	6	5	8
Flow systems and flow-injection analysis	3	4	9	15
Chromatographic detectors	4	1	7	11
Applications in biochemistry and drugs (organic species)	4	11	30	25
Applications in food, agriculture, air, water, and effluents	2	5	17	20

Table 25 (continued)
SUBJECT THEMES OF ELECTROANALYTICAL PAPERS ABSTRACTED IN *ANALYTICAL ABSTRACTS*

Theme	1969	1970	1979	1980
Applications in surfactant analysis	—	—	3	4
Oxygen electrodes (various)	4	9	9	7
Subtotals	39	44	96	102
Overall totals	237	264	373	409

* Many more abstracts fit into the headings of this section than are shown in the "Electrochemistry" section of *Analytical Abstracts*.

From Thomas, J. D. R., *Anal. Proc.*, 19, 60, 1982. With permission.

Table 26
PERCENTAGE OF ATOMIC ABSORPTION PAPERS ANALYZING VARIOUS MATRICES

Matrix	1961	1963	1965	1967	1969	1971
Agricultural	3.6	1.8	3.3	4.9	4.9	5.4
Biological	17.9	10.9	19.8	17.9	17.9	12.9
Food	—	—	1.1	4.6	2.0	3.2
Geochemical	3.6	3.6	10.9	9.8	10.4	10.2
Industrial	3.6	5.5	9.9	8.4	9.3	9.7
Instrumental	25.1	49.1	39.6	44.2	44.6	47.9
Metallurgical	14.2	20.0	7.7	7.7	9.2	9.8
Reviews	32.0	9.1	7.7	2.5	1.7	0.9

From Brooks, R. R. and Smythe, L. E., *Anal. Chim. Acta,* 74, 35, 1975. With permission.

Table 27
ATOMIC ABSORPTION PAPERS IN VARIOUS INSTRUMENTAL CATEGORIES EXPRESSED AS A PERCENTAGE OF THE TOTAL INSTRUMENTAL TOPIC

Instrumental category	1961	1963	1965	1967	1969	1971
Automation	—	—	3.6	4.0	5.2	3.0
Burners and flames	—	29.1	7.1	15.9	9.7	10.1
Electrodeless discharge tubes	—	—	—	—	7.8	3.0
Flameless atomic absorption (Hg)	—	—	—	4.7	0.6	9.3
Hollow cathodes	—	16.7	17.8	8.7	4.5	3.0
Instruments	50.0	29.2	21.4	15.8	7.1	2.6
Nebulization	—	8.3	3.6	5.5	3.2	1.1
Nonflame excitation (carbon rod, etc.)	—	—	—	2.4	5.2	9.7
Sources (other than above)	—	—	17.8	4.0	1.9	0.7
Techniques and theory	17.0	12.5	10.9	29.5	15.9	7.5
Other techniques	33.0	4.2	17.8	9.5	38.8	50.0

From Brooks, R. R. and Smythe, L. E., *Anal. Chim. Acta,* 74, 35, 1975. With permission.

Table 28
DISTRIBUTION OF ATOMIC ABSORPTION ANALYSIS PAPERS ACCORDING TO THE ANALYZED MATRICES

Matrix	Percent of papers
Mineral raw materials, ores, and concentrates	29.9
Metals and alloys	16.2
Technological liquors and electrolytes	11.1
High purity materials and semiconductors	10.3
Organic compounds	8.6
Reagents	7.7
Inorganic materials	6.8
Waste and natural water	5.1
Soil and plants	4.3

From Orient, I. M., Artemova, O. A., and Davidova, S. L., (in Russian), *Zavodsk. Lab.*, 43, 419, 1977; (English transl.), *Ind. Lab. USSR*, 43(4), 498, 1977. With permission.

It is interesting to compare these data with the results of Orient et al.,[30] who investigated the development and distribution of the literature on atomic absorption analysis on the basis of another data base, the 1965 and 1970 to 1975 volumes of *Referativnyj Zhurnal Khimiya* (Table 28).

Orient and co-workers[30] also report how, according to their measurements, the literature in the average of the 1966 to 1975 volumes of *Atomic Absorption Newsletter* is distributed according to analyses applied in various fields. The results are biology and medical sciences, 26.2%; miscellaneous industrial branches, 26.1%; metallurgy, 14.7%; geology and mining industry, 14.4%; agriculture, 9.3%; and protection of the biosphere, 9.3%. According to the various elements determined, the papers showed the following distribution: Al, 14.01%; Ag, 10.63%; Zn, 6.76%; Mg, 5.80%; Cu and Pd, 5.31% each; Fe and Pt, 4.83% each; Ni and Pb, 4.35% each; Co, 3.38%; Cd, K, Na, and Rb, 2.90% each; Mn and Ru, 2.42% each; Ca and Ir, 1.93% each; V, 1.45%; Mo, Sn, and Hg, 0.97% each; and other elements, 5.80%.

The statistical distribution of the literature on the analysis of mineral raw materials was investigated by Volkova and Kontsova[59] using sections 19 G and D of the 1968 to 1973 volumes of *Referativnyj Zhurnal Khimiya* as a data base. Their results are shown in Table 29. As can be seen from the table, both in the U.S.S.R. and abroad the majority of papers in 1973 was concerned with photometric, spectrometric, electroanalytical, volumetric and complexometric, atomic absorption, activation analysis, and chromatographic methods. However, in the U.S.S.R. the photometric and spectrometric methods (44%) were most widespread whereas in the other countries atomic absorption, photometric, and chromatographic methods (43.8%) were most widespread. The other methods have contributions of about 5% each, of which the most popular ones are radioanalytical and X-ray spectrometric methods (6% abroad, 3.3% in the U.S.S.R.). In the fields of photometric, electroanalytical, and spectrometric methods the Russian and foreign publication activities were at about the same level. Abroad, the greatest number of photometric papers were published in 1969 and 1972; in the U.S.S.R., in 1970. This is due to the fact that at that time some new organic reagents were thoroughly investigated from analytical aspects. The peak in 1970 to 1971 can be attributed to the consequences of the intense electroanalytical research carried out in the 1960s (amalgam, pulse, and film polarography). Papers dealing with the classical analytical methods (gravimetry, volumetry, and complexometry) applied to the analy-

Table 29
DISTRIBUTION OF PAPERS ON ANALYSIS OF MINERAL RAW MATERIALS

	1968		1969		1970		1971		1972		1973	
Method	U.S.S.R.	Rest of the world	U.S.S.R.	Rest of the world	U.S.S.R.	Rest of the world	U.S.S.R.	Rest of the world	U.S.S.R.	Rest of the world	U.S.S.R.	Rest of the world
Photometry (absorption)	23.4	19.4	25.9	21.5	27.7	14.5	25.8	17.7	20.7	22.9	23.1	16.1
Spectrometry (emission)	25.0	13.1	24.3	8.6	18.2	12.9	13.3	9.4	21.4	4.7	20.9	7.9
Electroanalytical	14.3	7.5	10.8	8.2	14.3	10.6	14.5	9.3	9.3	8.1	9.6	7.7
Volumetric	3.1	2.8	2.8	2.8	3.9	3.4	2.3	2.6	3.5	5.3	5.0	4.2
Complexometric	5.2	3.7	3.5	5.2	3.3	4.5	2.7	3.7	3.0	4.6	3.2	3.6
Atomic absorption	1.3	7.6	2.5	8.3	2.8	11.2	4.1	9.9	7.3	12.5	5.6	16.5
Neutron activation	12.8	14.5	8.4	11.3	4.9	9.4	5.9	19.6	12.1	8.6	5.4	8.9
Chromatography	2.3	8.5	3.5	7.5	5.0	9.9	4.5	6.2	2.7	7.6	5.2	11.2
Nuclear methods	1.0	0.9	1.2	0.8	1.6	1.3	1.8	2.0	0.3	3.9	1.7	1.7
X-ray spectrometry	1.9	7.4	1.4	5.3	3.7	3.7	2.2	3.6	2.0	2.2	1.6	4.3
Flame photometry	2.2	2.2	2.3	1.8	3.1	1.9	3.5	1.2	1.2	1.3	1.4	0.8
Gravimetry	1.2	2.5	1.9	2.9	1.5	2.2	1.3	2.2	1.8	3.6	0.9	2.2
Radioanalytical	0.4	1.1	1.0	1.6	0.4	1.1	0.3	0.9	0.7	0.4	0.4	0.7
Mass spectrometry	—	—	—	1.9	0.4	1.9	0.3	2.6	0.2	0.6	—	0.5
Other	5.9	9.8	10.5	12.3	9.2	11.6	17.5	9.4	13.8	15.7	16.0	13.7

From Volkova, G. A. and Kontsova, V. V., (in Russian), *Zavodsk. Lab.*, 42, 395, 1976. With permission.

sis of geological raw materials have approximately the same share in the U.S.S.R. and the rest of the world. The literature dealing with flame photometry is meager and has not changed during the past years (approximately 2%). This method is already well known, thoroughly investigated, and widely applied (mostly in the U.S.S.R.). Atomic absorption methods are the subject of mainly foreign publications: in 1973 three times more papers were published abroad than in the U.S.S.R. The number of papers dealing with activation methods suddenly increased between 1970 and 1972.

Melikhov and Berdonosova[33] investigated the distribution of the literature on inorganic sorbents according to various points of view. Subert and Blesova[60] subjected the information flows dealing with pharmaceutical analysis to detailed analysis. Futekov et al.[45] analyzed the distribution of the analytical subject literature of selenium and tellurium among the various instrumental methods in the 1970 to 1975 period. Finally, Kara-Murza[61] investigated the mechanism and speed of dissemination of gas and liquid chromatography through the statistical investigation of the information flows dealing with this subject.

V. MISCELLANEOUS

The application of mathematical methods greatly determines the exactness of scientific research work on various fields. In this respect it may be interesting to study the diffusion of mathematical methods into analytical chemical research. Such information can be practically acquired in two ways: by analyzing and carefully interpreting the papers published on the given fields of analytical chemistry or by purely formal scientometric methods. The latter appears to be more advantageous since it is less laborious and can be better formalized. For the scientometric method two possibilities are offered.

First, the bibliographic references of scientific publications can be analyzed and thus from the number of cited mathematical publications, for instance, the frequency of their use can be estimated. The authors, however, often cite textbooks, reviews, or bibliographies, and in such cases it cannot be determined what kind of mathematical method is concerned. Moreover, analytical chemical literature may also deal with mathematical methods, and thus the author applying a specific mathematical method may refer to this paper and not to a mathematics textbook.

Another possibility arises from the formal investigation of the mathematical terms found in analytical papers. From the analysis of their character, the changes over time in their uses can be traced. This latter method was applied by Preobrazhenskaya et al.[62] by investigating the frequency of terms used by mathematical statistics and analytical error estimation. The data base was comprised of the papers published on emission spectrometry during 1958 to 1968 in the periodicals *Zhurnal Analiticheskoi Khimii* (U.S.S.R.), *Zavodskaya Laboratoriya* (U.S.S.R.), and *Analytical Chemistry* (U.S.). The authors have shown that the papers investigated contained a very large number of statistical terms. According to the authors this large number is due to the lack of a universal terminology; analytical chemists apply several terms for expressing the very same concept. The nonuniqueness of terminology is also obvious. In *Zhurnal Analiticheskoi Khimii* 31 variants, in *Zavodskaya Laboratoriya* 26 variants, and in *Analytical Chemistry* 17 variants are used to express the concept of "error". The paper cited also discusses several other results of the statistical evaluation of mathematical terms.

Chapter 6

ANALYTICAL CHEMISTRY JOURNALS

I. THE SCATTER OF ANALYTICAL CHEMISTRY LITERATURE, BRADFORD'S LAW

The concentration or the scatter of scientific literature is regulated by objective "laws", the most general of which proved to be the Bradford law of literature scatter discovered in 1934 and again stated in 1948.[63] Bradford's original question was: "To what extent do different journals contribute papers to a given subject literature?" In general, a relation between a quantity (journals) and a yield (papers) was of interest. After ingeniously arranging the data he formulated the law as follows: " . . . if scientific journals are arranged in order of decreasing productivity of papers on a given subject, they may be divided into a nucleus of periodicals more particularly devoted to the subject and several groups or zones containing the same number of papers as the nucleus, where the number of periodicals in the nucleus and succeeding zones will be as $1:m:m^2$ (with usually $m = 5$). This indicates that the same number of papers is produced by a number of journals which increases from zone to zone in such a way that the ratio between the number of journals in the second and first zone is the same as between the third and second . . . The first group can be called "core journals". In other words, articles of interest to a specialist must occur not only in periodicals specializing in his subject, but also, from time to time, in other periodicals, which grow in number as the relation of their field to that of his subject lessens and the number of articles on his subject in each periodical diminishes. It is to be noted that during phases of rapid and vigorous growth of knowledge in a scientific field, articles of interest to that field appear in increasing numbers in periodicals distant from that field."

It has become apparent that Bradford's law is in fact a particular case of a very common statistical distribution. It was soon recognized that the division into three zones was not unique; the same distribution could be fitted to any number of zones. It has also been realized that Bradford's law is related to another well-known distribution, Zipf's law. Zipf, a linguist, was particularly interested in the frequencies with which different words are used. He demonstrated that if the words appearing in any reasonably long piece of text are counted and ranked in order of frequency of occurrence, this frequency is proportional to the rank order. For example, a word ranked tenth in terms of frequency of usage is employed one tenth as often as the word ranked first. According to Zipf,[64] this sort of rank-frequency relationship is obeyed by a wide range of social phenomena. It is a result, he argues, of a natural tendency to use more frequently those intellectual tools with which one is best acquainted and which are more flexible. The rank-order relationship, therefore, reflects the operation of some "principle of least effort". The relationship between Bradford's and Zipf's laws can be easily seen if we rephrase the latter slightly. Instead of relating frequency of word usage to rank order, we ask the question: how many words (N) occur exactly x times in a text? The answer follows from Zipf's law: $N = 1/x^2$. But Bradford's law can also be rewritten in a power law from the number of journals j containing exactly p articles on a specified subject which is given by $j = 1/p^2$.[65]

The graphical formulation of Bradford's law is a plot of the cumulative number of articles vs. the logarithm of the cumulative number of journals in which the articles appear. The plot, as currently understood, has an S-shape with a central straight section following Bradford's log-linear law. The upward-curving bottom of the curve represents the small nuclear zone of the most relevant journals. The upper end of the curve represents the peripheral zone where relevant references are widely scattered

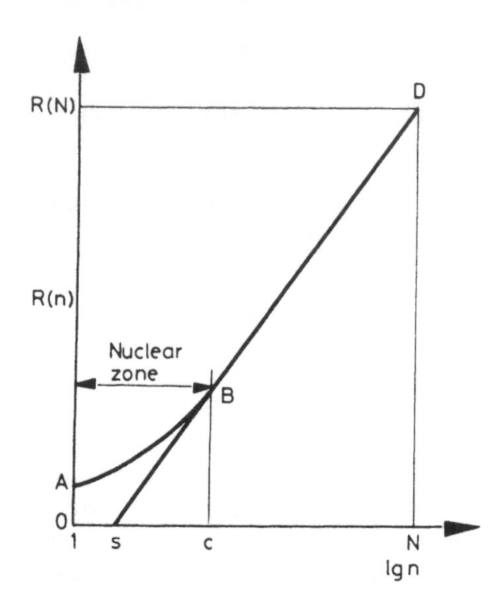

FIGURE 32. The idealized Bradford distribution. Along the horizontal axis are the ranked periodicals 1, 2, . . . , n in decreasing order of productivity; R(n) is the cumulative total of papers. The resulting graph begins with a rising curve AB which at some critical points B runs into a straight line BD. (From Brookes, B. C., *J. Doc.*, 24, 247, 1968. With permission.)

among a great number of journals (Figure 32).[66] How may the Bradford-Zipf distribution be conceptually modeled? Why, for example, are articles on a subject distributed in this way among journals? Brookes[66] suggests the following model. The first papers on a new topic are distributed at random among a set of N journals. Initially, the probabilities of a paper on the new topic being published in any journal of this set are equal to 1/N. But as soon as any journal of the set publishes its first paper on the topic, the probability of that journal publishing a second paper on the same topic increases from 1/N to 2/N. As soon as any journal publishes two papers on the topic, its probability of attracting a third paper increases to 3/N, and so on. Such a model leads to the straight-line portion of the distribution. But limitations of space and editorial decisions impose restrictions on the number of new papers entering the "core" or "nuclear zone", which therefore contains fewer papers than a straight-line relation would predict.

Suppose that we have an analytical chemistry bibliography, a special literature collection, or an abstract journal. If we analyze the items according to the frequency of occurrence of each journal title, we can plot the data as in Figure 32. The plot need only be continued far enough beyond point B to establish the slope of the line BD. Brookes[66] has shown that this slope R(n)/log n is equal to N, an estimate of the expected total number of journals that will contribute at least one paper to the field. N is marked on the log n axis, and the vertical ND drawn to cut the line BD. The horizontal from D then establishes R(N), the expected total number of papers in the field. This can be compared with the actual number of papers in the bibliography, collection, or abstract journal to give an estimate of its completeness.

An alternative to graphical estimation is to use the expression for the linearity:

$$R(n) = k \log n/s \qquad (c \leq n \leq N) \qquad (12)$$

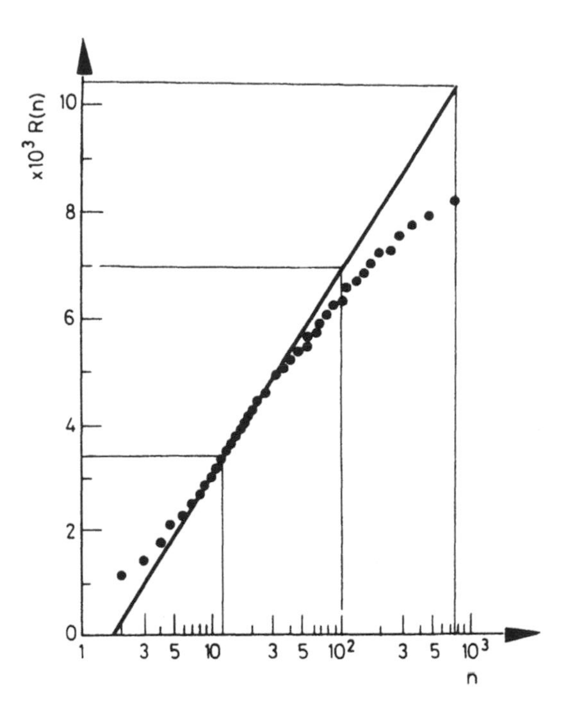

FIGURE 33. Scatter of analytical chemical literature. Bradford's distribution of the abstracts in *Analytical Abstracts*, 1977. R(n), cumulative total of papers contributed by the journals ranked 1 to n; n, the rank of journals. (From Braun, T., Bujdosó, E., and Lyon, W. S., *Anal. Chem.*, 52(6), 617A, 1980. With permission. Copyright 1980, American Chemical Society.)

where R(n) is the number of papers, k is the slope of the curve, n is the journal rank, and s is the intercept. The application of Bradford-type scattering analyses to analytical chemistry was done until now in only a very few cases. To determine scatter in the general analytical chemistry literature, Braun et al.[20] used the 1977 issues of *Analytical Abstracts* as a data base. There were 8311 papers distributed in 741 journals. The Bradford distribution (Figure 33) shows 12 core and 88 neighboring analytical journals. Table 30 shows data for the first 50 of these. The Bradford curve for *Analytical Abstracts* drops heavily at the end. A possible reason for this could be that the core and one half of the neighboring journals are completely (cover to cover) abstracted, but gathering scattered articles from other peripheral journals is somewhat haphazard.

A similar study has been made by Braun and Bujdosó[27] concerning the scatter of the radioanalytical literature. Their results are shown in Figure 34.

The characterization of periodicals either dealing with analytical chemistry or publishing papers of analytical character and the ranking of these periodicals had already been the subject of investigation before the Bradford-type studies. Using *Chemical Abstracts* and *Chemisches Zentralblatt* as data base, Boig and Howerton[44] attempted to rank analytical chemical periodicals for the period 1977 to 1950 on the basis of their productivity (Table 31).

Later, Brooks and Smythe,[12] investigating what they call the important analytical periodicals on the basis of the 1970 volumes of *Analytical Abstracts*, stated that some 26 journals accounted for 35% of the world total of papers on analytical chemistry in 1970. The major journals (each containing more than 1.0% of the world total) arranged alphabetically are as follows: *Analytica Chimica Acta, Analytical Biochemis-*

Table 30
THE FIRST 50 LEADING JOURNALS ON ANALYTICAL CHEMISTRY COMPUTED FROM *ANALYTICAL ABSTRACTS*, 1977, RANKED BY PRODUCTIVITY

Rank	Journal	Number of papers found
1	*J. Chromatogr.*	710
2	*Anal. Chem.*	511
3	*Anal. Chim. Acta*	338
4	*Zh. Anal. Khim.*	315
5	*Anal. Biochem.*	300
6	*Fresenius Z. Anal. Chem.*	212
7	*J. Assoc. Off. Anal. Chem.*	206
8	*Talanta*	198
9	*Bunseki Kagaku*	171
10	*Zavodsk. Lab.*	170
11	*Chem. Anal. (Warsaw)*	169
12	*Clin. Chem.*[a]	161
13	*J. Radioanal. Chem.*	147
14	*Analyst*	146
15	*J. Pharm. Sci.*	132
16	*Mikrochim. Acta*	116
17	*Clin. Chim. Acta*	109
18	*Anal. Lett.*	92
19	*Chromatographia*	91
20	*Radiochem. Radioanal. Lett.*	85
21	*J. Chromatogr. Sci.*	76
22	*Revista Chim. (Bucharest)*	75
23	*J. Agric. Food Chem.*	71
24	*Indian J. Chem. Sect. A*	60
25	*Farmatsiya (Moscow)*	52
26	*Rev. Sci. Instr.*	52
27	*Lab. Pract.*	51
28	*J. Clin. Chem. Clin. Biochem.*	51
29	*Pharmazia*	46
30	*Appl. Spectrosc.*	45
31	*Quim. Anal.*	40
32	*Ukr. Khim. Zh.*	37
33	*Environ. Sci. Technol.*	37
34	*J. Electroanal. Chem. Interfacial Electrochem.*	35
35	*Biochem. Med.*	34
36	*Appl. Opt.*	34
37	*J. Am. Oil Chem. Soc.*	33
38	*Bull. Environ. Contam. Toxicol.*	33
39	*Int. J. Appl. Radiat. Isot.*	32
40	*Z. Lebensm. Unters. Forsch.*	31
41	*X-ray Spectrom.*	31
42	*At. Absorpt. Newsl.*	31
43	*J. Phys. Sci. Instrum.*	30
44	*Nucl. Instrum. Methods*	29
45	*An. Quim.*	29
46	*Yukugaku Zasshi*	28
47	*Curr. Sci. (India)*	28
48	*Acta Pol. Pharm.*	28
49	*Z. Chem. (Leipzig)*	27
50	*Chem. Pharm. Bull. (Tokyo)*	27

[a] End of the core journals.

From Braun, T., Bujdosó, E., and Lyon, W. S., *Anal. Chem.*, 52(6), 617A, 1980. With permission. Copyright 1980, American Chemical Society.

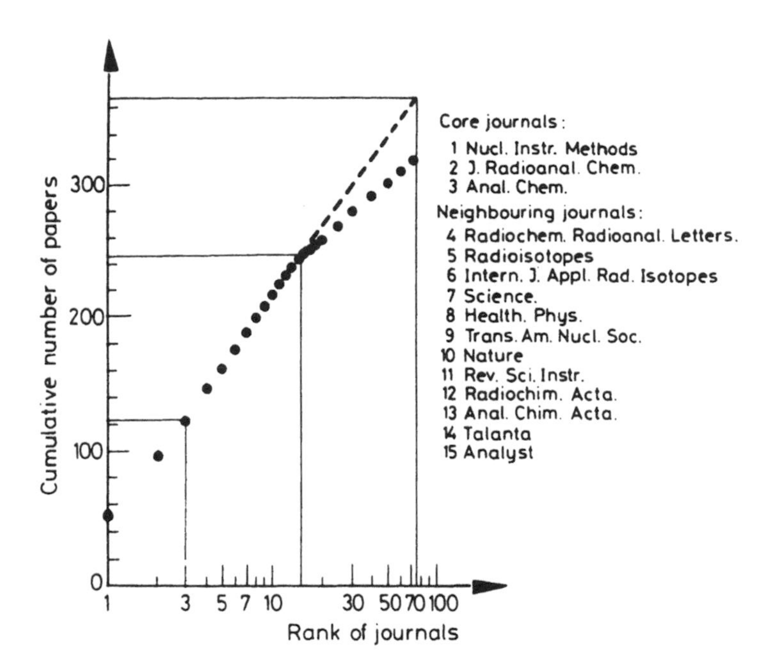

FIGURE 34. Bradford's distribution for the journals and articles cited by Lyon et al.[50,67 69] in the period 1972 to 1978. (From Braun, T. and Bujdosó, E., *J. Radioanal. Chem.*, 50, 9, 1979. With permission.)

*try, Japan Analyst, Journal of Electroanalytical Chemistry, Journal of the Association of Official Analytical Chemists, Mikrochimica Acta, Nucleonika**, *Talanta, Analyst, Zavodskaya Laboratoriya, Zeitschrift fur Analytische Chemie,* and *Zhurnal Analiticheskoi Khimii.* These 13 journals together accounted for about 30% of the world total.

Braun et al.,[20] on the other hand, showed that in 1977 (Figure 35) 50 and 90% of the papers could be found in only 3 and 36% of the journals, respectively.

On the basis of the 1975 volumes of *Referativnyj Zhurnal Khimiya,* Orient et al.[30] investigated the distribution of papers dealing with atomic absorption analysis among the various periodicals. Their results (Table 32) show that 50% of all papers are concentrated in eight periodicals (two U.S., two international, two Russian, one German, and one Japanese). Of these papers, 20% can be found in 13 very widespread analytical periodicals, and the remaining 30% are dispersed among the journals of various countries and "Sborniks" and "Trudys" published by various high schools, universities, and other institutions of the U.S.S.R.

As in the case of all very rapidly developing subfields, e.g., atomic absorption analysis, initially the supply of information in the leading periodicals is very high. Later, however, when the method has become routine technique, the emphasis is shifted from the leading analytical periodicals to more specialized ones. It is worth observing, as already mentioned in the general consequences of the previous section, that 45% of the Russian papers on atomic absorption analysis were not published in "regular" periodicals.

On the basis of *Atomic Absorption and Flame Emission Abstracts* and the bibliographies of *Atomic Absorption Newsletter,* Brooks and Smythe[47] also investigated the distribution of atomic absorption papers among the various journals in the 1950 to

* Seemingly this journal was included into the list by mistake.[13]

Table 31

PRODUCTIVITY RANKING OF PERIODICALS DEDICATED TO ANALYTICAL CHEMISTRY (1877 TO 1907 INCLUSIVE, *CHEMISCHES ZENTRALBLATT* AND 1917 TO 1950 INCLUSIVE, *CHEMICAL ABSTRACTS*)

Journal	Country	1877	1887	1897	1907	1917	1927	1937	1947	1948	1949	1950
Anal. Chem. (1947—)[a]	U.S.	—	—	—	—	—	—	—	1	1	1	1
Ind. Eng. Chem. (1909—1946)	U.S.	—	—	—	—	1	3	2	—	—	—	—
Zavodsk. Lab. (1935—)	U.S.S.R.	—	—	—	—	—	—	1	2	2	2	2
Zh. Anal. Khim. (1946—)	U.S.S.R.	—	—	—	—	—	—	—	18	—	3	3
Anal. Chim. Acta (1947—)	Holland	—	—	—	—	—	—	—	28	3	4	4
Analyst (1876—)	England	—	14	20	17	12	2	5	4	6	5	5
Bull. Soc. Chim. Fr. (1858—)	France	7	11	9	12	—	20	15	6	10	17	6
Chem. Listy (1907—), *Listy Chem.* (1875—1891)[b]	Czechoslovakia	—	19	—	—	—	—	—	—	—	—	7
Mikrochem. Ver. Mikrochim. Acta (1914—)[c]	Austria	—	—	—	—	—	19	6/7	8	8	11	8
Z. Anal. Chemie (1862—)	Germany	1	1	8	5	5	1	3	—	4	6	9
An. (Real.) Soc. Esp. Fis. Quim. (1902—)[d]	Spain	—	—	—	—	17	—	—	5	5	10	10
Nature (1869—)	England	—	—	—	—	—	—	—	26	—	14	11
Chim. Anal. (1919—), *Ann. Chim. Anal.* (1983—1919)[e]	France	—	—	—	3	7	17	—	10	7	7	12
J. Assoc. Off. Agr. Chem. (1918—)	U.S.	—	—	—	—	—	16	16	30	—	8	13
Metallurgia	England	—	—	—	—	—	—	—	15	9	13	14
Izv. Akad. Nauk (1925—)	U.S.S.R.	—	—	—	—	—	—	—	—	—	—	15
Izv. Sektora Platiny...(1923—)	U.S.S.R.	—	—	—	—	—	—	—	—	—	—	16
J. Chem. Soc. Jpn. (1880—)	Japan	—	—	—	—	—	—	12	3	—	—	—
C. R. (1835—)	France	4	15	3	13	14	12	18	9	13	18	—
Chemist-Analyst (1912—)	U.S.	—	—	—	—	3	10	10	11	—	—	—
Angew. Chem.[f]	Germany	—	6	4	8	20	8	8	—	—	—	—
J. Appl. Chem. USSR (1923—)	*U.S.S.R.*	—	—	—	—	—	—	*4*	—	—	—	—
J. Am. Chem. Soc. (1876—)	U.S.	—	—	2	2	2	9	9	—	15	—	—
J. Chem. Educ. (1924—)	U.S.	—	—	—	—	—	—	17	—	16	—	—
J. Soc. Chem. Ind. (1882—)	England	—	8	6	10	6	—	13	—	12	12	—
Chem. Ztg. (1877—)	Germany	—	2	1	1	4	5	—	—	—	—	—
Pharm. Zentralhalle (1860—)	Germany	8	3	1	11	—	—	—	—	—	—	—
Chem. News (1859—1932)	England	3	4	12	—	10	—	—	—	—	—	—
Chem. Ber. (1945—), *Berichte* (1867—1945)[g]	Germany	2	10	—	15	—	11	—	—	—	—	—

[a] *Anal. Chem.* (1947) was previously known as *Ind. Eng. Chem.* (1909—1947), *Analytical Edition.*

[b] *Chem. Listy* (1907—) may be a somewhat belated revision of *Listy Chem.*, published from 1875 to about 1891.

[c] *Mikrochemie* was combined with *Mikrochim. Acta* in 1937.

[d] *An. Soc. Esp. Fis. Quim.* was changed about 1941 to *An. Real. Soc. Esp. Fis. Quim.*

[e] *Ann. Chim. Anal.* was changed in 1919 to *Chim. Anal.*

[f] *Angew. Chem.*, the present name, was formerly known as *Z. Angew. Chem.* (1882—1932) and *Repertorium Anal. Chem.* (1881—1887).

[g] *Berichte* was changed in 1945 to *Chem. Ber.*

From Boig, F. S. and Howerton, P. W., *Science*, 115, 555, 1952. With permission.

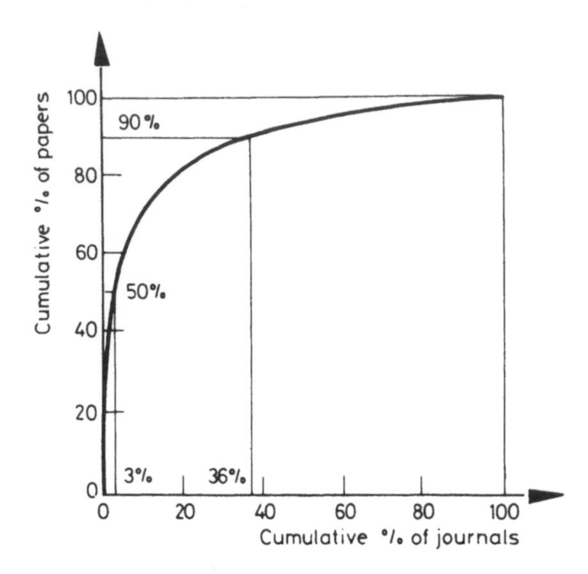

FIGURE 35. Concentration of the analytical chemical literature: 50% of the papers abstracted by *Analytical Abstracts* in 1977 were concentrated in only 3% of the journals. (From Braun, T., Bujdosó, E., and Lyon, W. S., *Anal. Chem.*, 52(6), 617A, 1980. With permission. Copyright 1980, American Chemical Society.)

1971 period. According to their results (Table 33), 21 journals published nearly 60% of the world total atomic absorption papers since 1955. *Spectrochimica Acta* and *Analyst* were the first in publishing atomic absorption papers to any extent before 1962.

II. CITATION ANALYSIS OF ANALYTICAL CHEMISTRY JOURNALS

Rankings based on the Bradford distribution use productivity as an indicator. It is interesting to see how rankings change when they are based on some journal "quality" indicator. Quality is, of course, quite difficult to define. Braun et al.[20] have used the so-called "impact factor" for ranking suggested by Garfield.[70] The impact factor is a spin-off from citation counts in the *Science Citation Index* data base and is a measure of the frequency with which the average cited article in a journal has been cited. Thus the 1978 impact factor is the number of 1978 citations of journal articles from 1976 and 1977 divided by the total number of articles the journal published in those same 2 years. Table 34 shows analytical journal rankings based on this indicator. By combination of the data from Tables 30 and 34, it was found[20] that 49% of the analytical citations in 1978 were coming from the 12 core journals of Table 30.

III. INFLUENCE AND INTERRELATIONSHIP OF ANALYTICAL CHEMISTRY JOURNALS

As seen, scientists engaged in analytical chemistry research disseminate their findings to the international scientific community by publishing in a wide variety of scientific journals. The range of subjects covered in a particular journal extends from the specific focus of a narrowly defined research specialty, e.g., the *Journal of Electroanalytical Chemistry*, to the broad analytical coverage of *Analytica Chimica Acta* or to the even broader cross-disciplinary coverage of the physical, chemical, and life sciences in interdisciplinary journals, i.e., *Science* and *Nature*. The process of grouping chemistry

Table 32
DISTRIBUTION OF PAPERS ON AAS IN DIFFERENT JOURNALS

Journal	Percent of all papers
Anal. Chem.	11.60
Anal. Chim. Acta	8.20
At. Abs. Newsl.	7.51
Zh. Anal. Khim.	6.48
Bunseki Kagaku	5.80
Talanta	5.12
Zavodsk. Lab.	4.44
Z. Anal. Chem.	3.75
Analysis	2.39
Zh. Prikl. Spektrosk.	2.05
Appl. Spectrosc.	1.71
Spectrochim. Acta, Part B	1.36
Analyst	1.36
SCAN	1.36
Anal. Lett.	1.02
Lab. Pract.	1.02
Mikrochim. Acta	1.02
Bull. Chem. Soc. Jpn.	1.02
Flame Emission At. Absorpt.	1.02
Erdöl Kohle	1.02
Chem. Anal. (Warsaw)	1.02
Other journals outside the U.S.S.R.	18.77
Various other publications in the U.S.S.R.	10.58

From Orient, I. M., Artemova, O. A., and Davidova, S. L., *Zavodsk. Lab.* (in Russian), 43, 419, 1977; (English transl.), *Ind. Lab. U.S.S.R.*, 43(4), 498, 1977.

or analytical chemistry journals into subject categories requires a decision as to the level of aggregation that will render the classification most useful for subsequent analyses. A useful criterion for designating a separate category is to require that the candidate group of journals serve as the primary publication outlet for the area of study. An example from the analytical chemistry literature will serve to illustrate this point. Journals dealing with chromatography and those dealing with electroanalytical chemistry represent subfields of analytical chemistry; however, a large amount of research in these subfields is published not in specialty journals but rather in general analytical chemistry or even general chemistry journals.

The journal impact factor introduced by Garfield[70] is a size-independent measure or indicator of some sort of quality for journals. This measure or indicator, e.g., the total number of citations, has no meaning on an absolute scale. In addition, the impact factor suffers from two significant limitations. First, although the size of a journal as reflected by the number of articles is corrected, the average length of individual papers appearing in the journal is not. Thus, journals publishing longer papers, i.e., review journals, tend to have higher impact factors. The second limitation is that the citations are unweighted, all citations being counted with equal weight regardless of the citing journal. It seems more reasonable to give higher weight to a citation from a prestigious journal than to a citation from a peripheral one.

To overcome these limitations three related influence measures were developed, each

Table 33
DISTRIBUTION OF PAPERS ON AAS IN DIFFERENT JOURNALS

Journal	1955—1959	1960—1961	1962—1963	1964—1965	1966—1967	1968—1969	1970—1971	Total	Percent of all papers
At. Abs. Newsl. (U.S.)	—	—	16	69	110	92	94	381	15.22
Anal. Chim. Acta (Holland)	—	4	11	10	30	67	72	194	7.75
Anal. Chem. (U.S.)	1	5	11	17	31	45	44	154	6.15
Spectrochim. Acta (U.K.)	5	12	5	8	17	37	31	115	4.59
Analyst (U.K.)	4	7	3	2	10	35	33	94	3.75
Appl. Spectrosc. (U.S.)	—	—	2	2	15	25	21	63	2.52
Bunseki Kagaku (Japan)	—	—	—	3	2	9	37	51	2.04
J. Assoc. Off. Anal. Chem. (U.S.)	—	—	—	2	10	16	22	50	2.00
Talanta (U.K.)	—	—	1	6	6	23	13	49	1.96
Zh. Prikl. Spektrosk. (U.S.S.R.)	—	—	—	2	10	18	17	47	1.88
Zh. Anal. Khim. (U.S.S.R.)	—	1	4	8	4	10	18	45	1.80
Z. Anal. Chem. (Germany)	1	1	2	5	6	12	11	37	1.48
Clin. Chem. (U.S.)	—	—	—	—	6	11	18	35	1.40
Zavodsk. Lab. (U.S.S.R.)	—	1	4	6	3	7	10	31	1.24
Nature (U.K.)	1	7	3	4	6	2	5	30	1.20
Clin. Chim. Acta (Holland)	—	—	—	—	—	12	6	18	0.72
Anal. Biochem. (U.S.)	—	—	1	1	6	6	4	16	0.64
Mikrochim. Acta (Austria)	—	2	2	2	1	—	6	13	0.52

Chim. Anal. (France)	—	—	—	1	—	5	6	12	0.48
Jpn. Analyst (Japan)	—	—	—	3	1	4	2	10	0.40
Chem. Listy (Czechoslovakia)	—	—	—	1	2	1	2	6	0.24
Total	11	40	65	152	276	435	472	1451	57.98

From Brooks, R. R. and Smythe, L. E., *Anal. Chim. Acta*, 74, 35, 1975. With permission.

Table 34
IMPACT FACTOR AND TOTAL NUMBER OF CITATIONS IN 1978 OF 50 LEADING JOURNALS ON ANALYTICAL CHEMISTRY RANKED BY IMPACT FACTOR

Rank	Journal	Impact factor	Total number of citations
1	*Clin. Chem.*	3.106	6,948
2	*Anal. Chem.*	3.058	19,507
3	*J. Chromatogr. Sci.*	2.586	2,142
4	*Anal. Biochem.*	2.309	15,131
5	*J. Chromatogr.*	2.302	11,770
6	*Appl. Spectrosc.*	2.161	1,263
7	*Chromatographia*	1.972	1,200
8	*Appl. Opt.*	1.934	6,136
9	*Clin. Chim. Acta*	1.676	8,120
10	*Analyst*	1.666	2,606
11	*J. Electroanal. Chem. Interfacial Electrochem.*	1.525	4,506
12	*J. Agric. Food Chem.*	1.503	4,665
13	*Environ. Sci. Technol.*	1.436	2,123
14	*Anal. Chim. Acta*	1.404	3,856
15	*Anal. Lett.*	1.244	980
16	*Talanta*	1.182	2,304
17	*J. Pharm. Sci.*	1.171	6,724
18	*J. Clin. Chem. Clin. Biochem.*	1.163	298
19	*Nucl. Instrum. Methods*	1.141	6,035
20	*Rev. Sci. Instrum.*	1.131	4,730
21	*X-ray Spectrom.*	1.118	212
22	*Chem. Pharm. Bull.*	0.979	4,497
23	*Fresenius Z. Anal. Chem.*	0.930	2,077
24	*J. Assoc. Off. Anal. Chem.*	0.916	2,373
25	*Int. J. Appl. Radiat. Isot.*	0.790	1,091
26	*J. Phys. Sci. Instrum.*	0.714	1,375
27	*Mikrochim. Acta*	0.697	808
28	*J. Radioanal. Chem.*	0.685	997
29	*Radiochem. Radioanal. Lett.*	0.639	674
30	*Z. Chem. (Leipzig)*	0.623	1,344
31	*Pharmazie*	0.483	881
32	*Z. Lebensm. Unters. Forsch.*	0.455	502
33	*Bunseki Kagaku*	0.410	641
34	*Zh. Anal. Khim.*	0.403	1,701
35	*An. Quim.*	0.393	552
36	*Indian J. Chem. Sect. A*	0.322	313
37	*J. Am. Oil Chem. Soc.*	0.278	2,616
38	*Acta Pol. Pharm.*	0.262	313
39	*Curr. Sci. (India)*	0.221	1,167
40	*Ukr. Khim. Zh.*	0.173	840
41	*Yukugaku Zasshi*	0.145	1,251
42	*Zavodsk. Lab.*	0.117	974
43	*Revista Chim. (Bucharest)*	0.082	106

Note: Data unavailable for *At. Absorpt. Newsl., Biochem. Med., Bull. Environ. Contam. Toxicol., Chem. Anal. (Warsaw), Farmatsiya (Moscow), Lab. Pract. (London),* and *Quim. Anal.*

of which measures one aspect of journal influence, with explicit recognition of the size factor. These measures are

1. The influence weight of the journal
2. The influence per publication for the journal
3. The total influence of the journal

The influence methodology and its statistical underpinning has been worked out by Narin and co-workers,[71,72] and our description of these topics relies heavily on their writings.

Influence weights for journals are derived from an analysis of the referencing interactions among members of the set of journals being considered. The influence weight is a weighted, normalized citation measure weighted by the influence of the referencing journal and normalized by the size of the journal.

The reference-citation matrix contains the information describing the flow of influence among journals. It has the form

$$C = \begin{vmatrix} C_{11} & C_{12} & \cdots & C_{1n} \\ C_{21} & C_{22} & \cdots & C_{2n} \\ \cdot & \cdot & \cdots & \cdot \\ \cdot & \cdot & \cdots & \cdot \\ C_{n1} & C_{n2} & \cdots & C_{nm} \end{vmatrix} \tag{13}$$

Here, the terms reference and citation have been used interchangeably but to avoid confusion in discussing influence weights, a specific role is assigned to each of these terms. The term reference will be used to designate the issuing unit, while the term citation will designate the receiving unit. A term C_{ij} in the reference-citation matrix indicates both the number of references unit i gives to unit j and the number of citations unit j receives from unit i. The time frame of a citation matrix must also be clearly understood in order that a measure derived from it may be properly interpreted. Suppose that the citation data are based on references issued in 1978. These may refer to papers published in any year, up to and including 1978. In general, the papers issuing the references will not be the same as those receiving the citations. Any conclusions drawn from such a matrix assume an ongoing, relatively constant nature for each of the units. It is assumed that the journals have not changed in size relative to each other, and that they represent a constant subject area. Journals in rapidly changing fields and new journals must therefore be treated with caution.

Starting with the reference-citation matrix, an algorithm was developed for the calculation of the "influence weight" for each journal. As a data base Narin et al.[71,72] used the citation tapes of the *Science Citation Index* data base for 1973. The reference-citation matrix may be thought of as an input-output matrix, with the medium of exchange being the citation. Each journal gives out references and receives citations; it is above average if it has a "positive citation balance", i.e., receives more than it gives out. This reasoning provides a first-order approximation to the weight of each journal, which is simply

$$w^{(1)} = \frac{\text{total number of citations to the journal from other journals}}{\text{total number of references from the journal to other journals}} \tag{14}$$

This ratio is the starting point for an iterative procedure for the calculation of the influence weights. In the first approximation, all citations to a journal were weighted equally; however, some of these citations come from peripheral journals. In the next order of approximation, a reference from any journal is weighted with the weight it received in the first approximation, yielding a set of second-order weights $w^{(2)}$. This process is continued and rapidly converges to a stable, self-consistent set of influence weights. This set of influence weights provides a size-independent measure for each journal. There is no tendency for a journal to have a higher or a lower weight due to its size, whether measured by the number of publications or by the average length of its publications.

There is the second influence measure, introduced by Narin et al.,[71,72] called "influence per publication". Two journals might have the same influence weights, but one may contain much longer articles, as would be the case for review journals. This journal would be expected to have a larger influence on a per publication basis. From the mathematical formulation of the problem it turns out that the two measures are related by

$$\text{(influence per publication)} = \text{(influence weight)} \times \text{(reference per publication)} \tag{15}$$

The influence per publication measure is particularly valuable in weighting counts of publications since it compensates for editorial policies which might affect the number of references in an article and because a count of publications is the natural starting point for comparative analysis.

This measure should not be confused with the third influence measure, the "total influence" of a journal. Two journals could have the same influence weight and the same influence per publication and yet have widely different total influence, solely due to the difference in the number of publications. The total influence of a journal is defined as

$$\text{(total influence)} = \text{(influence per publication)} \times \text{(number of publications)} \tag{16}$$

It should be emphasized that the influence measures could properly be called "citation influence measures". There are numerous factors which are relevant to a journal that contribute to these *de facto* measures: true merit, journal circulation, availability, degree of specialization, country of origin, language, etc.

Some influence measures for analytical journals are presented in Table 35. The analytical journals were considered as the basic units for an analysis resulting in an influence map for these journals as seen in Figure 36. The following conventions apply to this map.

1. A solid rectangle is used to represent journals within the field and subfields. The area of a rectangle is proportional to the size of a journal as measured by the number of articles, notes, and reviews in the corporate index of the *Science Citation Index* in 1973.
2. The vertical scale shows influence per publication for each journal on a logarithmic scale where weights for a set of units tend to be distributed uniformly, i.e., less crowding for the lower weight units.
3. The horizontal direction is used to separate either different subfields appearing on the map or journals with different specific focuses. Journals in the same column tend to be more similar to each other than to journals in neighboring columns.
4. Arrows are directed from a journal to the other journals to which it refers most frequently, exclusive of itself. Usually, two arrows are drawn from each journal showing the two other journals that are most frequently referenced; occasionally three are given if the number of references to the second and third are close, or

Table 35
INFLUENCE MEASURES FOR SOME ANALYTICAL JOURNALS

Journal	Influence weight	References/ publication	Influence/ publication	Publications	Total influence
Anal. Lett.	0.40	7.2	2.9	123	353
Analusis	0.12	8.5	1.1	85	90
Analyst	0.82	9.7	7.9	131	1039
Anal. Chem.	0.75	20.9	15.6	603	3401
Anal. Chim. Acta	0.51	8.9	4.5	344	1555
J. Assoc. Off. Anal. Chem.	1.04	5.1	5.3	321	1692
J. Chromatogr. Sci.	0.94	11.9	11.2	118	1317
J. Chromatogr.	0.53	10.7	5.6	631	3553
J. Radioanal. Chem.	0.20	6.4	1.3	143	186
J. Therm. Anal.	0.34	9.7	3.3	45	147
Jpn. Analyst (Bunseki Kagaku)	0.02	25.9	0.5	231	119
Microchem. J.	0.31	7.5	2.3	86	201
Microchim. Acta	0.43	9.9	4.2	124	525
Talanta	0.49	10.1	4.9	155	763
Z. Anal. Chem.	0.76	6.4	4.9	249	1210
Zh. Anal. Khim.	0.15	8.4	1.3	463	602

From Pinski, G., *J. Chem. Inf. Comp. Sci.*, 17, 67, 1977. With permission.

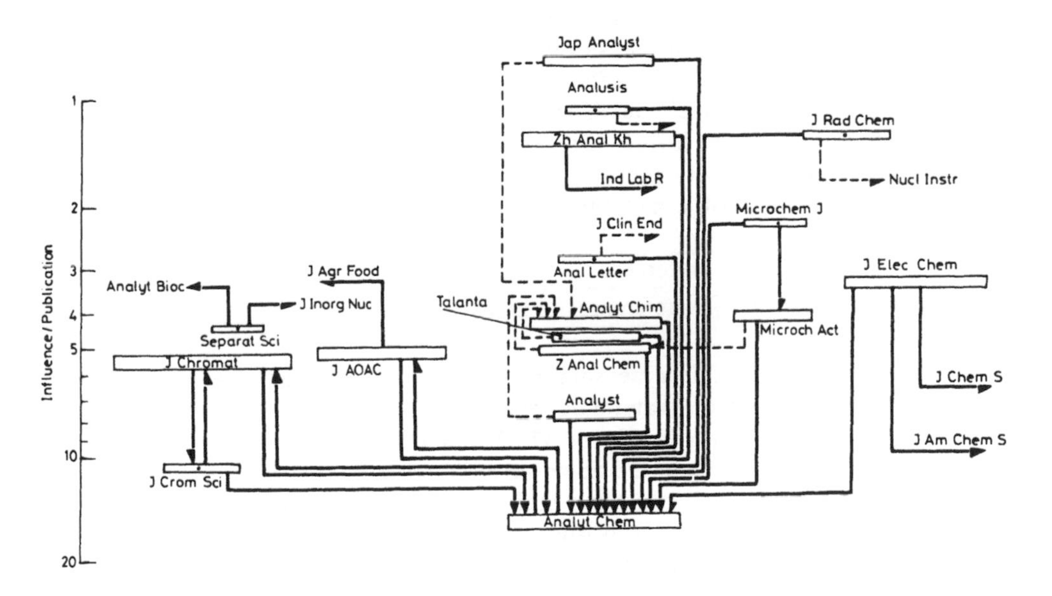

FIGURE 36. Influence map for analytical chemistry journals. See text for information. (From Pinski, G., *J. Chem. Inf. Comp. Sci.*, 17, 67, 1977. With permission.)

there may be only one if a single arrow best characterizes the referencing priority of the journal. A full arrowhead is used for a first arrow (largest number of references), while half an arrowhead is used for a second or third arrow. A dotted arrow is used for a secondary arrow, which is considerably weaker than the primary arrow.

5. Arrows directed out of the field to journals which are not of central importance for the field are generally short ones leading to the unenclosed journal name. In this case there is no significance to the vertical placement of the cited journal.

Classification of the chemistry journals into subfields on the basis of influence measures enabled Pinski[73] to analyze the flow of influence among subfields. In the overall flow of influence through the fields of science, there is a hierarchy which can be represented by the scheme

$$\text{biology} \rightarrow \text{chemistry} \rightarrow \text{physics} \rightarrow \text{mathematics}$$

The subfields of chemistry may be thought of as spread out in a spectrum from biology to physics. The interface between chemistry and biology has itself developed into a major field, i.e., biochemistry. Although the *Journal of Biochemistry,* for example, is published by the American Chemical Society, it is one of the central journals to the literature of biochemistry, and only a small portion of its references are directed at chemistry journals. The interface at the other end of the spectrum presents a different situation. Physical chemistry has remained within chemistry, and as a subfield is not a borderline area between chemistry and physics. Chemical physics does provide an interface linking chemistry and physics. Although the direct physics-chemistry linking is too weak to establish a strong hierarchical relationship, the inclusion of chemical physics provides a strong connection with both chemistry and physics, establishing the strong hierarchical relationship

$$\text{chemistry} \rightarrow \text{chemical physics} \rightarrow \text{physics}$$

This means that chemistry refers to chemical physics more often than it is cited by chemical physics, and chemical physics refers to physics more often than it is cited by physics. Pinski[73] aggregated the influence data for journals from nine subfields of chemistry including analytical chemistry. A 9 × 9 reference-citation matrix was then constructed and the subfield influence measures were derived. These are given in Table 36. The different subfields of chemistry interact to a very different degree with the two "endpoints" of the chemistry spectrum, biochemistry and chemical physics. The ratio

$$\frac{\text{references to chemical physics}}{\text{references to biochemistry}} \tag{17}$$

calculated for each subfield is a measure of the physical to biochemical orientation of each field. This suggests a diagrammatic representation of the chemical literature as a generalized hierarchy shown in Figure 37. The vertical coordinate is the influence weight with values increasing in the downward direction, while the horizontal coordinate is the above ratio. A logarithmic scale is used for both coordinates. Physical orientation of subfields increases towards the right. As a subfield, analytical chemistry is of medium influence and size at about one half the distance between biochemistry and physical chemistry.

The interdependency, interaction, and cluster trees for the major broad-based journals (i.e., those dealing with all aspects of analytical chemistry and publishing research papers from theory through analytical operations until data processing[74]) and specialty analytical chemistry journals were investigated by Bujdosó et al.[75] on the basis of the reference-citation matrices. The expectation analysis for the interdependency of journals showed diagonal strength values between 1.9 and 4.4 for *Analytical Chemistry, Analytica Chimica Acta, Analyst, Analytical Letters,* and *Talanta,* while *Fresenius Zeitschrift fuer Analytische Chemie* and *Analusis* were between 10.9 and 13.8 and *Zavodskaya Laboratoriya,* had an especially high value, 43.4. These values indicate how much more a journal cites itself than would be supposed from the group behavior. The analysis also showed a tightly linked subcluster of journals, *Zavodskaya Labora-*

Table 36
INFLUENCE MEASURES AND PUBLICATION DATA FOR THE SUBFIELDS OF CHEMISTRY

Subfield	Influence weight	References/ publication	Influence/ publication	Publications	Total influence
General chemistry	1.043	15.72	16.40	16,089	263,809
Biochemistry	0.747	20.65	15.42	12,049	185,749
Analytical chemistry	0.737	11.78	8.69	4,243	36,860
Organic chemistry	0.598	14.24	8.52	5,968	50,855
Inorganic chemistry	0.776	14.47	11.23	2,515	28,244
Applied chemistry	0.846	12.22	10.34	2,762	28,571
Polymer chemistry	0.562	11.37	6.39	3,224	20,614
Physical chemistry	1.340	12.37	16.58	7,188	119,206
Chemical physics	3.218	15.71	50.54	3,220	162,732

From Pinski, G., *J. Chem. Inf. Comp. Sci.*, 17, 67, 1977. With permission.

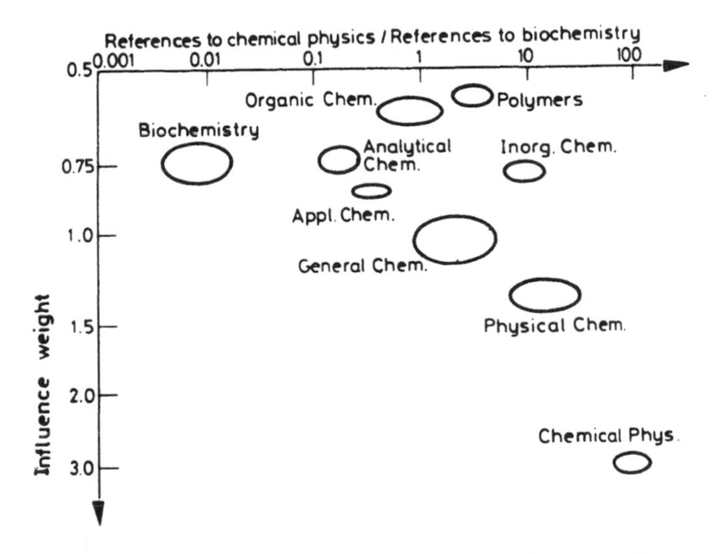

FIGURE 37. Influence structure for the chemical literature. (From Pinski, G., *J. Chem. Inf. Comp. Sci.*, 17, 67, 1977. With permission.)

toriya and *Zhurnal Analiticheskoi Khimii*, that is little related to others. There are especially tight links between the following pairs: *Analusis* and *Analytical Letters*, *Analytica Chimica Acta* and *Analyst*, and *Bunseki Kagaku* and *Talanta*.

The cluster trees with the amalgamation distances for the group of broad-based and specialty journals and for their combinations are shown in Figures 38 to 40. Amalgamation distance should be considered as a distance in a 2 m dimensional space in the case of an m × m matrix between two journals or between the center of the "masses" of subgroups represented by their weighted citation values as coordinates.

In the cluster tree for the group of broad-based journals (Figure 38), the Russian subcluster (1, 9, 10), the *Analusis — Bunseki Kagaku — Analytical Letters* subcluster (3, 6, 5), and the *Talanta — Analyst — Analytica Chimica Acta* subcluster (8, 4, 2) are clearly visible. In the groups of specialty analytical journals (Figure 39), the subclusters of the chromatographic journals (1, 6, 2) and the two microchemical journals (9, 10) are significant. The cluster tree of all the 22 analytical journals (Figure 40) shows new interconnections and also the separation of some specialty journals (22, 13, 14, 17, and 21) and the two "big" journals *Analytical Chemistry* and *Journal of Chromatography*.

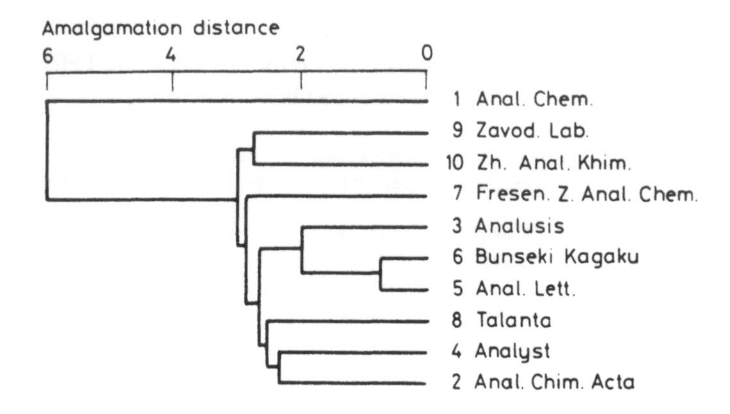

FIGURE 38. Cluster tree of the group of the main broad-based analytical journals.

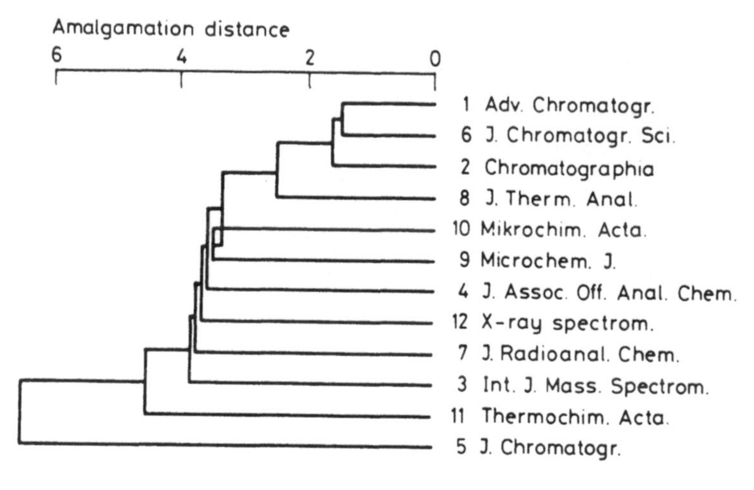

FIGURE 39. Cluster tree of the group of specialty analytical journals.

IV. INFLOW-OUTFLOW OF ANALYTICAL CHEMICAL INFORMATION

Analytical chemistry as a subfield of chemistry is characterized by the prompt response to results and methods of other fields applicable for its purposes, urged by its own inner activity and the growing needs of science and technology for analytical results. This activity manifests itself in a strong information transfer with subfields and fields of other disciplines. This information transfer was investigated by Bujdosó et al.[75] by citation analysis using the 1978 *Journal Citation Reports*.[70] The extent to which authors in one journal reference the work of scientists in another journal can be taken to be a good measure of cross-journal information flow. A study of the flow of these information quanta through the citation network connecting journals and also the fields they represent can map the interdependency of basic research in analytical chemistry.

As representing the subfield of analytical chemistry, a group of the major broad-based analytical journals was selected according to Petruzzi.[74] References from one of the broad-based journals were taken as *inflow*, and citations to one of them as *outflow*

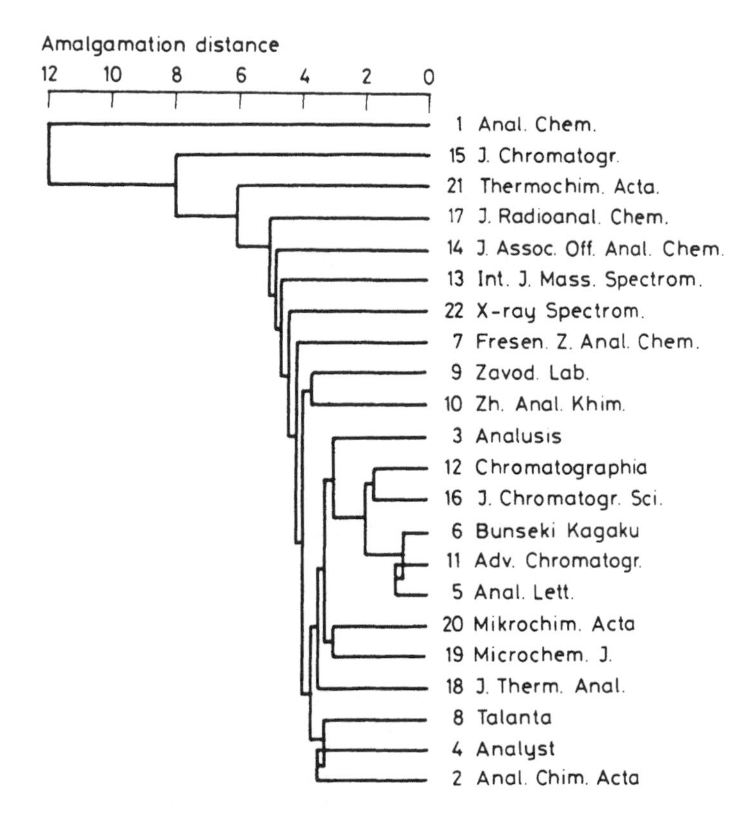

FIGURE 40. Cluster tree of the analytical journals.

to or from analytical chemistry. In order to be able to measure the flow of information among the fields of science, journals were classified into subfields according to Narin.[71]

The information flow among the group of main broad-based and specialty analytical journals as well as journals of all other disciplines is shown in Figure 41. Figure 42 shows the information flow map for chemistry subfields drawn from the broad-based journal data base including the group of specialty analytical chemistry journals. The in- and outflow data are in percentages of the total analytical information flow in the field of chemistry. The area of circles representing the subfields is proportional to their virtual size, i.e., the number of published items multiplied by the appropriate impact factor of the journals, then summed for the whole subfield.

According to Figure 42, the main information sources of analytical chemistry within overall chemistry are the specialty analytical, general, and physical chemistry journals. The rate of in- and outflow changes in every subfield; for applied, general, inorganic, and nuclear chemistry, analytical chemistry is an emitter of information, i.e., all these fields are drawing information from analytical chemistry.

Figure 43 shows the information flow between analytical chemistry and other disciplines. The main information sources of analytical chemistry are physics, clinical medicine, and earth and space sciences, i.e., analytical chemistry absorbs information from all these disciplines.

V. PEER REVIEW IN ANALYTICAL CHEMISTRY JOURNALS

De Solla Price[76] has suggested that the number of published scientific articles that are "good" increases with the square root of the total number of published articles.

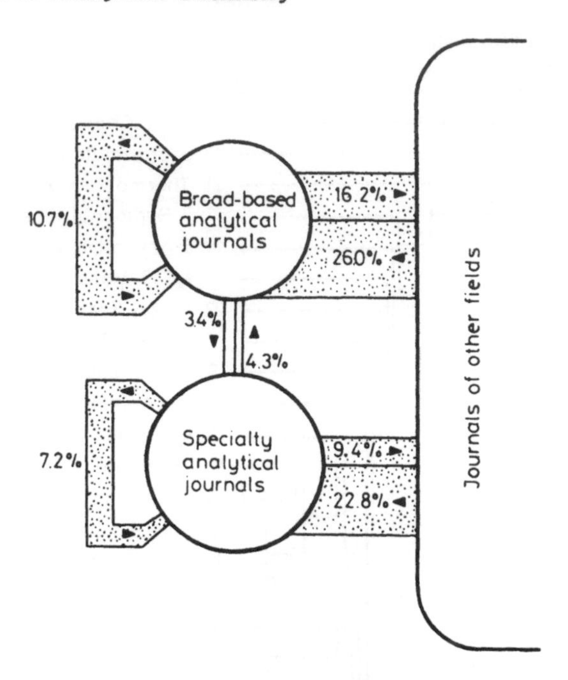

FIGURE 41. The information flow among the group of main broad-based and specialty analytical journals as well as the journals of all other disciplines. Data are in percentages of whole information traffic. (From Bujdosó, E., Braun, T., and Lyon, W. S., *Trends Anal. Chem.*, 1, 268, 1982. With permission.)

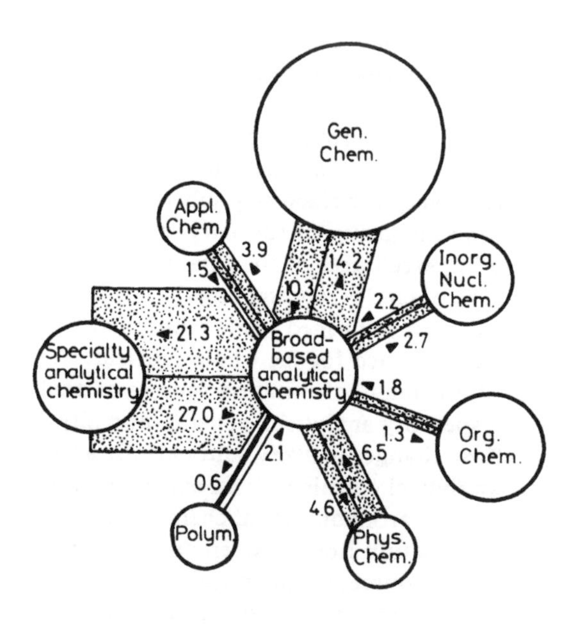

FIGURE 42. The information flow among the group of main broad-based analytical journals and the subfields of chemistry in percentages of the total analytical information flow in the field of chemistry. (From Bujdosó, E., Braun, T., and Lyon, W. S., *Trends Anal. Chem.*, 1, 268, 1982. With permission.)

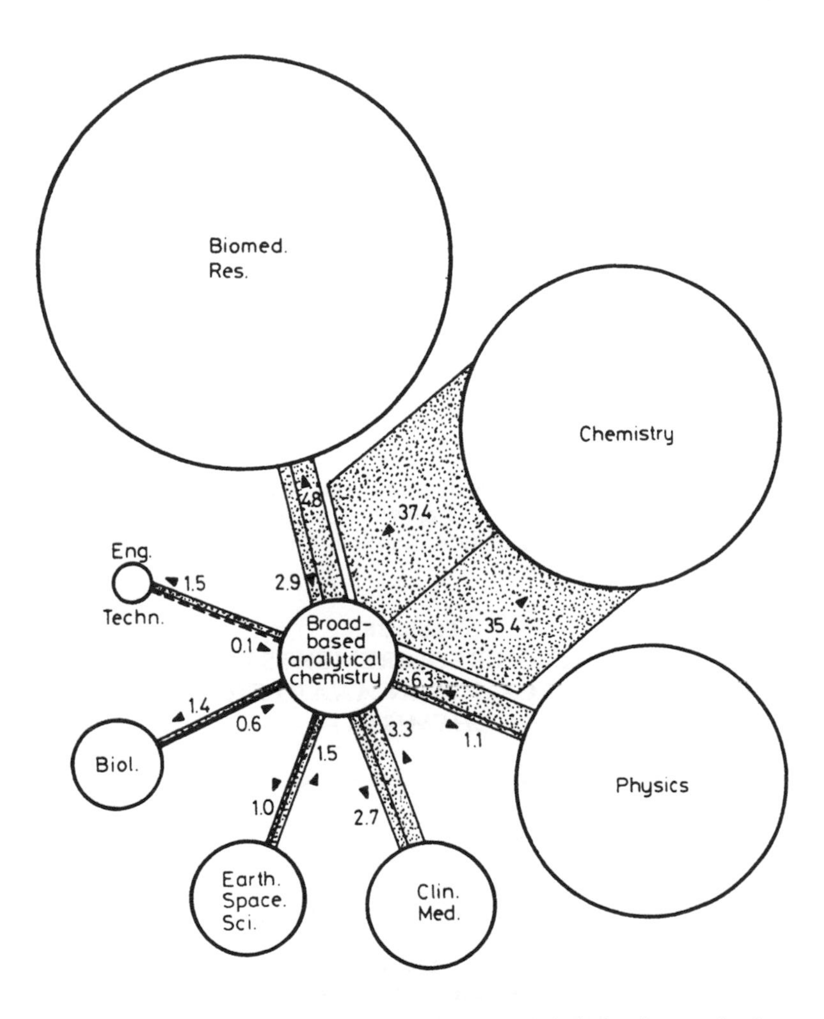

FIGURE 43. The information flow among analytical chemistry and other fields of science in percentages of the total analytical information flow. (From Bujdosó, E., Braun, T., and Lyon, W. S., *Trends Anal. Chem.*, 1, 268, 1982. With permission.)

This implies that if total publication volume continues to increase, the percentage of those that are "good" decreases. The burden on readers to sort "wheat from chaff" would increase. This might increase the need for lower acceptance rates, resulting in an increased quality of the manuscripts accepted for publication. The present acceptance rate is about 55% for journals receiving at least 300 manuscripts per year and about 70% for smaller journals.[77] The acceptance rate is also higher for papers in the "hard" (i.e., physical) sciences than in the "soft" ones. It can be as low as 10% in some social science journals and as high as 90% in mathematics and physics.[78] The systems used by research journals to evaluate and select manuscripts for publication are seen to be of vital importance to science. Not surprisingly, one finds the behavior of individuals taking part in these systems, especially the behavior of anonymous referees consulted by journal editors, to be an issue of great sensitivity and heated debate. Within this debate, accusations of deliberate referee bias are sometimes presented, usually by aggrieved authors. Such deliberate bias is not, however, thought to occur very often, and it can usually be identified by the editor, with or without a second or third referee's report or author's letter of appeal. Its influence upon the overall pattern of growth of the literature of a discipline can be considered minimal. A very thorough analysis of this quite controversial issue was published in 1982.[79]

Table 37
EDITORIAL DECISION CONCERNING MANUSCRIPTS IN THE JOURNAL *ANALYTICAL CHEMISTRY*

Decision	Number	Percent
Accepted	299	68.5
Rejected	107	24.5
Inactivated	15	3.4
In process	12	2.7
Withdrawn	4	0.9

From Petruzzi, J. M., *Anal. Chem.*, 48, 875A, 1976. With permission. Copyright 1976, American Chemical Society.

Table 38
REVISION REQUIREMENTS FOR MANUSCRIPTS IN *ANALYTICAL CHEMISTRY*

Action requested	Number	Percent
Minor revision	177	40.5
Major revision	130	29.7
Accepted without change	19	4.3
Rejected or withdrawn	111	25.4

From Petruzzi, J. M., *Anal. Chem.*, 48, 875A, 1976. With permission. Copyright 1976, American Chemical Society.

In the field of analytical chemistry the peer review process — as far as we know — has been investigated in some detail only for the journal *Analytical Chemistry*. Thus, Petruzzi[80] followed the fate of 437 manuscripts submitted for publication from February 21 through August 20, 1975. The results of this study are shown in Table 37. The "in process" manuscripts represent papers either out for second review or in the hands of the authors for revision. The revision requirements for manuscripts are shown in Table 38.

Two or more reviewers reviewed 95% of the manuscripts, 16% were sent to referees (a term reserved for persons consulted when the two reviews did not agree), and 33.3% of the rejected manuscripts were seen by a referee. Data collected on the reasons for the rejection of manuscripts showed that 52% of the rejections could be attributed to reasons such as "insufficient new information", "already published", "lack of originality", and "not relevant". As other reasons, insufficient supporting data (13%) and questions regarding scientific merit and validity of data (12%) are mentioned.[80]

VI. GATEKEEPING PATTERNS IN THE PUBLICATION OF ANALYTICAL CHEMISTRY RESEARCH

The invention of a mechanism for the systematic publication of scientific work may well have been the key event in the history of modern science.[81] The main channel through which this publication flows is provided by the scientific journals.

Thus Gordon[82] states: "Publication of papers in primary research journals is widely accepted as having a central role to play in the continuance of science as an intellectual and social activity. In particular it is recognized as being both a means by which researchers are able to establish and advance themselves professionally, and the medium through which contributions are made to a discipline's body of *ratified knowledge*. Consequently, journal editors, in controlling systems of manuscript evaluation and selection, occupy *powerful strategic* positions in the collective activity of their discipline. The practices and preferences which they adopt in their roles as editors are therefore of considerable significance."

In an earlier paper[83] the same author had said, "editors and referees who control the access to the coveted pages of scientific journals, particularly those who 'gatekeep' for the more prestigious publications, hold *vital strategic* positions in the orchestration of science."

There are three main groups of questions considered of paramount importance in the whole complex problem of editorial gatekeeping in journals.

1. How does this gatekeeping system function and on what criteria do journal editorial board members base their decisions?
2. What is the structure of the powerful body of journal gatekeepers? In other words, who is chosen to perform gatekeepings tasks and to which countries do they belong?
3. How can the evaluators be evaluated? In other words, what special characteristics give these individuals the right to sit in judgment?

It is useful to concentrate here on gatekeeping in analytical chemistry publications to try to find answers to the last two questions. The first question is not dealt with as there is no answer to it that is specific to analytical chemistry. The gatekeepers of analytical chemistry journals use criteria similar to those used by science journal gatekeepers[84] in general, and these criteria have been quite thoroughly investigated.[82]

To find answers to the other two questions, Braun and Bujdosó[85] have analyzed the national composition of gatekeeping boards of analytical chemistry journals and sought correlations between the number of gatekeepers, their citation rates, and the number of analytical papers published by scientists from the country in question. A comparison has been made of the citation rates of the gatekeepers of organic, inorganic, and analytical chemistry journals, and the citation data for the gatekeepers of analytical chemistry journals have been scrutinized.

As a data base, 14 analytical, 9 organic, and 4 inorganic chemistry journals — considered among the most significant in their respective fields — were chosen. The group of analytical chemistry journals was further divided into a subgroup of seven broad-based analytical journals that deal with all branches of analytical chemistry[74] and seven specialty journals. The interrelationship between these two groups of journals was discussed in Section III of this chapter.

The broad-based journals were *Analytical Chemistry, Analytical Letters A and B, Analusis, Analyst, Analytica Chimica Acta, Mikrochimica Acta,* and *Talanta.* The specialty journals were *Chromatography, Journal of Chromatography, Journal of Radioanalytical Chemistry, Journal of Thermal Analysis, Radiochemical and Radioanalytical Letters, Spectrochimica Acta, Part A,* and *Spectrochimica Acta, Part B.*

These journals were examined with respect to the nationality of their gatekeepers. Braun and Bujdosó considered as gatekeepers the editor(s)-in-chief, the editor(s), the managing editor(s), and the members of the editorial and advisory boards, but not the technical editor(s).[85] For the characterization of publication activities of the various countries in the field of analytical chemistry, papers published in the 14 analytical chemistry journals in 1978 were counted and grouped according to countries. In that year 1560 papers were published in the 7 broad-based journals and 3610 in the whole group of 14 analytical chemistry journals considered.

As a measure of "effectiveness", "eminence", "impact", "importance", "influence", "quality", "significance", or "utilization" of the scientific work of the gatekeepers,[86,87] the number of citations was considered. As a data base the 1970 to 1974 cumulative volumes of the *Science Citation Index* published by the Institute for Scientific Information in Philadelphia were chosen, and the citations under the gatekeepers' names were counted.

Table 39 shows the national distribution and citation counts of the gatekeepers for the chosen analytical journals. The number of gatekeepers from various countries and their specific citation rates vary between wide limits. About one half of the gatekeepers for analytical chemistry journals originate from only four countries (U.S., U.K., France, and West Germany).

In co-opting scientists for journal gatekeeping functions, many points of view are probably taken into account. Here the attention is limited to only two factors affecting the "visibility" of an individual with regard to his/her selection as a potential gatekeeper, i.e., publication productivity in some broad-based or specialized analytical field and the impact of the research.

Accordingly, correlations were sought, on the one hand, between the number of gatekeepers from a given country and the number of papers published yearly in the two groups of journals (broad-based and specialty) from that country and, on the other, between the number of gatekeepers and their citation rates. The results are shown in Figure 44 as log-log plots; r and m represent the correlation coefficient and the slope, respectively. The overall correlation coefficient is $r = 0.8$. It appears that the two factors examined have an equal effect upon the selection of the gatekeepers.

The value of m in the relationship $y = ax^m$, i.e., the exponent of publication productivity of quality, is usually below 1.0, its mean value being 0.71. This shows that the relationship is nonlinear. In other words, to increase the number of gatekeepers from a given country, a progressively larger effort is necessary.

Along with the regression lines the standard deviation limits are also shown. Those cases that fall outside these limits are regarded as deviating significantly from the general group behavior. For instance, taking the broad-based analytical journals as an example (Figure 44 b), the U.S., U.K., France, Belgium, Switzerland, and Denmark give more gatekeepers than would be expected from their publication activity. In the citation rates of the gatekeepers of the same journals, it is again the U.S., U.K., France, and Belgium that figure foremost, along with Canada, the Netherlands, and Italy. On the editorial boards of the broad-based analytical chemistry journals, India, South Africa, and Israel are relatively underrepresented.

The impact factors of chemistry journals differ over about the same relative range as the citation rates of their gatekeepers. Do the scientific quality and distinction of the gatekeepers have a repercussion upon their gatekeeping activities?

An answer to this question can be given by comparing the impact factors of the journals with the citation rates of their gatekeepers. The data were taken from Zsindely et al.[88] Tables 40 to 42 contain data for organic, inorganic, and analytical chemistry journals, respectively.[85]

The citation frequencies of the gatekeepers are roughly 3:2:1 for the organic, inorganic, and analytical chemistry journals, whereas the average impact factors are almost the same for the organic and inorganic journals, and that for the analytical journals is

Table 39

NATIONAL DISTRIBUTION AND CITATION RATES OF GATEKEEPERS OF BROAD-BASED AND SPECIALTY ANALYTICAL CHEMISTRY JOURNALS

		Analytical chemistry journals		Broad-based analytical chemistry journals		Specialty analytical chemistry journals	
Rank[a]	Country	Number of gatekeepers	Citation rate	Number of gatekeepers	Citation rate	Number of gatekeepers	Citation rate
1	U.S.	154	220	81	260	73	180
2	U.K.	75	240	49	165	26	370
3	France	59	90	40	85	19	90
4	West Germany	31	230	13	120	18	305
5	Hungary	30	120	5	415	25	65
6	Czechoslovakia	24	120	2	310	22	95
7	U.S.S.R.	23	375	6	685	17	255
8	Japan	19	325	9	285	10	360
9	Canada	19	220	10	105	9	360
10	Belgium	18	75	12	80	6	60
11	Italy	17	95	4	30	13	115
12	Switzerland	15	165	8	170	7	160
	Austria	15	120	6	115	9	130
14	The Netherlands	14	90	6	50	8	120
15	Sweden	12	300	8	320	4	260
16	Australia	9	290	7	345	2	120
17	Poland	8	150	4	235	4	65
18	Denmark	7	345	6	345	1	345
19	East Germany	6	85	2	120	4	70
	Israel	6	390	2	495	4	345
	South Africa	6	330	2	845	4	70
	Yugoslavia	6	140	2	50	4	180
23	Brazil	5	65	2	55	3	90
	India	5	260	1	140	4	290
	Romania	5	300	3	200	2	440
26	Mexico	3	1	2	1	1	1
	Norway	3	67	1	35	2	85
28	Greece	2	36	2	35	—	—
	New Zealand	2	123	1	110	1	135
	Spain	2	15	—	—	2	15
31	Egypt	1	—	1	—	—	—
	Argentina	1	7	—	—	1	7
	Other	6	15	5	20	1	2
Total		608		302		306	
Average			193		200		162

Note: Citation rates are given as the average number of citations per gatekeeper over a 5-year period (1970—1974) and were rounded off by the program used.

[a] According to the number connected with all 14 analytical chemical journals.

From Braun, T. and Bujdosó, E., *Talanta*, 30, 161, 1983. With permission.

only about 25% lower (Table 43). These differences in impact factor are not significant. Figure 45 shows a plot of the data. Between the specific citation rates of the gatekeepers and the impact factors of their journals there is a significant correlation (r = 0.6). The slope of the regression line is 0.4, which means that the prestige of journals is only slightly raised by increasing the prestige of the gatekeepers.

The distribution of the gatekeepers of various countries and their citation rates is uneven, just as is the distribution of the scientific productivity,[89] area, and national

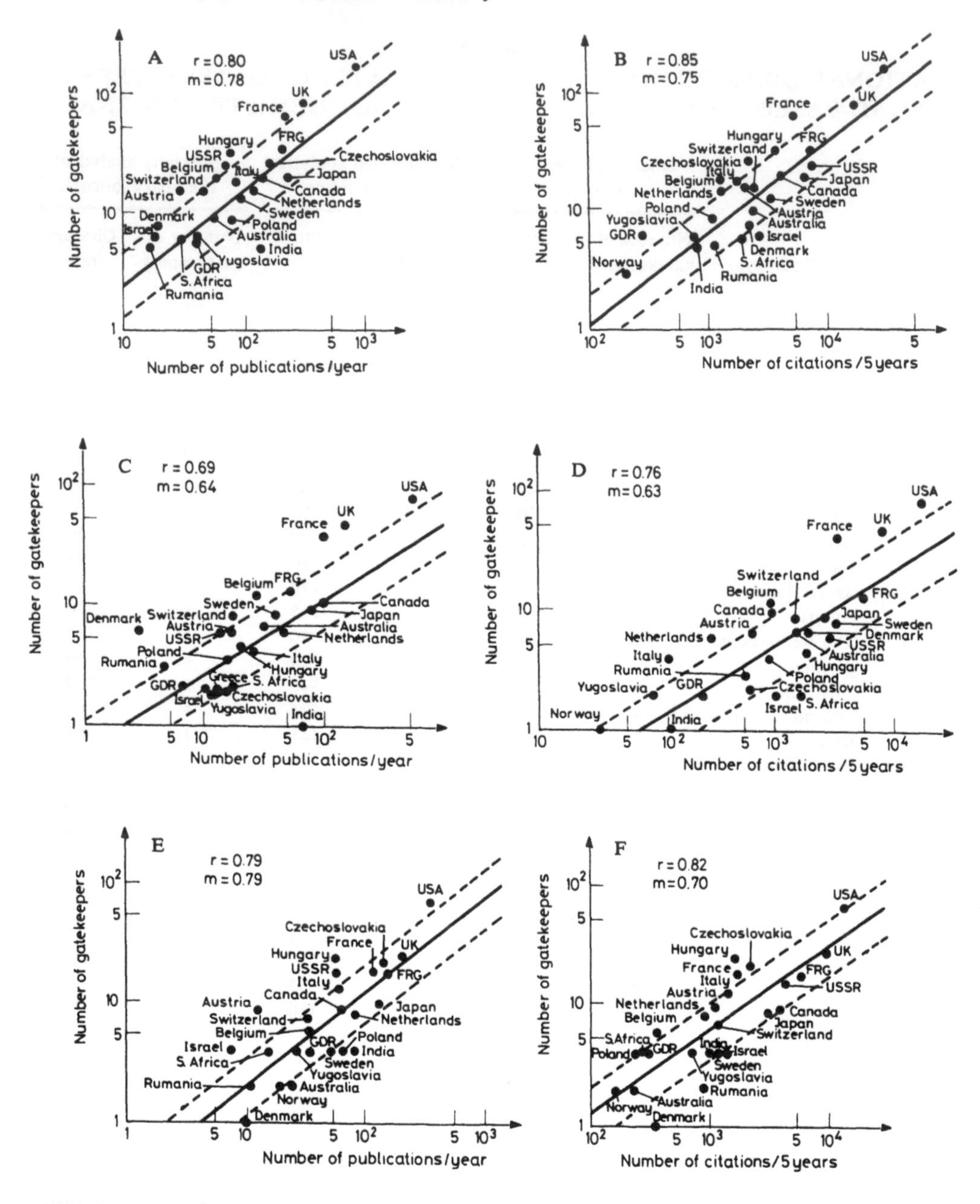

FIGURE 44. Relationships between the number of gatekeepers and the number of publications for a given country (A to C) and between the number of gatekeepers and their citation rates (D to F) for 14 analytical chemistry journals (A and D), 7 analytical chemistry journals of broad-based character (B and E), and 7 specialty analytical chemistry journals (C and F). (From Braun, T. and Bujdosó, E., *Talanta*, 30, 161, 1983. With permission.)

wealth of these countries. The Lorenz curve is a graphical presentation of the concentration, i.e., the inequality of distribution, of various items over a population. A point on the Lorenz curve shows what percentage of the countries examined are endowed with a given percentage of the item plotted on the vertical axis. For example, in Figure 46 we see that 25 (i.e., 60%) of the 42 countries* dealt with can muster between them

* There are 42 countries represented in the editorial boards of the international chemistry journals.[88] Our data also refer to 42 countries in the case of analytical, inorganic, and organic chemistry journals.

Table 40
IMPACT FACTORS OF ORGANIC CHEMISTRY JOURNALS AND CITATION DATA FOR THEIR GATEKEEPERS

		Gatekeepers		
Journal	Impact factor	Number	Total citations	Citations per capita
Carbohydr. Res.	1.431	53	6,638	125
J. Organomet. Chem.	2.331	7	9,888	1,413
Monatsch. Chem.	0.831	38	13,584	357
Org. Magn. Resonance	1.379	39	16,553	424
Org. Mass. Spectrom.	1.253	37	15,178	410
Synthesis	1.758	24	27,026	1,126
Synth. Commun.	1.178	30	18,360	612
Tetrahedron	1.745	71	60,285	849
Tetrahedron Lett.	2.114	65	60,097	925

From Braun, T. and Bujdosó, E., *Talanta*, 30, 161, 1983. With permission.

Table 41
IMPACT FACTORS OF INORGANIC CHEMISTRY JOURNALS AND CITATION DATA FOR THEIR GATEKEEPERS

		Gatekeepers		
Journal	Impact factor	Number	Total citations	Citations per capita
Inorg. Chim. Acta	2.859	79	42,130	533
Inorg. Nucl. Chem. Lett.	1.141	26	14,441	555
J. Inorg. Nucl. Chem.	1.017	73	28,635	392
Z. Anorg. Allg. Chem.	1.333	38	15,220	400

From Braun, T. and Bujdosó, E., *Talanta*, 30, 161, 1983. With permission.

only 8% of the gatekeepers of broad-based analytical journals, the remaining countries having the other 92% of the gatekeepers. In this way Figure 46 a to d tells us that 72, 80, and 83% of the editors and 74, 90, and 96% of the gatekeeper citations of analytical, inorganic, and organic chemistry journals, respectively, stem from only 8 countries (20%).

In Lorenz-type graphs an even distribution is represented by the diagonal. The divergence of the Lorenz curve from the diagonal is reflected in the Gini index, which is a measure of the normalized area between the diagonal and the Lorenz curve. It ranges from zero, i.e., complete equality, to unity, i.e., total inequality.

Upon comparing the Gini indexes shown in Figure 46 it becomes clear that the greatest inequality in the distribution of the citation rates of the gatekeepers appears for the inorganic and organic chemistry journals, $G = 0.81$ and 0.86, respectively. On the other hand, the national distribution of the gatekeepers of specialty analytical chemistry journals is the most even ($G = 0.62$). For comparison, the Gini index of world scientific publication productivity is $G = 0.91$; the indexes for the distribution of total national production and of population are $G = 0.85$ and 0.75, respectively.[90]

Participation in gatekeeping for some scientific journals represents a form of reward

Table 42
IMPACT FACTORS OF ANALYTICAL CHEMISTRY JOURNALS AND CITATION DATA FOR THEIR GATEKEEPERS

Journal	Impact factor	Gatekeepers: Number	Total citations	Citations per capita
Anal. Chem.	2.803	17	3,193	188
Anal. Lett. Part A and Part B	0.884	62	15,471	250
Analusis	0.774	50	6,169	123
Analyst	1.702	42	8,664	206
Anal. Chim. Acta	1.488	40	7,795	195
Chromatographia	1.394	33	8,978	272
J. Chromatogr.	1.846	46	11,543	251
J. Radioanal. Chem.	0.890	49	4,535	93
J. Thermal Anal.	0.506	34	3,625	107
Mikrochim. Acta	0.779	42	8,830	210
Radiochem. Radioanal. Lett.	0.515	74	6,546	88
Spectrochim. Acta, Part A	1.023	34	5,589	164
Spectrochim. Acta, Part B	1.621	33	15,527	471
Talanta	0.907	51	10,831	212

From Braun, T. and Bujdosó, E., *Talanta*, 30, 161, 1983. With permission.

Table 43
COMPARISON OF ORGANIC, INORGANIC, AND ANALYTICAL CHEMISTRY JOURNALS (MEAN AND STANDARD DEVIATION)

Characteristics	Organic chemistry	Inorganic chemistry	Analytical chemistry
Average impact factor	1.56 ± 0.47	1.59 ± 0.85	1.22 ± 0.62
Average number of gatekeepers per journal	40 ± 20	54 ± 26	43 ± 14
Average citations per gatekeeper	693 ± 415	470 ± 85	202 ± 97

From Braun, T. and Bujdosó, E., *Talanta*, 30, 161, 1983. With permission.

for the person involved. Participation in many journals is naturally a cumulated reward, and in such cases no doubt the "Matthew effect" is at work[91] (see also Chapter 7, Section II). It has been shown that scientists who are already known, i.e., more "visible", are given more reward than others who may have similar scientific achievements but are less visible and/or less widely known.

Among the 608 gatekeepers of the 14 analytical chemistry journals considered, 61 are members of 2 editorial boards, and 19 participate in 3 or more (Table 44).

The citation rate of gatekeepers of analytical chemistry journals can be well described by a logarithmic normal distribution curve (Figure 47). The median corresponds to M = 100 citations per 5 years; in other words, 50% of the gatekeepers receive over 20 citations per year, whereas 68% of them get between 3 and 100 yearly citations ($M \pm \sigma$) (Table 45).

The results of this chapter can be summarized as follows.

1. In the case of analytical chemistry journals, whether broad-based or specialized in character, a correlation has been shown to exist between the number of gatekeepers of a given nationality and the number of analytical papers published in these groups of journals by scientists in the country concerned.

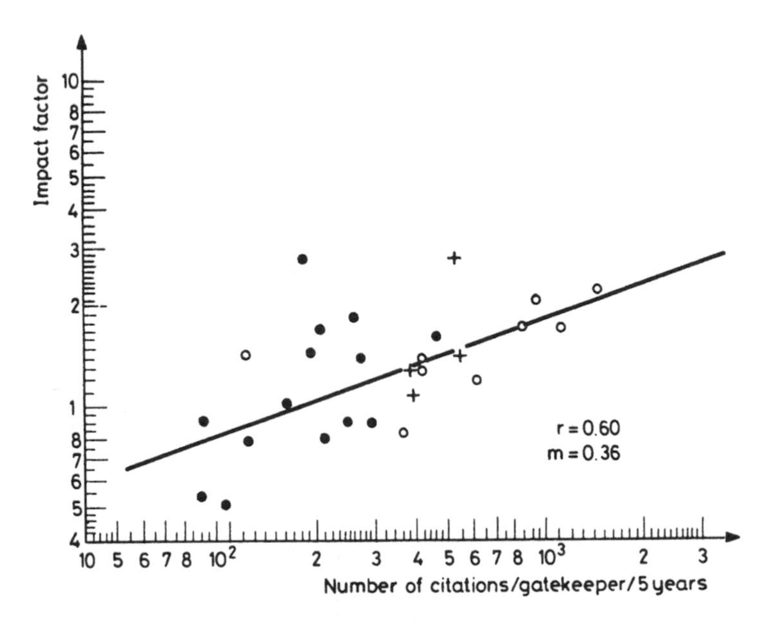

FIGURE 45. Correlation between the citation rate of gatekeepers and journal impact factors: • analytical chemistry, O organic chemistry, and + inorganic chemistry. (From Braun, T. and Bujdosó, E., *Talanta*, 30, 161, 1983. With permission.)

2. For the journals of analytical chemistry a correlation also exists between a number of gatekeepers and their citation rate. This correlation is of about the same strength for broad-based and specialized analytical chemistry journals.
3. The relationship between the number of gatekeepers (n) and their publication productivity (i.e., their citation rate, N) is $n \sim aN^m$, where m shows values between 0.6 and 0.8. In other words, for the journals mentioned so far, the effort needed for a country to increase its number of gatekeepers by 1, e.g., from 50 to 51 or from 100 to 101, would be 2 and 3 times, respectively, as large as that necessary to effect an increase from 10 to 11.
4. There is yet another correlation between the impact factors of the journals and the citation rates of their gatekeepers. In the relationship $n \sim b\,(\text{impact factor})^m$, the exponent $m = 0.4$ is smaller than in the corresponding relationship involving the number of gatekeepers. The citation rate of the gatekeepers is, therefore, reflected in the impact factors of the journals.
5. The citation rates of the gatekeepers of organic and inorganic chemistry journals are 3 and 2.5 times, respectively, those for the gatekeepers of the analytical chemistry journals, and the impact factor of the latter journals is about 0.3 below that for the other 2 types of journal.
6. Of the 608 gatekeepers of analytical chemistry journals, 237 have an average citation rate of more than 20 per year, 113 have over 50 citations per year, and 58 are cited more than 100 times a year. The quality or impact of their research has an immediate effect on the prestige (impact factor) of the journals. Among the 608 editors, 61 are members of more than 1 board; 19 of more than 2; and 9 of more than 3 boards of analytical chemistry journals.

The results show that 75% of the positions of power influencing the publication of new results in almost all areas of analytical chemistry are concentrated into the hands of scientists from no more than ten countries of the world.

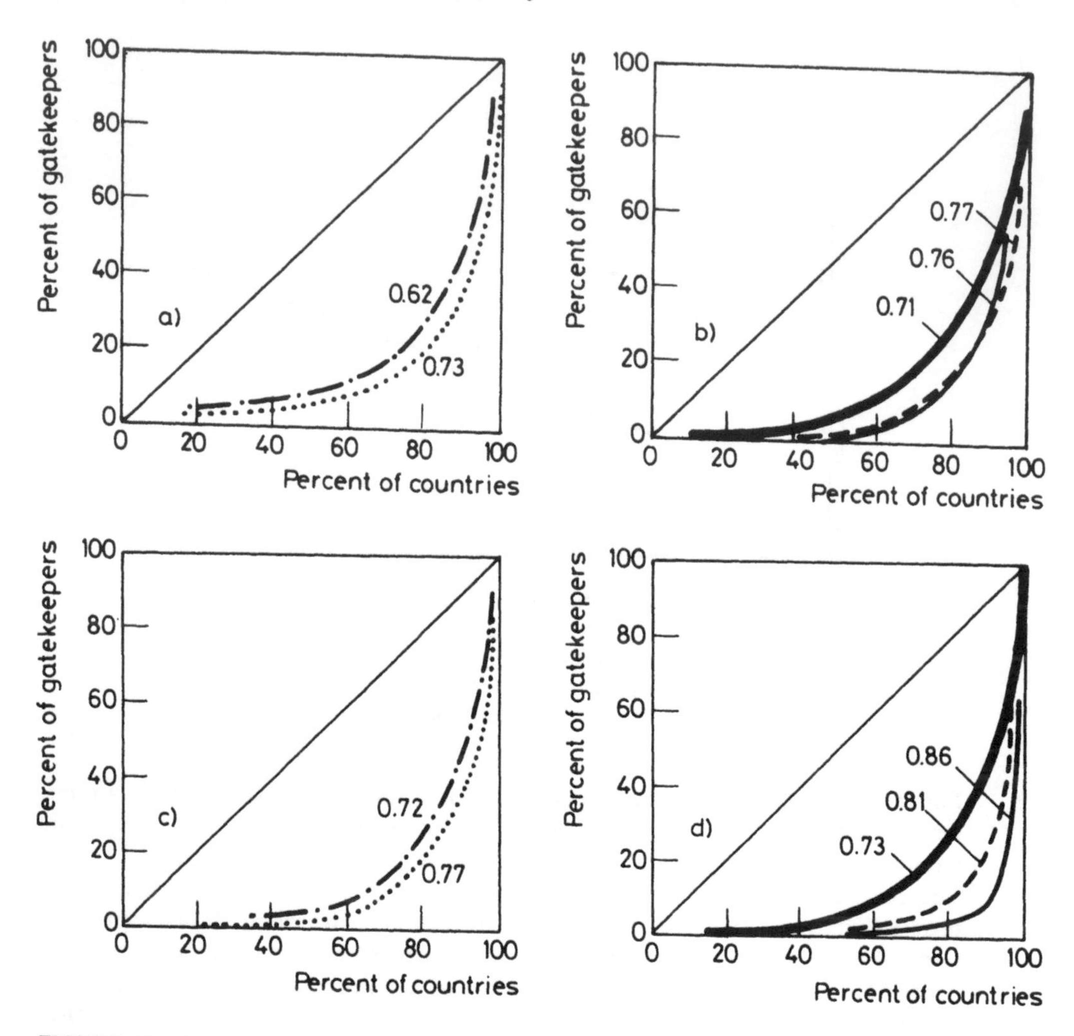

FIGURE 46. Lorenz curves for the national distribution of gatekeepers of various groups of analytical chemistry journals (a and b) and their citation rate (c and d). ••••••• broad-based, —·—·— specialty, ▬▬▬ analytical chemistry (the sum of the first two groups), ——— organic chemistry, and — — — inorganic chemistry. The corresponding Gini indexes are also shown. (From Braun, T. and Bujdosó, E., *Talanta*, 30, 161, 1983. With permission.)

VII. PUBLICATION SPEED (TIME) OF PAPERS IN ANALYTICAL CHEMISTRY JOURNALS

Modern science has developed a particular mechanism of communication which began with the appearance of the first scientific journals in the 17th century and which has remained basically the same ever since. Briefly this mechanism is based on the selective publication of fragments rather than complete treatises. It is this selective concern with fragments of knowledge, represented primarily by journal articles, that enables science to function effectively and is responsible for its phenomenal growth and development. It follows from this that the primary literature represents the only genuine record of scientific achievement.

J. M. Ziman[81] has said, "The aim of the scientist is to create, criticise, or contribute to a rational consensus of ideas and information. If you accept this as a general notion, you will agree that the results of research only become completely scientific when they are published." Implicit in this assumption is the crucial role of communication in the

Table 44
LIST OF SCIENTISTS PARTICIPATING IN THE EDITORIAL ACTIVITIES OF MORE THAN TWO ANALYTICAL CHEMISTRY JOURNALS

Rank	Name	Number of journals	Journal
1	Hoste, J.	6	*Analusis*
			Analyst
			Anal. Chim. Acta
			Mikrochim. Acta
			J. Radioanal. Chem.
			Radiochem. Radioanal. Lett.
2—4	Duyckaerts, G.	5	*Analusis*
			Anal. Chim. Acta
			J. Chromatogr.
			Spectrochim. Acta, Part A
	Guiochon, G.	5	*Anal. Chem.*
			Analusis
			Chromatographia
			J. Chromatogr.
			Mikrochim. Acta
	Pungor, E.	5	*Analyst*
			Anal. Chim. Acta
			Anal. Lett. A
			Mikrochim. Acta
			Talanta
5—8	Alimarin, I. P.	4	*Zh. Anal. Khim.*
			J. Radioanal. Chem.
			Radiochem. Radioanal. Lett.
			Talanta
	Belcher, R.	4	*Analyst*
			Anal. Lett. A
			Mikrochim. Acta
			Talanta
	Ruzicka, J.	4	*Analyst*
			Anal. Chim. Acta
			Mikrochim. Acta
			J. Radioanal. Chem.
	West, T. S.	4	*Analusis*
			Analyst
			Anal. Chim. Acta
			Mikrochim. Acta
9—16	Boumans, P. W. J. M.	3	*Analusis*
			Spectrochim. Acta, Part A
			Spectrochim. Acta, Part B
	Dyrssen, D.	3	*Analyst*
			Anal. Chim. Acta
			Anal. Lett. A
	Haerdi, W.	3	*Analusis*
			Anal. Chim. Acta
			Anal. Lett. A
	Huber, J. F. K.	3	*Chromatographia*
			J. Chromatogr.
			Mikrochim. Acta
	Marcus, Y.	3	*Analusis*
			Anal. Lett. A
			J. Chromatogr.

Table 44 (continued)
LIST OF SCIENTISTS PARTICIPATING IN THE EDITORIAL ACTIVITIES OF MORE THAN TWO ANALYTICAL CHEMISTRY JOURNALS

Rank	Name	Number of journals	Journal
	Kosta, L.	3	*J. Radioanal. Chem.*
			Mikrochim. Acta
			Radiochem. Radioanal. Lett.
	Simon, W.	3	*Anal. Chim. Acta*
			Chromatographia
			Mikrochim. Acta
	Zuman, P.	3	*Analyst*
			Anal. Chim. Acta
			Anal. Lett. A

From Braun, T. and Bujdosó, E., *Talanta*, 30, 161, 1983. With permission.

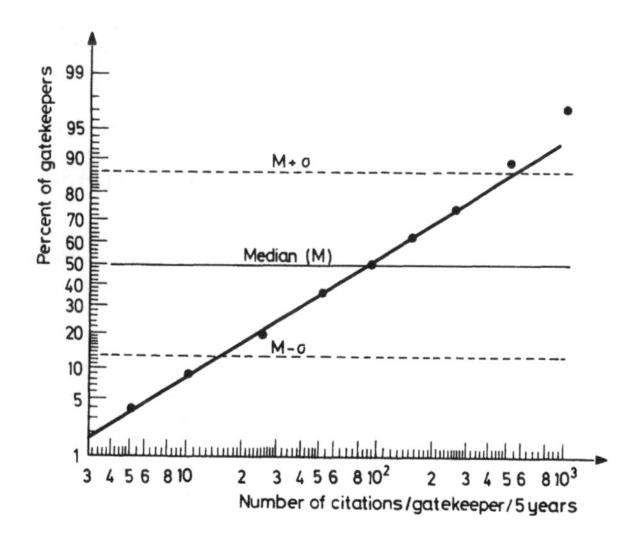

FIGURE 47. Distribution of the number of citations for gatekeepers of journals in analytical chemistry plotted on Gauss paper with logarithmic abscissa. (From Braun, T. and Bujdosó, E., *Talanta*, 30, 161, 1983. With permission.)

growth of science. Ziman[89] has stated that "the communication system of contemporary science is the product of an organic evolution, functionally inseparable from the evolution of the scientific activity that it serves." This can be supplemented by the accepted fact that publication in a professional journal is a critical factor in determining a scientist's visibility and mobility within the social stratification system of science.

The crucial role primary journals play in the scientific communication process involves the acute need to evaluate how well are they fulfilling their role. Attempts have been made in the past to do this using ratings. However, more recently scientometric methods have been devised which are altogether simpler and more objective.[92] Authors of research papers usually complain about how much time it takes to get their manuscripts published and how outdated the research findings are at publication as a consequence of lengthy publication times. As this problem is of paramount importance to the whole communication process, Braun and Nagydiósi-Kocsis[93] have investigated one

Table 45
MOST CITED SCIENTISTS PARTICIPATING IN THE EDITORIAL ACTIVITIES OF ANALYTICAL CHEMISTRY JOURNALS

Rank	Name	Country	Journal	First author citations (5 years, 1970—1974)
1	Bellamy, L. J.	U.K.	*Spectrochim. Acta, Part A*	3341
2	Stahl, E.	West Germany	*Chromatographia*	1932
3	Gillespie, R. J.	Canada	*Spectrochim. Acta, Part A*	1694
4	Irving, H. M. N. H.	South Africa	*Analyst*	1645
5	Ito, M.	Japan	*Spectrochim. Acta, Part A*	1578
6	Kiselev, A. V.	U.S.S.R.	*Chromatographia*	1520
7	Poster, G.	U.S.	*Spectrochim. Acta, Part A*	1160
8	Bieman, K.	U.S.	*Mikrochim. Acta*	1078
9	Perrin, D. D.	Australia	*Anal. Lett. A*	1048
10	Busev, A. J.	U.S.S.R.	*Anal. Lett. A*	1007
11	Rao, C. N. D.	India	*Spectrochim. Acta, Part A*	1006
12	Guilbault, G. G.	U.S.	*Anal. Lett. A, B*	962
13	Shimagouchi, T.	Japan	*Spectrochim. Acta, Part A*	841
14	Tanaka, M.	Japan	*Analusis, Mikrochim. Acta*	834
15	Miller, F. A.	U.S.	*Spectrochim. Acta*	788
16	Gray, P.	U.K.	*Analyst*	780
17	Snyder, L. R.	U.K.	*J. Chromatogr.*	777
18	Belcher, R.	U.K.	*Anal. Lett. A, Analyst, Mikrochim. Acta, Talanta*	773
19	Zuman, P.	U.S.	*Analyst, Anal. Acta, Anal. Lett. A*	769
20	Meites, L.	U.S.	*Anal. Lett. A*	756
21	Clark, D. T.	U.S.	*Spectrochim. Acta, Part A*	736
22	Bayer, E.	West Germany	*Chromatographia*	724
23	Horning, E. C.	U.S.	*Anal. Lett. B, J. Chromatogr.*	717
24	Wunderlich, B.	U.S.	*Thermal Anal.*	704
25	James, A. T.	U.K.	*J. Chromatogr.*	690
26	Price, W. C.	U.K.	*Spectrochim. Acta, Part A*	686
27	Balaban, A. T.	Romania	*Radiochem. Radioanal. Lett.*	677
28	Kemula, W.	Poland	*Analusis*	625
29	Sawicki, E.	U.S.	*Anal. Lett. A, Mikrochim. Acta*	618
30	Alimarin, I. P.	U.S.S.R.	*J. Radioanal. Chem., Radiochem. Radioanal. Lett., Talanta*	616
31	Nicholson, R. S.	U.S.	*Anal. Chem.*	607
32	Pungor, E.	Hungary	*Analyst, Anal. Chim. Acta, Anal. Lett. A, Mikrochim. Acta, Talanta*	603
33	Savvin, S. B.	U.S.S.R.	*Analusis*	602
34	Charlot, G.	France	*Analusis*	590
35	Fritz, J. S.	U.S.	*Talanta*	586
36	Willis, J. B.	Australia	*Anal. Chim. Acta*	585
37	Gehrke, C. W.	U.S.	*J. Chromatogr.*	567
38	Murray Royce, W.	U.S.	*Anal. Chem.*	564
39	Lord, R. C.	U.S.	*Spectrochim. Acta, Part A*	563
40	Hadzi, D.	Yugoslavia	*Spectrochim. Acta, Part A*	548
41	Zolotov, Yu. A.	U.S.S.R.	*Anal. Chim. Acta, Mikrochim. Acta*	544
42	Bowen, H. J. M.	U.K.	*Radiochem. Radioanal. Lett.*	533
43	Wendland, W. W.	U.S.	*Thermal Anal.*	519
44	Zlatkis, A.	U.S.	*J. Chromatogr., Chromatographia*	517
45	Sjovell, J.	Sweden	*Anal. Lett. B*	511
46	Grant, D. M.	U.S.	*Spectrochim. Acta, Part A*	499
47	Marcus, Y.	Israel	*Analusis, Anal. Lett. A, J. Chromatogr.*	494

Table 45 (continued)
MOST CITED SCIENTISTS PARTICIPATING IN THE EDITORIAL ACTIVITIES OF ANALYTICAL CHEMISTRY JOURNALS

Rank	Name	Country	Journal	First author citations (5 years, 1970—1974)
48	Samuelson, O.	Sweden	*J. Chromatogr.*	486
49	Schwab, G. M.	West Germany	*J. Chromatogr.*	483

of the steps — that of journal publication time — in the long way leading to the publication of a paper in analytical chemistry journals. Namely, between the hypothesis and publication of research results in analytical chemistry, time is needed to design and complete the research leading to the paper, to write it, to edit it, etc. All these steps precede publication (journal handling times). A rough estimate indicates that publication lapse (PL), i.e., the time between submission of a manuscript to a journal and its ultimate publication, is only about 25% of the total.[94] Psychologically, the interval seems to be much longer, for once the scientist has completed his work, time hangs heavy. Authors are not only eager to see their own results in print but also may be concerned that similar results from someone else might be published before theirs. Authors generally do not recognize that the major lapses in time are spent in setting up the investigation, doing it, and writing the manuscript. Even if editing and journal processing were instantaneous, the research idea would be on average many years old by the time it was read.

This chapter reports some data about the PL of manuscripts in some analytical journals. PL is generally and automatically defined as the time that elapses from the acceptance of the manuscript until the date of appearance of the issue in which it is published. Median PL has been defined as the time during which the first 50% of the body of manuscripts in a given issue is getting through the publication process. Likewise characteristic is the length and location on the abscissa (on the cumulative publication lapse curves) of the time interval during which the main cohort, i.e., the central one half of between 25 to 75%, of the manuscripts is getting through the process (interquartile range). PLs in various journals may range from several months up to about 1 year with 6 to 12 months as an average.[20,95]

Figures 48 and 49 show the cumulative PL curves for papers in *Analytical Chemistry* (U.S.) and *Journal of Radioanalytical Chemistry* (Switzerland and Hungary). Table 46 reports some comparative data for PLs of papers in four leading analytical chemistry journals.

Figure 50 shows the PL curves and histograms for papers in *Radiochemical and Radioanalytical Letters* for each year from 1969 to 1981. Table 47 presents the so-called 90% PL, i.e., the time necessary for 90% of the papers to be published in *Radiochemical and Radioanalytical Letters.* You will note that in contrast to the data for the journals of Figures 48 and 49 and Table 46, for which all PLs are given in months, the data for *Radiochemical and Radioanalytical Letters* are in weeks.

Another approach of the publication time problem could be the following. As shown in Chapter 4, the results of research and the publication itself become obsolete with time. Let us suppose that a countdown begins when the experimental work and its evaluation by intellectual processes are completed, i.e., a manuscript is ready for publication. It is reasonable to allow a maximum of 10% delay during the publication time, i.e., the advising and editing, printing, the author's galley proofreading as well as mailing, etc. of a paper. Taking into account the 4-year half-life of a paper in analytical chemistry,[28] this time can be at maximum 8 months.[20] The leading analytical journals seem to be trying to keep their publication times under this value.

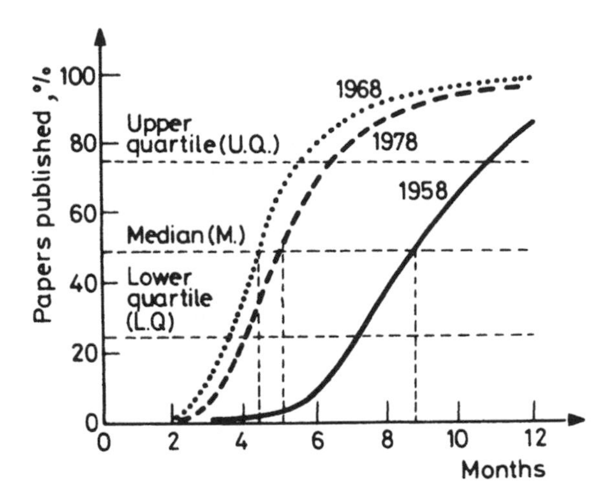

FIGURE 48. PL of papers in *Analytical Chemistry.* (From Braun, T. and Nagydiósi-Kocsis, Gy., *Radiochem. Radioanal. Lett.*, 52, 327, 1982. With permission.)

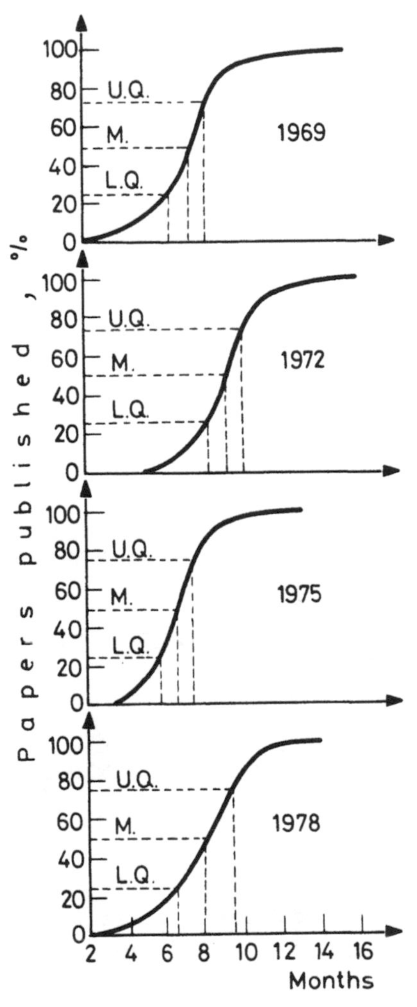

FIGURE 49. PL of papers in *Radioanalytical Chemistry.* V. Q., M., and L. Q. are upper quartile, median, and lower quartile, respectively. (From Braun, T. and Nagydiósi-Kocsis, Gy., *Radiochem. Radioanal. Lett.*, 52, 327, 1982. With permission.)

Table 46
PUBLICATION TIMES IN MONTHS OF PAPERS IN FOUR LEADING ANALYTICAL JOURNALS

Journal	Year	Median	Interquartile range	Range
Anal. Chem.	1958	8.7	7.2—10.9	1—26
	1968	4.4	4.6—5.4	2—14
	1978	5.1	3.9—6.4	3—17
Anal. Chim. Acta	1958	6.8	5.8—7.6	3—11
	1968	5.2	4.7—7.6	1—21
	1978	5.7	4.7—7.0	1—17
Analyst	1958	—	—	—
	1968	6.2	5.0—7.3	4—40
	1978	5.6	4.2—6.8	3—19
Talanta	1958	—	—	—
	1968	7.1	5.9—8.6	2—18
	1978	8.1	6.1—10.3	2—22

From Braun, T., Bujdosó, E., and Lyon, W. S., *Anal. Chem.*, 52(6), 617A, 1980. With permission. Copyright 1980, American Chemical Society.

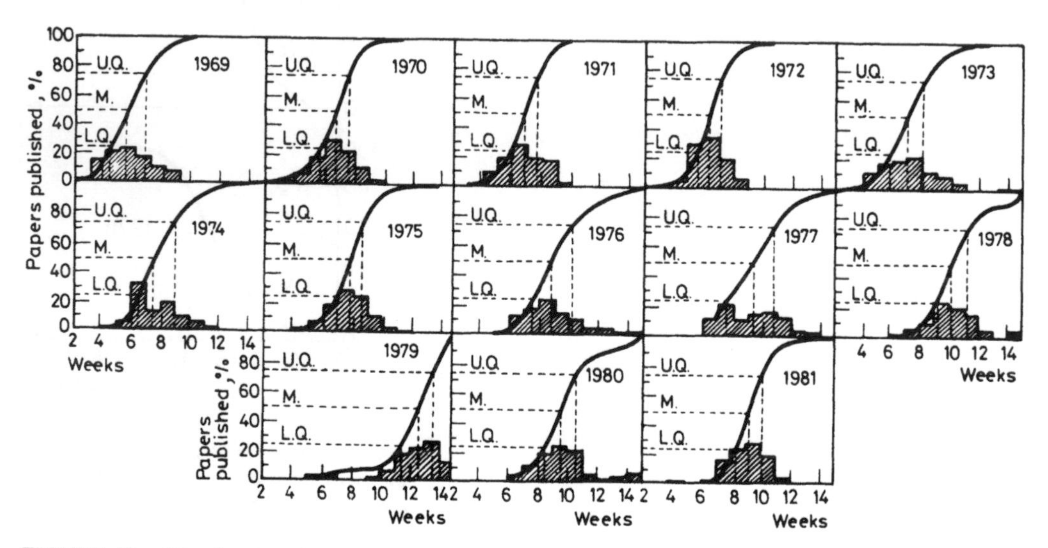

FIGURE 50. PL of papers in *Radiochemical and Radioanalytical Letters* for the years 1969 to 1981. V. Q., M., and L. Q. are upper quartile, median, and lower quartile, respectively. (From Braun, T. and Nagydiosi-Kocsis, Gy., *Radiochem. Radioanal. Lett.*, 52, 327, 1982. With permission.)

Table 47
PLs OF PAPERS IN *RADIOCHEMICAL AND RADIOANALYTICAL LETTERS*

Year	Publication delay (weeks)	Year	Publication delay (weeks)
1969	7.8	1976	12.0
1970	8.4	1977	12.2
1971	8.4	1978	12.6
1972	7.9	1979	14.4
1973	9.4	1980	13.0
1974	10.4	1981	10.4
1975	9.6	1969—1981	12.0

From Braun, T. and Nagydiósi-Kocsis, Gy., *Radiochem. Radioanal. Lett.*, 52, 327, 1982. With permission.

Chapter 7

EPIDEMIOLOGY AND ORIGIN OF A REVOLUTION WITHIN A REVOLUTION: FLOW INJECTION ANALYSIS

I. INTRODUCTION

Analytical chemistry may well have been the locus of two coupled scientific revolutions within the past 25 years, one revolution within the other. Continuous flow analysis (CFA), invented in 1957, ushered in the era of automated analysis. Flow injection analysis (FIA), described in 1975, changed the concept of CFA and made possible very rapid simplified analysis. These developments can be viewed within the context of certain historical and philosophical ideas. For example, normal science as defined by Kuhn[96] means "research firmly based upon one or more past scientific achievements, that some particular community acknowledges for a time as supplying the foundation for its further practice." These achievements, referred to as paradigms, share two characteristics. They are "sufficiently unprecedented to attract an evolving group of adherents away from competing modes of scientific activity" and "sufficiently open ended to leave all sorts of problems for the redefined group of practitioners to resolve." Transformations of paradigms are scientific revolutions and "the successive transition from one paradigm to another via revolution is the usual developmental pattern of mature science."

What is the mechanism whereby the new paradigm is substituted for the old? One rather provocative hypothesis, proposed by Goffman and Newill,[97] is to treat such events as epidemics. The epidemic model thus patterns the transmission of ideas after the transmission of disease. "These principles are a part of epidemiology," Goffman says. The elements necessary for the spread of an infectious disease are a specified population and an exposure to infectious material. The members of the population belong to one of three basic classes at a given instant in time: (1) infectives, those members who are host to the infectious material; (2) susceptibles, those who can become infectives given contact with infectious material; and (3) removals, those who have been removed for one of a variety of reasons, i.e., death, immunity, hospitalization, etc. These latter members may have been either susceptibles or infectives at the time of their removal.

Within the field of analytical chemistry the invention of CFA in 1957 resulted in the overthrow of an old paradigm and the substitution of a new. But again in 1975 the development of FIA seems to have resulted in a second revolution, with modification and change of the new paradigm. Though our emphasis is on FIA, a brief background in the history of CFA will be a useful preliminary to the study of FIA.

This chapter will explore briefly the development of CFA, which led to one of the most widely used automated analytical systems, and FIA, which opened CFA to rapid, simplified methods. The discussion will consider these processes in the light of scientific revolution and epidemic theory. An attempt will be made to delineate some of the forces that act upon an applied discipline (i.e., analytical chemistry); it is these forces that bring about revolution and change.

II. CFA AND FIA: BACKGROUND AND DEFINITIONS

Three excellent reviews of CFA and FIA have recently appeared. Snyder[98] discusses mainly the technical aspects of the two techniques, Stewart[99] attempts to review the early history of FIA (and to some extent CFA), and Mottola[100] in a literate and fair

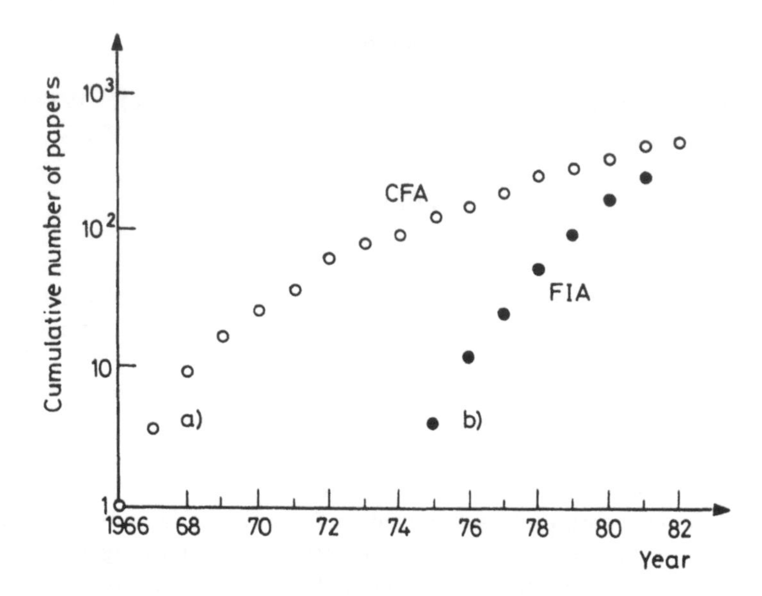

FIGURE 51. Growth of publications in (a) CFA and (b) FIA.

appraisal shows the history of two techniques, a history that as far as FIA is concerned has not been without controversy and some acrimony regarding the sequence of events leading to its discovery.[101,102] Yes, even scientific revolutions show seeds of discord, charges of deviationism, and reams of revisionist writings.

CFA as defined here is the technique of processing large numbers of liquid samples through an essentially human-free automated pipeline flow system in which the samples are segmented, usually by air. Contemporary thought held that air segmentation was necessary to prevent sample mixing; such segmentation severely limits the scope and speed of automated analysis and, of course, adds a complicating step. FIA we define as the unsegmented technique and, bowing to the will of the originators of the term, restrict it to those methods which meet the following three requirements: sample injection, controlled dispersion, and reproducible timing.[103]

III. CFA

Mottola[100] notes in his review that CFA rests, as do all techniques, upon a continuous background of analytical work. But as Mottola points out, Skeggs,[104] in developing a modular continuous flow concept of analysis, "thought of what nobody else thought before. He proposed a novel manner of sample reagent, mixing and transport to detection, and emphasized the utility inherent in all modular set-ups." Skeggs, in 1957, literally revolutionized the industrial and clinical laboratory by enunciating the flow concept; commercial automated instruments soon followed.

Spackman et al.[105] in the following year published in *Analytical Chemistry* a detailed account of their automated amino acid analyzer, and this paved the way for rapid clinical determinations. As Goffman[97] might say, the infection had begun to spread, and Figure 51 shows the growth curves for world publication of papers on CFA. A citation count taken as of the end of 1982 shows total cumulative citations to Skeggs of 356, whereas to Spackman et al. the total citations are an astounding 3611.

IV. FIA

FIA represents a revolution within a revolution; but it is in a sense only an advance in CFA. As most reviewers concede, advancements in CFA presaged the discovery of FIA; what was needed was someone to consolidate the information found in numerous papers (e.g., Nagy et al.,[106] Stewart et al.,[107] and Eswara Dutt and Mottola,[108] to make possible automated analysis without air segmentation. Ruzicka and Hansen[109] in their classic 1975 paper coined the term "flow injection analysis" and clearly outlined the basis of the method. Their position as revolutionary rulers was strengthened and consolidated in a series of experimental and theoretical papers in subsequent years. Thus at the end of 1982 their 1975 paper had received 124 citations, almost a factor of 10 more than any other paper that reported nonsegmented automated analysis during the middle 1970s. That an epidemic was again rampant is indicated in Figure 51 which shows the growth curve for the world population of FIA.

One FIA paper stands out in the citation picture: the Ruzicka and Hansen 1975 paper[109] has 124 citations; the closest rival for priority claim, Stewart has but 53 split between 15 to a 1974 oral presentation[107] and 38 to a 1976 paper in a specialty journal.[110] Ruzicka and co-workers completely dominated the early publication picture for FIA: between 1975 to 1978 they published 12 papers in major journals that described the method, developed theory, and broke new ground in applications. There are 478 total citations to these papers.

Stewart, working with the U.S. Department of Agriculture, chose to publish in his usual manner. His articles appeared in analytically peripheral specialty journals. He did not follow up, as Ruzicka and co-workers did, with explanatory and expansionary papers. Ruzicka seized the opportunity and exploited it. Ruzicka was also within the establishment and published in an established nonspecialty journal with a reasonably high impact factor. Publication in specialty journals restricts the audience. FIA (and CFA also) beautifully demonstrates the danger of publishing broad-based research in specialty journals. The number of citations per year to the first Ruzicka and Hansen article is 15, almost 10 times the journal average of 1.5.

Being within or a part of an establishment always helps in getting general acceptance for new ideas. Ruzicka and Hansen were at the Chemistry Department of the Technical University of Denmark, Copenhagen at the time FIA was developed. Ruzicka was well known to the world analytical chemistry community by two previous remarkable achievements on substoichiometric analysis and ion-selective electrodes. But since the Ruzicka-Hansen article is the most frequently cited publication by the members of the Chemistry Department of the Technical University of Denmark, the prominence of the authors is again insufficient by itself to account completely for its popularity. Another possible reason for the large number of citations is that the paper has serious methodological implications. "Methodology papers often appear among articles that are most-cited, but not all methodology papers are heavily cited."[111] A closer analysis of citations to the 1975 journal article by Ruzicka and Hansen demonstrates two major concepts that have to be taken into account here: the *invisible colleges* and the *Matthew effect.*

The concept of invisible college was first used in the 17th century to refer to the collection of scientists who eventually formed the Royal Society of London. Such a group, like the faculty of a college, represented a collection of intellectuals with a sense of allegiance to each other, frequently interacting both professionally and socially. The adjective "invisible" was used because the membership of the group was not confined to a particular academic setting and was not obvious to persons who had little knowledge about 17th century science.

In 1963, De Solla Price[1] reintroduced the term invisible college to describe the existence of such groups in modern sciences. These groups are collections of scientists who live in disparate geographical locations, but who often attend the same conferences, publish in the same journals, invite each other to give presentations at their home institutions, and share preprints of their research endeavors. It is through the political power of such colleges that many of the changes in a science are made.

The concept of invisible colleges does not connote a conspiratorial or subversive process in which power groups engage in Machiavellian struggles for control of a science. Invisible colleges exist in every scientific topical field, and virtually every researcher/scientist can be identified as belonging to some particular invisible college.[112] De Solla Price,[1] in introducing the term, described how scientists must form into reasonably small, homogeneous groups to protect themselves from the pressures of "big science", the "publish or perish" ethic, and the tremendous growth of scientific literature. Such forces are sufficiently powerful to overwhelm virtually any scientist who attempts to remain isolated and yet make significant advances in his or her field. It is through the formation of small, informal collegial groups that individual scientists can gain sufficient support and power to become effective.

The citation analysis of the Ruzicka and Hansen article demonstrates the existence of an invisible college within the FIA field. Of the citations to the Ruzicka-Hansen article, 64% have been made by researchers who are or had been associated with the authors, suggesting that there exists an invisible college connected with the Technical University of Denmark group's approach to FIA. Gradually this college expanded to include analytical chemists at many institutions in Brazil, the U.K., the U.S., and elsewhere.

This college had a major impact on analytical thinking about FIA during the second one half of the 1970s. First it was the members of this college who cited and promoted the Ruzicka-Hansen article. In addition at the end of the 1970s some industrial links were created and the promotion began to be supplemented by an instrument marketing activity.

The second important concept which can be used to understand the dramatic number of citations to the Ruzicka-Hansen article is the Matthew effect. The effect was first described by Merton[91] and is named for a verse in the Gospel according to Matthew which says "For unto every one that hath shall be given, and he shall hath abundance; but from him that hath not shall be taken away even that which he hath" (Matthew 25:29).

One of the manifestations of the Matthew effect is that most persons who publish journal articles receive very few citations to their articles (they have not). There are, however, a few who publish in scientific journals whose articles attract a large number of citations (they have abundance). The latter are researchers of established reputations who are likely to serve on editorial boards, to receive grants, to travel and lecture, and who have solid connections to industrial firms that manufacture products derived from their research. Initially FIA was promoted by the members of the invisible college, and the number of citations to the Ruzicka-Hansen article showed a rapid growth. A likely contributing reason to this rapid growth of citations is the Matthew effect. The article had one co-author whose publications had already been demonstrated to "hath abundance". In addition, the relatively large number of early citations to this article increased the probability that the article would attract the attention of other researchers. In short, the initial promotion of the Ruzicka-Hansen article by the FIA invisible college was enhanced by the Matthew effect. A high number of citations resulted.

As a conclusion to this chapter we can say that two main types of forces can be used to account for changes within any science. The first is the *cognitive* (internal, scientific) force which concerns the quality and amount of evidence that supports an innovation in a science before it is adopted. It is the cognitive aspect which researchers are trained to emphasize. Theories, hypotheses, methodologies, and statistics — these are the cognitive aspects of science which supposedly determine if a piece of research is to be accepted. The second force involved in scientific change is the *noncognitive* (external, extra-scientific) which pertains to socioeconomic, personal career, and reward-regulated issues and internal politics.

Good science does not always result in rapid acceptance and personal recognition; the cognitive force is a necessary but not a sufficient condition. Noncognitive forces also play their part in technical accomplishment. Here we have looked at both cognitive and noncognitive forces; indeed they are usually closely entwined and not easily separated. Such approaches are always incomplete; the *ex post facto*, outside observer can never know all the contributing factors that produce a result. But such attempts are often useful because they can contribute to a better understanding of the issues that surround a skyrocketing[31] field such as FIA.[113]

Chapter 8

AUTHORS OF ANALYTICAL CHEMISTRY PAPERS

I. PRODUCTIVITY OF ANALYTICAL AUTHORS, LOTKA'S LAW

Authors are central to the cumulative literature of any subject. No matter what editors and publishers add to or subtract from manuscripts, the published results remain unequivocally the authors' contributions to knowledge. Studies analyzing the characteristics of subject literatures have increasingly focused attention on the quantity and rates at which authors publish in their fields.

Of course, this measure is far from perfect. One paper may report an epoch-making discovery while another describes only some relatively trivial redetermination of certain experimental values or variations in the conditions for a reaction. There is a difficulty also over multiple authorship, which is increasing as a proportion of all work. Because publications are used as a basis to help in deciding appointments and promotions, there may be a tendency to try to spin out a given amount of work into more papers than are really necessary. Whatever it is that is measured by a paper count is perhaps better described as productivity than as quality. But at least it is something measurable by which one can compare the output of different scientists. Though there are flagrant anomalies in individual cases, there does seem on the whole to be reasonably good correlation between productivity in terms of papers and eminence as judged in other ways.

Using paper counts, one can investigate the frequency distribution of scientific productivity. What kind of distribution would one expect? Would one guess that the productivity of most scientists clusters around some average number of papers, with only exceptionally good or bad scientists producing many more or many fewer? In fact, one finds quite a different kind of distribution which is nothing like the sort of normal distribution that one gets for chance events. In Figure 52, plotted as a graph, curve 1 does not resemble the familiar bell shape of curve 2. The great majority of people who publish at all do not publish more than a few papers. Many manage to produce one or two as research students, but never any more. Only a selected minority produce more than a few.

In 1926 Lotka[114] proposed an inverse-square law relating authors of scientific papers to the number of papers written by each author. Lotka was interested in determining "if possible, the part which men of different calibre contribute to the progress of science." He counted the number of authors and the number of contributions made by each of them in the decennial index of *Chemical Abstracts*, 1907 to 1916; only the letters A and B were covered. Similar data were also collected from *Geschichtstafeln der Physik* (J. A. Barth, Leipzig, 1910). Lotka plotted on a logarithmic scale the number of authors against the number of contributions made by each author and found that in each case the points were closely scattered about a straight line having a slope of approximately two. On the basis of these data, Lotka deduced the general equation

$$mn^c = k \qquad (18)$$

where m is the fraction of authors making n contributions each, and c and k are constants. For the special case of c = 2 (inverse-square law of scientific productivity), the value of the constant k was 0.6079. This meant that the proportion of authors who contribute a single item should be 0.6079, or just over 60% of the total number of authors. The observed figures for the proportion of authors making 1 contribution

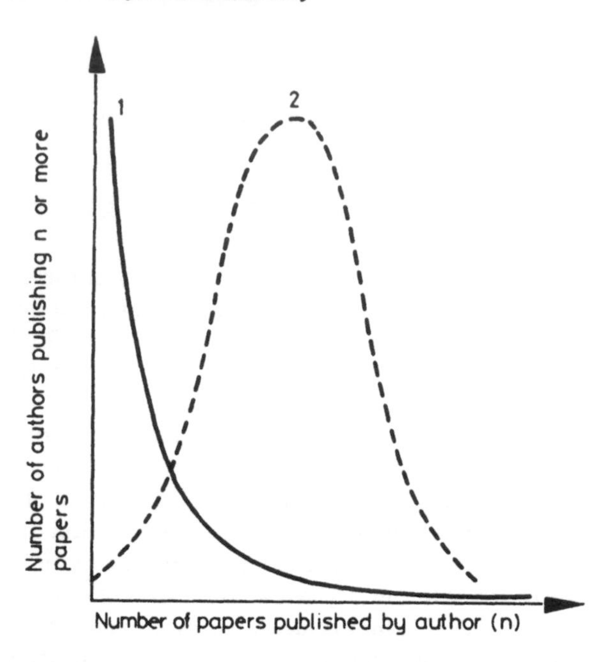

FIGURE 52. Productivity of authors of scientific papers. Curve 1, real distribution; curve 2, hypothetical distribution.

each were 57.9% for the *Chemical Abstracts* data and 59.2% for the Auerbach data. Lotka[114] summarized his findings: "In the cases examined, it is found that the number of persons making 2 contributions is about one fourth of those making one; the number making 3 contributions is about one ninth, etc; the number making n contributions is about $1/n^2$ of those making one; and the proportion of all contributors, that make a single contribution, is about 60 percent."

In other words, for every 100 authors contributing 1 article each, there would be 25 others contributing 2 articles each ($100/2^2 = 25$), about 11 contributing 3 articles each ($100/3^2 = 11.1$), about 6 contributing 4 articles each ($100/4^2 = 6.25$), and so on.

Lotka's law can also be expressed by the equation

$$a_n = a_1/n^2, \qquad n = 1, 2, 3.... \tag{19}$$

where a_n is the number of authors contributing n papers each. De Solla Price[1] has suggested that one half of all the scientific papers is contributed by the square root of the total number of scientific authors.

In recent years several attempts have been made to verify Lotka's law and to apply it to the literature of various fields.[115] For example, in some cases an exponent different from two was required to provide a good fit with empirical data. The range of exponents extended from 1.5 to 4.0.

Coming now to analytical chemistry there were only a very few attempts to verify Lotka's law. Vlachy[116] found in the author indexes of *Analytical Abstracts*, 1954 and 1970 (January to December) and 1971 and 1974 (July to December), exponents ranging from 3.5 to 4.0. Braun et al.[28] found an exponent around 2.0 working with a world activation analysis bibliography for the period 1936 to 1970.[117]

Though Lotka's law of scientific productivity may be applicable with or without modification to the literature of analytical chemistry or to that of its subfields, it seems obvious that more thorough investigations are necessary.

In some studies on Lotka's law, when considering a paper written by a number of authors, it was assumed each co-author contributed one paper. Coile[118] has pointed out that such an assumption can lead to erroneous conclusions. Clearly, contributing part of a paper as a co-author in a team effort is not the same as writing a paper by oneself. It is extremely difficult, if not impossible, to ascertain the extent of contribution of each co-author. In Lotka's original study, joint contributions were credited to the senior author only.

II. CITATION ANALYSIS IN ANALYTICAL CHEMISTRY

It is emphasized by Nalimov and Mulchenko[6] that if the progress of science is attempted to be interpreted through an information model, particular attention must be paid to the analysis of the specific coding language of scientific references and citations within the flow of scientific information. The references found in a scientific publication can be regarded as the particular, specific language of scientific information. All scientific papers are based on the multitude of already expressed ideas. These ideas may be rather new and unknown for the reader of a paper. Despite this fact, the author of the paper does not deal with the explanation of these ideas at any length; instead he refers to the previous papers in which these ideas were first explained or stated. The system of scientific references and citations is the code that enables the author to write concise papers without repetitions. The theories on which the research is based are conveyed in this language. According to the statement of Kessler,[119] the references reflect the spiritual atmosphere in which the publication was born. Many scientists understand and apply this language to such an extent that they are able to form an idea about a paper without actually reading it. Let us suppose that we look through a mathematical periodical and bump into a lengthy and hardly digestible paper with a short abstract that is also indigestible. Then, we will turn our attention to the references. If we find familiar names or publications, immediately an idea can be formed about the subject of the paper.

The contents of scientific publications are, therefore, coded or indexed by the author in the references. In a certain sense this coding reflects the richness in ideas of a publication much better than other methods of coding, e.g., with descriptors. Consequently, the system of bibliographic references can be regarded as a special information language. Its statistical analysis has proved to be an efficient means of investigating the development of scientific information flow.[120] By applying this method, the progress and trends of scientific fields and the infiltration of new ideas or methods into frontier areas can be followed. Also the impact of the work of scientists can be evaluated by starting from their impact on the information flow of scientific literature.

Based mainly on the above considerations, there is a tendency in the scientometric literature[121] to use citations in order to inject a quality factor into the evaluation of scientific publications.[6] The claim is that such a factor can be determined by counting not the number of publications but the number of citations a publication of a set of publications receives in the scientific literature. The image underlying this claim is that the impact of a scientific paper is in its influence on subsequent research papers, and each instance of such influence will manifest itself by the influenced paper referring to the influencing paper.

Although the authors of the present monograph believe in the utility of citation analysis as a research tool, they also have to mention that the above philosophy is not free of deficiencies, and neither is the practical application of it. In fact, citation measures have been blamed on various grounds, including that they simply describe and do not explain, that they are dubious until we understand the psychology and sociology

Table 48
AVERAGE LEVEL OF CITABILITY OF ANALYTICAL CHEMISTRY PAPERS PUBLISHED BY SOVIET AUTHORS IN 1965

			From the total	
Number of authors	Number of authors cited in *SCI*	Total number of citations	Citations in world analytical literature	Citations in Soviet analytical literature
119	90	1098	508	590

Note: Average level of citability is 1098:90 or 12.2; in world literature, 503:90 or 5.6; and in Soviet literature, 590:90 or 6.6.

From Orient, I. M., *Zavodsk. Lab.*, 33, 1383, 1967. With permission.

of who cites whom and why, that the *Science Citation Index (SCI)* is not a true reflection of the citation structure, and that citations are not all of equal value and intent.[122,123]

Nevertheless, citations are used with increasing frequency as both practical and conceptual tools.[6,121] It is a question of matching a given purpose with a tool of appropriate reliability and not expecting exceedingly much from a given application.

According to Orient and Markusova,[25] all scientific publications can be regarded as the result of two factors. The first factor includes the ideas taken by the author from the papers of others and coded in the references. The second contains the new ideas of the author produced in connection with the notions known before. When the paper of a given author is cited by future papers, just these new ideas are utilized. Therefore, by investigating citations the dissemination of new ideas and methods can be followed; the interrelationship of the development of ideas and the internal structure of scientific research can be revealed. Citation measures the contribution of the cited author to the information flow. However, it must be mentioned that this criterion has substantial limitations. While the great impact of a paper witnesses its efficiency or perhaps its value, the lack of impact does not necessarily indicate the worthlessness of the given paper. In investigating the impact of a paper it must be taken into account that this depends also on the "popularity" of the periodical in which the paper was published, the speed of the research front in the given field, the citation habits of the given field or country, etc.

The most-used data base for citation analysis is the *SCI* compiled by the Institute for Scientific Information, Philadelphia, which is available from 1964 both in printed form and on magnetic tape. Concerning all aspects and details of the data base, citation indexing, and analysis, we refer to Garfield's[120] recent monograph.

Orient[124] carried out elaborate investigations on the basis of the *SCI* in order to determine in what measure the papers of Russian analysts were cited in the world chemical literature during 1965. As the subject of the investigations, a group of authors was chosen who had published not less than 2 papers yearly for 3 years in *Zhurnal Analyticheskoi Khimii* and *Zavodskaya Laboratoriya.* Of the 119 authors chosen, 90 were cited 1098 times. The data given in Table 48 were subjected to further analyses. In the 1965 volume of the *SCI,* 29 of the above 119 authors were not cited at all. Of the authors cited, 1 appeared more than 100 times; 2 more than 70 times; 2 more than 40 times; 4 more than 30 times; 9 more than 20 times; 6 more than 10 times; 18 more than 5 times; and 38 were cited less than 5 times. The distribution of citations referring to the various subfields of analytical chemistry are given in Table 49.

Finally, an attempt was made to determine the "quality level" of Russian analytical

Table 49
DISTRIBUTION OF CITATION OF ANALYTICAL PAPERS ACCORDING TO SUBFIELDS

Methods	Total citation	In foreign literature	In Soviet literature
Electrochemistry	296	99	197
Polarography	197	57	140
Amperometry	66	26	40
Other	33	16	17
Photometry	265	126	139
Fluorometry	30	11	19
Extraction	47	31	16
Other	188	84	104
New organic reagents	158	99	59
Determination of constants and theoretical work	111	67	44
Titrimetry	75	27	48
Nonaqueous	26	8	18
Other	49	19	30
Gas chromatography	58	34	24
Phase analysis of gases in minerals, ores, alloys, and steel	52	7	45
Kinetic methods	43	21	22
Radiochemical methods	40	28	12
Total	1098	508	590

From Orient, I. M., *Zavodsk. Lab.*, 33, 1383, 1967. With permission.

chemical research on the basis of five different criteria.[124] The data are given in Table 50.

Since citation is used as a measure of the impact of publications and since — as we mentioned — citation is the language of science which codes the information published previously in a given paper, it is worth investigating the logical bases of citation in papers published on analytical chemistry. Orient[124] has subjected the experimental analytical papers published in *Zhurnal Analyticheskoi Khimii* and *Zavodskaya Laboratoriya* to conceptual analysis. Before starting the analysis no classification was fixed; this developed during the process in which all citations received their appropriate meaning. The classification developed after the analysis of 1200 references, and the corresponding distribution is shown in Table 51. A part of the reference has a "soft" character and reflects the individual citation habits of the author, i.e., references to previous works, critical judgment, etc. The number of such references is usually very different in various papers, and the references reflect the traditions of the journal or sometimes the lack of generalizing or refer to review papers. The other part of the references is "hard", i.e., handbook data, methods of the evaluation of results, description of syntheses, preparation of reagents, or hints to the procedure followed. Hard references, as it appears from Table 51, amount to nearly 38% of all references.

It is obvious that the citations concerning the level of "contribution" are not equal in value, but to assign different weights to them in order to take into account citability as a criterion would be complicated and unsound in practice. If results are compared to the citations found in metallurgical papers,[125] it can be seen that in analytical chemistry the weight of methodological references is significantly higher.

Special attention may be devoted to the work of Kara-Murza[126] in which the birth

Table 50
RANKINGS OF SOVIET ANALYTICAL CHEMISTRY RESEARCH IN FIVE CATEGORIES

Methods	Ranking I	II	III	IV	V
Electrochemistry	2	1	1—3	4—5	6
New organic reagents	6—8	3	1—3	2	2
Determination of constants, theoretical papers	10	4	4	1	1
Photometry	1	2	1—3	8—9	8
Gas chromatography	4	6	5	8—9	7
Radiochemical methods	9	8—10	7—8	4—5	4
Extraction	6—8	8—10	6	7	5
Kinetic methods	11	8—10	9	3	3
Titrimetry	3	5	7—8	10	9
Phase analysis of alloys and ores	6—8	7	10	6	10
Liquid chromatography	5	11	11	11	11

Note: Rank based on I, number of papers; II, number of citations; III, number of citations in foreign literature; IV, specific citation (citations per papers); and V, specific citation in foreign literature.

From Orient, I. M., *Zavodsk. Lab.*, 33, 1383, 1967. With permission.

Table 51
CATEGORIZATION OF REFERENCES IN ANALYTICAL CHEMISTRY PAPERS

Type of reference	Percent of total number of references
Stated in preceding papers (reviews)	40
Reference to apparatus used, method, preparation of reagent, calculation, and synthesis ("methodological")	20
Reference to the use of ideas, methods, and theory ("idea")	15
Reference to earlier established facts and effects ("phenomonological")	13
Critical discussions	4
Comparison of results with literature data	3
References to handbook data	2
Historical	1
Not identified	2

Note: Number of references: 1200.

From Orient, I. M., *Zavodsk. Lab.*, 41(9), 1071, 1975; (English transl.), *Ind. Lab. USSR*, 41, 1327, 1975. With permission.

and spread of affinity chromatography is studied by means of citation analysis. Affinity chromatography was established almost simultaneously in Sweden[127,128] and in the U.S.[129] Kara-Murza counted the number of citations to the fundamental papers of the authors and to their later papers. The data about the publication activity of the authors concerning the method and the extent of scientific relations and links were obtained from the 1968 to 1976 volumes of the *SCI.*

It was found in analyzing the "diffusion" of the method that it obeys a logistic model, i.e., in the initial stage of dissemination the dynamics of diffusion speed up

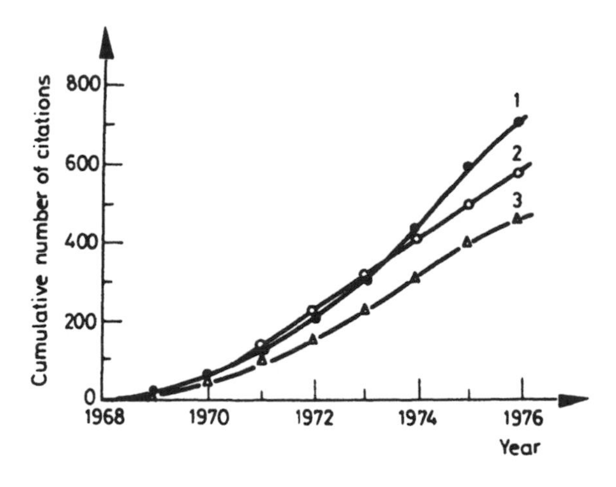

FIGURE 53. Growth of the number of citations to the first three papers on affinity chromatography. Curves: 1, Axén et al.;[127] 2, Cuatrecasas et al.;[129] 3, Porath et al.[128] (From Kara-Murza, S. G., *Nauchno Tekh. Inf. Ser. 1*, 1, 7, 1979. With permission.)

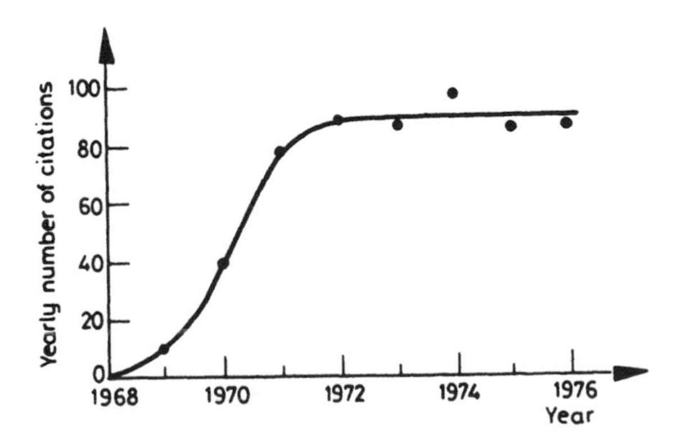

FIGURE 54. Yearly citation rate of the papers by Cuatrecasas et al.[129] (From Kara-Murza, S. G., *Nauchno Tekh. Inf. Ser. 1*, 1, 7, 1979. With permission.)

and can be expressed by an exponential curve. The accumulation of citations to the first three publications on affinity chromatography shows this relationship (Figure 53). However the exponential curves turn quite rapidly into straight lines, i.e., the growth rate of the number of citations becomes nearly constant, which is seen in the annual citation curve of the paper of Cuatrecasas et al.[129] (Figure 54). This can be attributed to the fact that the scientists who applied this method referred later to those papers which use affinity chromatography in their specific topic. The original papers are cited only by those who recommend the method for new topics.

The citation dynamics of the first papers on affinity chromatography show already after the first 3 to 4 years how rapidly this method was introduced into the research labs. No doubt that the discoverers became citation recorders. However, the rate of spreading of affinity chromatography can be estimated more thoroughly from the dynamics of the number of citations to all papers of Cuatrecasas published after the appearance of the first one (Figure 55).[126]

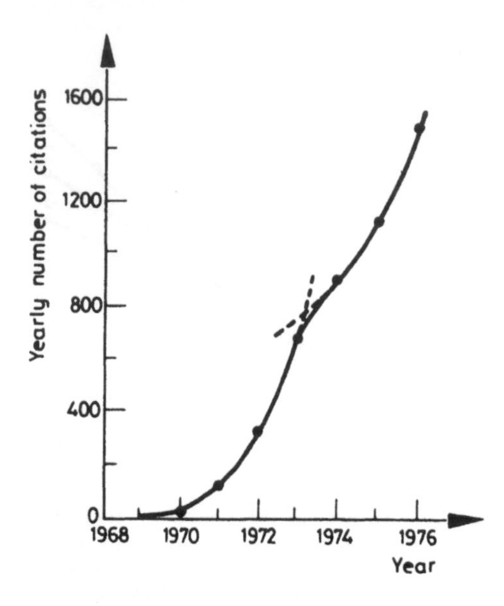

FIGURE 55. Yearly citation rate of papers published by Cuatrecasas after the paper of Reference 129. (From Kara-Murza, S. G., *Nauchno Tekh. Inf. Ser. 1*, 1, 7, 1979. With permission.)

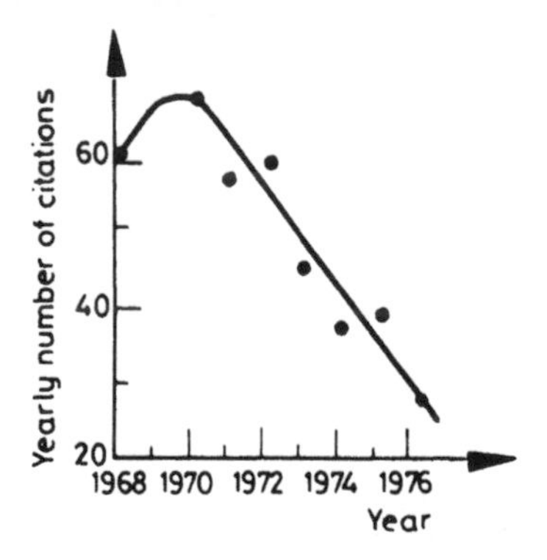

FIGURE 56. Dynamics of citations to papers published by Cuatrecasas[129] until 1968. (From Kara-Murza, S. G., *Nauchno Tekh. Inf. Ser. 1*, 1, 7, 1979. With permission.)

No doubt a certain fraction of citations can be found in the papers of those authors who do not utilize affinity chromatography but have not forgotten the important results discussed by Cuatrecasas in his publications. However, the difference is not great and can be calculated since the main activity of Cuatrecasas after 1968 was the application of affinity chromatography to various fields, which affected his results on his *par excellence* field of research (enzymology).

Figure 56 shows the dynamics of citation of the papers of Cuatrecasas published until 1968.[126] After 5 years it dropped to one half, roughly corresponding to the general laws pertaining to the obsolescence of "normal" scientific literature. The citation maximum of papers leaving the press between 1961 and 1967 appeared in 1970 with 67

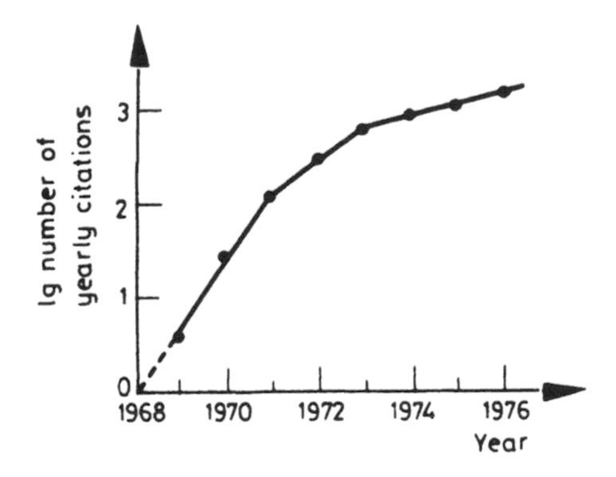

FIGURE 57. Growth of the number of citations in papers by Cuatrecasas published after 1968. (From Kara-Murza, S. G., *Nauchno Tekh. Inf. Ser. 1*, 1, 7, 1979. With permission.)

citations, i.e., the papers published during 1 year (in the interval investigated) received an average of 10 citations at the time of maximum citation. It is obvious that if this value were subtracted from the number of citations given in Figure 55, this would not affect the parameters of the curve significantly. At any rate, the error is certainly smaller than the number of citations connected exclusively with the application of affinity chromatography. This number drops to zero because in several later publications Cuatrecasas is not the first author, and thus he is not recorded in the *SCI.*

The citation curve shown in Figure 55 (after subtracting self-citations) consists of two exponential sections which can be observed particularly well in Figure 57 where the curve is plotted in logarithmic form. While the number of citations decreases in 1972, citation does not stop. Consequently, the growth rate of publications dealing with the application of affinity chromatography does not show a stabilization tendency. By the end of 1976, the number of citations to the papers in which Cuatrecasas was the first author amounted to more than 5500 whereas the number of those to the paper of Axén et al.[127] was 8 times less! To the paper of Porath et al.,[128] 467 citations were made by the end of 1976. This comparison does not show that the citation rate of the papers by Swedish authors is not high. In spite of this, the absolute number of citations to the two papers[127,128] is tremendous. The growth of citations reflecting the applications of the two variants of the method is shown in Figure 58.

In the first stage of the spread of affinity chromatography, when the authors cited the original papers, most of the citations referred to the authors of both variants. Consequently, in order to determine the true popularity of one or the other variant, one half of the citations should be subtracted from the scores of both variants. It is easy to see that this would increase the relative difference between the citation levels of the U.S. and Swedish authors.

The diffusion of the Swedish variant of the method is characterized in Figure 58 by the citations to Reference 127, not including the further papers of the authors. This is motivated by the fact that the scientific attention of the authors, if it is judged correctly from the titles of their post-1968 papers, had shifted to another field (immobilized enzymes), and a large number of citations refer to papers that are not related to affinity chromatography. Even if these citations were added, the parameters of curve 2 of Figure 58 would not change substantially. The works of Axén published after Reference 127 received a total of 365 citations in the whole period up to 1976.

Then what is the explanation of this enormous difference between the propagation of these two so similar variants? The difference, as it appears, is even more surprising

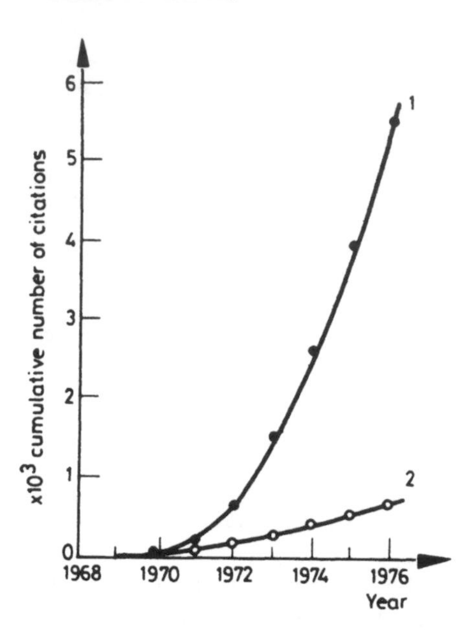

FIGURE 58. Growth rate of citations to the papers of Cuatrecasas (curve 1) and Axén (curve 2). (From Kara-Murza, S. G., *Nauchno Tekn. Inf. Ser. 1*, 1, 7, 1979. With permission.)

since the paper of the Swedish authors was published 1 year earlier in a first-class periodical, and one of its authors was Porath, who had become known worldwide for being one of the creators of the gel-filtration method. As can be seen from Figure 53, the papers[127,128] were widely cited immediately after their publication. The differences appearing later can be explained by the fact that Cuatrecasas, unlike the Swedish authors, took part personally and very actively in the use of affinity chromatography for solving a wide variety of problems. He assisted several researchers working in various fields of chemistry and biochemistry in solving the difficulties of learning the method that were mostly psychological in nature. He demonstrated the efficiency of the method and organized centers which accelerated the diffusion of the new method. Cuatrecasas played the role of "innovation champion", the key figure indispensable in the introduction of a new method. The necessity of such a figure was postulated by Schön[130] and proved by an extensive investigation of scientific-technical innovation processes. The *SCI* makes it possible to assess the proportions of the scientific links of Cuatrecasas. Figure 59 shows the "contact map" of the authors of the original paper, not only those researchers with whom Cuatrecasas, Anfinsen, and Wilchek co-authored a paper, but also some prominent co-authors of the co-authors. The basic concept of the map is that the co-author of a co-author is already in the field of scientific contact of a scientist since they may easily be contacted personally through their common acquaintances. It can be seen in Figure 59 that among the Cuatrecasas co-authors, excellent scientists of very diverse research fields can be found. This follows from the data of Table 52, in which the specialties of these scientists and the number of papers they published in 1968 are given. This "co-author map" contains important pieces of information on the connection of the two variants of affinity chromatography. Among the 1968 co-authors with Wilchek one can find Witkopf, an expert of bioorganic and organic chemistry, who was in 1966 and 1967 co-author with Porath, Axén, and Ernback, the Swedish discoverers.

The 1968 data shown in Figure 59 and Table 52 are enough to characterize the re-

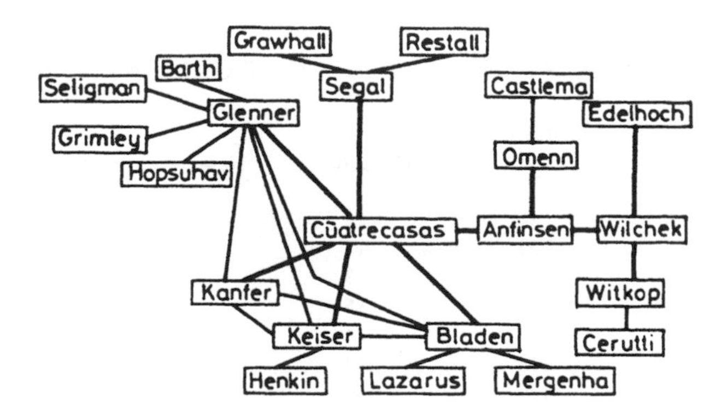

FIGURE 59. Map of scientific connections of Cuatrecasas et al.[129] in 1968. Thick line connects direct co-authors; thin line binds the co-authors of Cuatrecasas, Wilchek, and Anfinsen. (From Kara-Murza, S. G., *Nauchno Tekh. Inf. Ser. 1*, 1, 7, 1979. With permission.)

Table 52
SCIENTISTS ON THE PROFESSIONAL CONNECTION MAP OF THE AUTHORS OF THE PAPER IN REFERENCE 127

Name	Subject field	Number of co-authored papers (1968)
Glenner, G.	Histochemistry	21
Keiser, H.	Antibiotics	6
Bladen, H.	Bioorganic chemistry	4
Kaufer, J.	Lipids	5
Segal, S.	Transport processes	15
Crawhall, J.	Medicine	7
Restall, C.	Anesthesiology	4
Barth, W.	Connective tissues	6
Seligman, A.	Histochemistry	16
Grimley, P.	Citology and virology	8
Hopsuhav, V.	Enzymology	10
Henkin, R.	Endocrinology (corticosteroids)	9
Lazarus, G.	Connective tissues	6
Mergenha, S.	Polysaccharides (endotoxins)	12
Castlema, B.	Medicine	49
Witkop, B.	Bioorganic and organic chemistry	27
Cerutti, P.	Bioorganic chemistry	6
Edelhoch, H.	Bioorganic chemistry	6

From Kara-Murza, S. G., *Nauchno Tekh. Inf. Ser. 1*, 1, 7, 1979. With permission.

search style of Cuatrecasas, who maintained intense contact with a large number of research groups, which in itself promotes the dissemination of a method. In 1967 he published 13 papers with 15 co-authors. Nevertheless, his activity after 1968 was directed to the diffusion of affinity chromatography in cooperation with several research groups and the publication of numerous papers, with co-authors frequently changed.

Table 53 shows the amount, co-authors, and new co-authors of the publications of Cuatrecasas. Figure 60 shows the growth of papers and new co-authors after 1969.

Table 53
PUBLICATION ACTIVITY OF CUATRECASAS AFTER 1969

Year	Papers	Total after 1969	Co-authors	New co-authors[a]	Total number of new co-authors
1970	10	10	12	4	4
1971	21	31	17	6	10
1972	15	46	13	8	18
1973	28	74	15	10	28
1974	22	96	21	9	37
1975	23	119	14	5	42
1976	16	135	16	3	45

[a] Co-authors with whom Cuatrecasas did not co-author papers after 1968.

From Kara-Murza, S. G., *Nauchno Tekh. Inf. Ser. 1*, 1, 7, 1979. With permission.

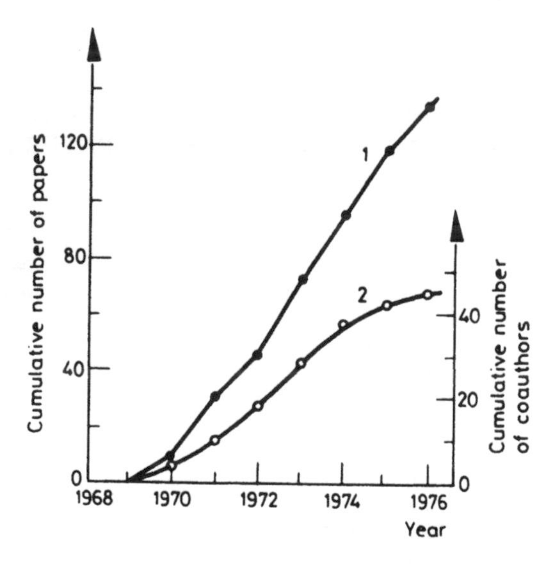

FIGURE 60. Publication activity and scientific links of Cuatrecasas after 1969. Curve 1, papers; curve 2, co-authors. (From Kara-Murza, S. G., *Nauchno Tekh. Inf. Ser. 1*, 1, 7, 1979. With permission.)

Although the number of Cuatrecasas' new co-authors is large, of course it is too low to be processed statistically. The curve shown in Figure 61 reflects the tendency to believe that the number of new co-authors has passed its maximum and in the last years it is decreasing. This is easy to explain; by now the method can be regarded as well known in all basic fields of application, it is described in detail in handbooks, and it is even in the program of undergraduate laboratory practice. The author already has nothing to do with its diffusion.

The activity of the Swedish authors developed in a completely different way. Their publication activity does not change after 1967. During 9 years (1967 to 1976) Axén published 22 papers, many of them in co-authorship with Porath. The latter published 50 papers, but only 16 of them deal with affinity chromatography; the rest of his work is directed mostly towards finding new ways for "anchoring" macromolecules to carriers.

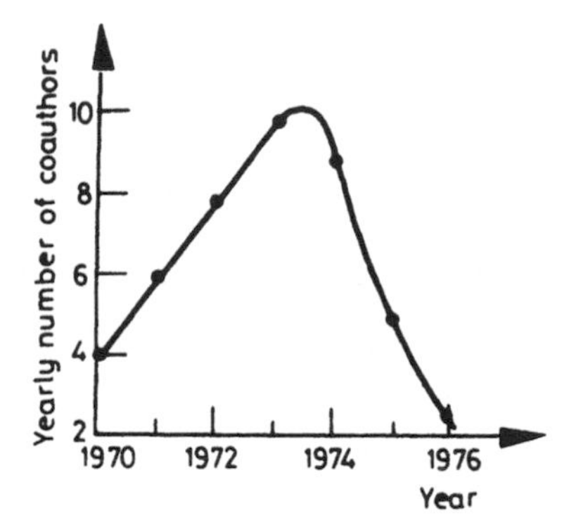

FIGURE 61. Dynamics of Cuatrecasas' scientific links between 1970 and 1976. (From Kara-Murza, S. G., *Nauchno Tekh. Inf. Ser. 1*, 1, 7, 1979. With permission.)

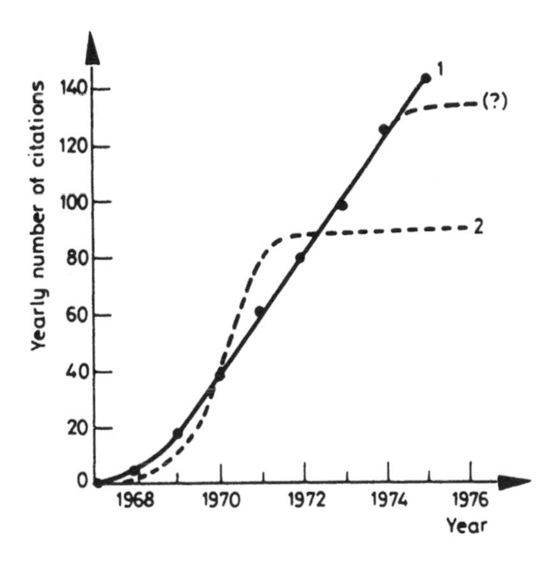

FIGURE 62. Citation dynamics of the first two papers on affinity chromatography. Curve 1, Axén et al.;[127] curve 2, Cuatrecasas et al.[129] (From Kara-Murza, S. G., *Nauchno Tekh. Inf. Ser. 1*, 1, 7, 1979. With permission.)

The fact that Axén, Porath, and Ernback published so little on the extension of the application sphere of the method explains why the citation dynamics of their original paper (Reference 127) are so different from the pattern shown in Figure 62 and the characteristics of the papers of the American authors. The citation level of this paper increases almost linearly up to 1975 (Figure 62). This is in relation with the fact that all scientists who learned and apply the given variant of the method may cite the paper of Axén et al. only, since there is no paper on the application of the Swedish variant of the method on fields closer to one or another research group, unlike the American variant.

It can be assumed that if a scientist with a broad research profile has introduced the Swedish variant of the method to various fields, the citation of the original paper would have reached saturation. In this case the "propagation agency" of the innovation takes the relay baton from the creators of the method.

It is noted by Laitinen and Ewing[131] that "a method's utilization needs commercial instrumentation for success, regardless of the method's inherent capabilities." This is extended by us with the remark that this not only holds for instrumentation but also for the special reagents, adsorbents necessary for a method. Had the column filling of affinity chromatography not been produced and marketed commercially, the method would not have been applied so widely.

It is worth noting here two other investigations. In one of them Orient[132] deals with citation analysis by the relative efficiency of books dedicated to analytical chemistry. In the other,[133] the same author investigates the scientific legacy of the prominent late Soviet analyst A. K. Babko by scientometrics, primarily citation analysis.

III. CITATION CLASSICS*

In a 1978 paper, Belcher[134] stated that "at least four chemists have been awarded the Nobel prize for the development of analytical techniques, but it is significant that in each instance the particular technique chosen accelerated the progress of biochemistry."

The correctness of this statement also transpires from two studies by Garfield[135,136] in which a list is compiled of the 100 papers most cited in the *SCI* for 1961 to 1972. The list, broken into 2 parts containing 50 papers each, can be seen in Tables 54 and 55. Papers related to analytical chemistry are marked with circles in the lists. We consider it extremely significant that although the list is completely interdisciplinary, i.e., pertains to the entire field of science, 47% of the papers are on the subject of analytical chemistry. We cannot imagine better proof for the support of viability and importance of this subject field.

* As per the definition of the Institute for Scientific Information, Philadelphia, a "citation classic" is a highly cited publication, as indicated by the *SCI*. Of course, citation rates differ for each discipline. The number of citations which would make a publication a classic in botany, a small field, might be much lower than the number required to make a classic in a large field, i.e., biochemistry. However, one can appreciate the relative impact of each classic by considering that the average article published in a journal covered by the *SCI* in 1973 received 5.7 citations, 1973 to 1976.

Table 54
MOST-CITED JOURNAL ARTICLES 1961 TO 1972, NUMBERS 1 TO 50

Rank	Times cited	Bibliographical data
1	29,655	Lowry, O. H., Rosebrough, N. J., Farr, A. L., and Randall, R. J., Protein measurement with the Folin phenol reagent, *J. Biol. Chem.*, 19(3), 265, 1951.
2	6,281	Reynolds, E. S., The use of lead citrate at high pH as an electron opaque stain in electron microscopy, *J. Cell Biol.*, 17, 208, 1963.
3	5,825	Fiske, C. H. and Subbarow, Y., The colorimetric determination of phosphorus, *J. Biol. Chem.*, 66, 375, 1925.
4	5,273	Luft, J. H., Improvements in epoxy resin embedding methods, *J. Biophys. Biochem. Cytol.*, 9, 409, 1961.
5	5,054	Folch, J., Lees, M., and Sloane Stanley, G. H., A simple method for the isolation and purification of total lipids from animal tissues, *J. Biol. Chem.*, 226, 497, 1957.
6	4,932	Bray, G. A., A simple efficient liquid scintillator for counting aqueous solutions in a liquid scintillation counter, *Anal. Biochem.*, 1, 279, 1960.
7	4,376	Spackman, D. H., Stein, W. H., and Moore, S., Automatic recording apparatus for use in the chromatography of amino acids, *Anal. Chem.*, 30, 1190, 1958.
8	4,367	Sabatini, D. D., Bensch, K., and Barrnett, R. J., Cytochemistry and electron microscopy; the preservation of cellular ultrastructure and enzymatic activity by aldehyde fixation, *J. Cell Biol.*, 17, 19, 1963.
9	3,967	Gornall, A. G., Bardawill, C. J., and David, M. M., Determination of serum proteins by means of the biuret reaction, *J. Biol. Chem.*, 78, 751, 1949.
10	3,621	Lineweaver, H. and Burk, D., The determination of enzyme dissociation constants, *J. Am. Chem. Soc.*, 56, 658, 1934.
11	3,464	Davis, B. J., Disc electrophoresis. II. Method and application to human serum proteins, *Ann. N.Y. Acad. Sci.*, 121, 404, 1964.
12	3,406	Burton, K., A study of the conditions and mechanism of the diphenylamine reaction for the colorimetric estimation of deoxyribonucleic acid, *Biochem. J.*, 62, 315, 1956.
13	2,903	Scheidegger, J. J., Une micro-methode de l'immuno-electrophorese, *Int. Arch. Allergy*, 7, 103, 1955.
14	2,770	Duncan, D. B., Multiple range and multiple *F* tests, *Biometrics*, 11, 1, 1955.
15	2,740	Nelson, N., A photometric adaptation of the Somogyi method for the determination of glucose, *J. Biol. Chem.*, 153, 375, 1944.
16	2,620	Reed, L. J. and Muench, H., A simple method of estimating fifty per cent endpoints, *Am. J. Hyg.*, 27, 493, 1938.
17	2,293	Dole, V. P., A relation between non-esterified fatty acids in plasma and the metabolism of glucose, *J. Clin. Invest.*, 35, 150, 1956.
18	2,230	Marmur, J., A procedure for the isolation of deoxyribonucleic acid from micro-organisms, *J. Mol. Biol.*, 3, 208, 1961.
19	2,226	Moorhead, P. S., Nowell, P. C., Mellman, W. J., Battips, D. M., and Hungerford, D. A., Chromosome preparations of leukocytes cultured from human peripheral blood, *Exp. Cell Res.*, 20, 613, 1960.
20	2,054	Warburg, O. and Christian, W., Isolierung und Kristallisation des Garungsferments Enolase, *Biochem. Z.*, 310, 384, 1941.
21	1,976	Jacob, F. and Monod, J., Genetic regulatory mechanisms in the synthesis of proteins, *J. Mol. Biol.*, 3, 318, 1961.
22	1,905	Martin, R. G. and Ames, B. N., A method for determining the sedimentation behavior of enzymes: application to protein mixtures, *J. Biol. Chem.*, 236, 1372, 1961.
23	1,887	Watson, M. L., Staining of tissue sections for electron microscopy with heavy metals, *J. Biophys. Biochem. Cytol.*, 4, 475, 1958.
24	1,885	Bartlett, G. R., Phosphorus assay in column chromatography, *J. Biol. Chem.*, 234, 466, 1959.
25	1,849	Palade, G. E., A study of fixation for electron microscopy, *J. Exp. Med.*, 95, 285, 1952.

Table 54 (continued)
MOST-CITED JOURNAL ARTICLES 1961 TO 1972, NUMBERS 1 TO 50

Rank	Times cited	Bibliographical data
26	1,841	Smithies, O., Zone electrophoresis in starch gels: group variations in the serum proteins of normal human adults, *Biochem. J.*, 61, 629, 1955.
(27)	1,814	Barker, S. B. and Summerson, W. H., The colorimetric determination of lactic acid in biological material, *J. Biol. Chem.*, 138, 535, 1941.
(28)	1,767	Warren, L., The thiobarbituric acid assay of sialic acids, *J. Biol. Chem.*, 234, 1971, 1959.
(29)	1,737	Trevelyan, W. E., Procter, D. P., and Harrison, J. S., Detection of sugars on paper chromatograms, *Nature*, 166, 444, 1950.
(30)	1,695	Dubois, M., Gilles, K. A., Hamilton, J. K., Rebers, P. A., and Smith, F., Colorimetric method for determination of sugars and related substances, *Anal. Chem.*, 28, 350, 1956.
31	1,662	Eagle, H., Amino acid metabolism in mammalian cell cultures, *Science*, 130, 432, 1959.
32	1,628	Litchfield, J. T., Jr. and Wilcoxon, F., A simplified method of evaluating dose-effect experiments, *J. Pharmacol. Exp. Ther.*, 96, 99, 1949.
33	1,398	Ellman, G. L., Tissue sulfhydryl groups, *Arch. Biochem. Biophys.*, 82, 70, 1959.
34	1,387	Bardeen, J., Cooper, L. N., and Schrieffer, J. R., Theory of superconductivity, *Phys. Rev.*, 108, 1175, 1957.
35	1,384	Andrews, P., Estimation of the molecular weights of proteins by Sephadex gel filtration, *Biochem. J.*, 91, 222, 1964.
(36)	1,384	Schmidt, G. and Thannhauser, S. J., A method for the determination of deoxyribonucleic acid, ribonucleic acid, and phosphoproteins in animal tissues, *Biochem. J.*, 161, 83, 1945.
37	1,344	Jaffe, H. H., A reexamination of the Hammett equation, *Chem. Rev.*, 53, 191, 1953.
38	1,333	Venable, J. H. and Coggeshall, R., A simplified lead citrate stain for use in electron microscopy, *J. Cell Biol.*, 25, 407, 1965.
39	1,324	Karnovsky, M. J., Simple method for staining with lead at high pH in electronmicroscopy, *J. Biophys. Biochem. Cytol.*, 11, 729, 1961.
40	1,317	Karplus, M., Contact electron-spin coupling of nuclear magnetic moments, *J. Chem. Phys.*, 30, 11, 1959.
41	1,305	Gell-Mann, M., Symmetries of baryons and mesons, *Phys. Rev.*, 125, 1067, 1962.
(42)	1,297	Ornstein, L., Disc electrophoresis. I. Background and theory, *Ann. N.Y. Acad. Sci.*, 121, 321, 1964.
(43)	1,294	Chen, P. S., Jr., Toribara, T. Y., and Warner, H., Microdetermination of phosphorus, *Anal. Chem.*, 28, 1756, 1956.
(44)	1,292	Moore, S., Spackman, D. H., and Stein, W. H., Chromatography of amino acids on sulfonated polystyrene resins, *Anal. Chem.*, 30, 1185, 1958.
(45)	1,278	Schneider, W. C., Phosphorus compounds in animal tissues. I. Extraction and estimation of deoxypentose nucleic acid and of pentose nucleic acid, *J. Biol. Chem.*, 161, 293, 1945.
46	1,239	Mancini, G., Carbonara, A. O., and Heremans, J. F., Immunochemical quantitation of antigens by single radial immunodiffusion, *Immunochemistry*, 2, 235, 1965.
47	1,226	Yphantis, D. A., Equilibrium ultracentrifugation of dilute solutions. *Biochemistry*, 3, 297, 1964.
48	1,214	Dulbecco, R. and Vogt, M., Plaque formation and isolation of pure lines with poliomyelitis viruses, *J. Exp. Med.*, 99, 167, 1954.
49	1,209	Weber, K. and Osborn, M., The reliability of molecular weight determinations by dodecyl sulfate-polyacrylamide gel electrophoresis, *J. Biol. Chem.*, 244, 4406, 1969.
(50)	1,207	Mandell, J. D. and Hershey, A. D., A fractionating column for analysis of nucleic acids, *Anal. Biochem.*, 1, 66, 1960.

Table 55
MOST-CITED JOURNAL ARTICLES 1961 TO 1972, NUMBERS 51 TO 100

Rank	Times cited	Bibliographical data
(51)	1204	Smithies, O., An improved procedure for starch-gel electrophoresis: further variations in the serum proteins of normal individuals, *Biochem. J.*, 71, 585, 1959.
52[a]	1171	Arnon, D. I., Copper enzymes in isolated chloroplasts: polyphenoloxidase in *Beta vulgaris*, *Plant Physiol.*, 24, 1, 1949.
(53)	1158	Moore, S. and Stein, W. H., A modified ninhydrin reagent for the photometric determination of amino acids and related compounds, *J. Biol. Chem.*, 211, 907, 1954.
(54[b])	1156	Somogyi, M., Notes on sugar determination, *J. Biol Chem.*, 195, 19, 1952.
(55)	1152	Hanes, C. S. and Isherwood, F. A., Separation of the phosphoric esters on the filter paper chromatogram, *Nature*, 164, 1107, 1949.
(56[b])	1149	Van Handel, E. and Zilversmit, D. B., Micromethod for the direct determination of serum triglycerides, *J. Lab. Clin. Med.*, 50, 152, 1957.
57	1145	Monod, J., Wyman, J., and Changeux, J. P., On the nature of allosteric transitions: a plausible model, *J. Mol. Biol.*, 12, 88, 1965.
(58)	1130	Friedemann, T. E. and Haugen, G. E., Pyruvic acid. II. The determination of keto acids in blood and urine, *J. Biol. Chem.*, 147, 415, 1943.
59[a]	1116	Millonig, G., Advantages of a phosphage buffer for OsO_4 solutions in fixation, *J. Appl. Phys.*, 32, 1637, 1961.
(60)	1103	Dische, Z., A new specific color reaction of hexuronic acids, *J. Biol. Chem.*, 167, 189, 1947.
(61)	1100	Hoffman, W. S., A rapid photoelectric method for the determination of glucose in blood and urine, *J. Biol. Chem.*, 120, 51, 1937.
(62[a])	1087	Bligh, E. G. and Dyer, W. J., A rapid method of total lipid extraction and purification, *Can. J. Biochem. Physiol.*, 37, 911, 1959.
(63)	1066	Poulik, M. D., Starch gel electrophoresis in a discontinuous system of buffers, *Nature*, 180, 1477, 1957.
64	1063	DeDuve, C., Pressman, B. C., Gianetto, R., Warriaux, R., and Appelmans, F., Tissue fractionation studies. VI. Intracellular distribution patterns of enzymes in rat-liver tissue, *Biochem. J.*, 60, 604, 1955.
(65)	1061	Bratton, A. C. and Marshall, E. K., Jr., A new coupling component for sulfanilamide determination, *J. Biol. Chem.*, 128, 537, 1939.
66	1059	Boyden, S. V., The adsorption of proteins on erythrocytes treated with tannic acid and subsequent hemagglutination by antiprotein sera, *J. Exp. Med.*, 93, 107, 1951.
(67)	1044	Van Slyke, D. D. and Neill, J. M., The determination of gases in blood and other solutions by vacuum extraction and manometric measurement, *J. Biol. Chem.*, 61, 523, 1924.
68	1029	Millonig, G., A modified procedure for lead staining of thin sections, *J. Biophys. Biochem. Cytol.*, 11, 736, 1961.
69[a]	1023	Shapiro, A. L., Vinuela, E., and Maizel, J. V., Jr., Molecular weight estimation of polypeptide chains by electrophoresis in SDS-polyacrylamide gels, *Biochem. Biophys. Res. Commun.*, 28, 815, 1967.
70	1022	Coons, A. H. and Kaplan, M. H., Localization of antigen in tissue cells. II. Improvements in a method for the detection of antigen by means of fluorescent antibody, *J. Exp. Med.*, 91, 1, 1950.
71	1012	Caulfield, J. B., Effects of varying the vehicle for OsO_4 in tissue fixation, *J. Biophys. Biochem. Cytol.*, 3(5), 827, 1957.
72	1010	Sever, J. L., Application of a microtechnique to viral serological investigations, *J. Immunol.*, 88, 320, 1962.
73	1006	Ahlquist, R. P., A study of the adrenotropic receptors, *Am. J. Physiol.*, 153, 586, 1948.
74	1001	Porter, R. R., The hydrolysis of rabbit gamma-globulin and antibodies with crystalline papain, *Biochem. J.*, 73, 119, 1959.
(75)	994	Dole, V. P. and Meinertz, H., Microdetermination of long-chain fatty acids in plasma and tissues, *J. Biol. Chem.*, 235, 2595, 1960.
(76)	990	Sperry, W. M. and Webb, M., A revision of the Schoenheimer-Sperry method for cholesterol determination, *J. Biol. Chem.*, 187, 97, 1950.

Table 55 (continued)
MOST-CITED JOURNAL ARTICLES 1961 TO 1972, NUMBERS 51 TO 100

Rank	Times cited	Bibliographical data
77	984	Boyer, P. D., Spectrophotometric study of the reaction of protein sulfhydryl groups with organic mercurials, *J. Am. Chem. Soc.*, 76, 4331, 1954.
78	968	Seldinger, S. I., Catheter replacement of the needle in percutaneous arteriography: a new technique, *Acta Radiol.*, 39, 368, 1953.
79	961	Abell, L. L., Levy, B. B., Brodie, B. B., and Kendall, F. E., A simplified method for the estimation of total cholesterol in serum and demonstration of its specificity, *J. Biol. Chem.*, 195, 357, 1952.
80	960	Boas, N. F., Method for the determination of hexosamines in tissues, *J. Biol. Chem.*, 204, 553, 1953.
81	955	Nirenberg, M. W. and Matthaei, J. H., The dependence of cell-free protein synthesis in *E. coli* upon naturally occurring or synthetic polyribonucleotides, *Proc. Natl. Acad. Sci. U.S.A.*, 47, 1588, 1961.
82	949	Bowden, K., Heilbron, I. M., Jones, E. R. H., and Weedon, B. C. L., Researches on acetylenic compounds. I. The preparation of acetylenic ketones by oxidation of acetylenic carbinols and glycols, *J. Chem. Soc.*, 38, 1946.
83	944	Hirs, C. H. W., The oxidation of ribonuclease with performic acid, *J. Biol. Chem.*, 219, 611, 1956.
84	928	Sweeley, C. C., Bentley, R., Makita, M., and Wells, W. W., Gas-liquid chromatography of trimethylsilyl derivatives of sugars and related substances, *J. Am. Chem. Soc.*, 85, 2497, 1963.
85	924	Bloembergen, N., Purcell, E. M., and Pound, R. V., Relaxation effects in nuclear magnetic resonance absorption, *Phys. Rev.*, 73, 679, 1948.
86	919	Allen, R. J. L., The estimation of phosphorus, *Biochem. J.*, 34, 858, 1940.
87	915	Reitman, S. and Frankel, S., A colormetric method for the determination of serum glutamic oxalacetic and glutamic pyruvic transaminases, *Am. J. Clin. Pathol.*, 28, 56, 1957.
88[b]	908	Cromer, D. T. and Waber, J. T., Scattering factors computed from relativistic Dirac-Slater wave functions, *Acta Cryst.*, 18, 104, 1965.
89[a]	906	Pariser, R. and Parr, R. G., A semi-empirical theory of the electronic spectra and electronic structure of complex unsaturated molecules, *J. Chem. Phys.*, 21, 767, 1953.
90[b]	903	Karnovsky, M. J., A formaldehyde-glutaraldehyde fixative of high osmolality for use in electron microscopy, *J. Cell Biol.*, 27, A137, 1965.
91[a]	896	Higgins, G. M. and Anderson, R. M., Experimental pathology of the liver. I. Restoration of the liver of the white rat following partial surgical removal, *Arch. Pathol.*, 12, 186, 1931.
92[a]	896	Hoffmann, R., An extended Huckel theory. I. Hydrocarbons, *J. Chem. Phys.*, 39, 1397, 1963.
93	894	Clarke, D. H. and Casals, J., Techniques for hemagglutination and hemagglutination-inhibitions with arthropod-borne viruses, *Am. J. Trop. Med. Hyg.*, 7, 561, 1958.
94[a]	894	Davis, B. D. and Mingioli, E. S., Mutants of *Escherichia coli* requiring methionine or vitamin B_{12}, *J. Bacteriol.*, 60, 17, 1950.
95[a]	890	Hales, C. N. and Randle, P. J., Immunoassay of insulin with insulin-antibody precipitate, *Biochem. J.*, 88, 137, 1963.
96	875	Peterson, E. A. and Sober, H. A., Chromatography of proteins. I. Cellulose ion-exchange adsorbents, *J. Am. Chem. Soc.*, 78, 751, 1956.
97[a]	871	Roothaan, C. C. J., New developments in molecular orbital theory, *Rev. Mod. Phys.*, 23, 69, 1951.
98	869	Bush, I. E., Methods of paper chromatography of steroids applicable to the study of steroids in mammalian blood and tissues, *Biochem. J.*, 50, 370, 1952.
99	867	Farquhar, M. G. and Palade, G. E., Junctional complexes in various epithelia, *J. Cell Biol.*, 17, 375, 1963.
100	864	Yalow, R. S. and Berson, S. A., Immunoassay of endogenous plasma insulin in man, *J. Clin. Invest.*, 39, 1157, 1960.

Table 55 (continued)
MOST-CITED JOURNAL ARTICLES 1961 TO 1972, NUMBERS 51 TO 100

[a] These articles achieved their highest citation rate in 1971.
[b] These articles achieved their highest citation rate in 1972.

Chapter 9

TRENDS AND PATTERNS OF THE LITERATURE OF PROMPT NUCLEAR ANALYSIS — A CASE STUDY

I. INTRODUCTION

In this chapter the publication trends and patterns of the subfield of prompt nuclear analysis,[137] i.e., the use of prompt radiation accompanying a nuclear reaction to measure elemental or isotopic concentrations, will be investigated.

Two bibliographies and a review paper by Bird et al.[138-140] were used as a data base. Original and review papers as well as conference proceeding papers related to nuclear reactions induced by positive ions, gamma rays, and neutrons as well as Rutherford backscattering and channeling were included in the data base. Papers describing applications of incident ion energy lower than 100 keV were not taken into account. Accordingly, the data base excludes papers on analysis by proton-, ion-, and high energy, heavy ion-induced X-ray emission.

Bibliographical entries used were those published up to 1976. In this way the eventual bias coming from the incompleteness at the end of an up-to-date bibliographical data base was eliminated. Citations were extracted from the *Science Citation Index (SCI)* from the Institute for Scientific Information, Philadelphia.

II. GROWTH OF THE LITERATURE ON PROMPT NUCLEAR ANALYSIS

Similar to science at large, a scientific subject field can be subjected to a sort of input (resources, manpower, etc.) and output (papers published, number of elements analyzed, etc.) analysis. Let us now examine the publication output of prompt nuclear analysis research.

The composition of the prompt nuclear analysis data base for the period 1949 to 1977 was the following:

Journal papers	1030
Papers published in conference proceedings and multiauthored books	452
Reports	149
Theses	38
Other items	16
Total	1685

A more detailed distribution is given in Table 56.

The total volume of literature in a given field must be corrected to take into account the fact that about 66% of papers appearing in conference proceedings eventually make their way into primary research journals.[94,141] Thus the estimated volume of the primary journal literature on prompt nuclear analysis amounts to 1250 research papers after deducting the roughly 110 review papers written during this period. This represents 2.5% of the activation analysis and about 0.3% of the total analytical chemistry primary journal literature by the end of 1977.

The time distribution of papers on prompt nuclear analysis is shown in Figure 63. As shown, in 25 years the yearly output has increased by more than two orders of magnitude. The proportion of primary journal papers, theses, and patents was con-

Table 56
DISTRIBUTION OF THE PROMPT NUCLEAR ANALYSIS DATA BASE AMONG VARIOUS DOCUMENT TYPES

	Percentage of publications				
Type of document	—1960	1961—1965	1966—1970	1971—1975	1976—
Journal papers	69.0	61.8	65.1	62.4	61.1
Conference proceedings[a]	4.8	9.1	14.1	25.0	26.8
Report	21.4	24.5	15.8	9.0	8.8
Thesis	4.8	3.6	3.4	2.3	2.2
Patent	—	0.9	1.4	1.2	1.0

[a] Including multiauthored books.

Bujdosó, E., Lyon, W. S., and Noszlopy, I., *J. Radioanal. Chem.*, 74, 197, 1982. With permission.

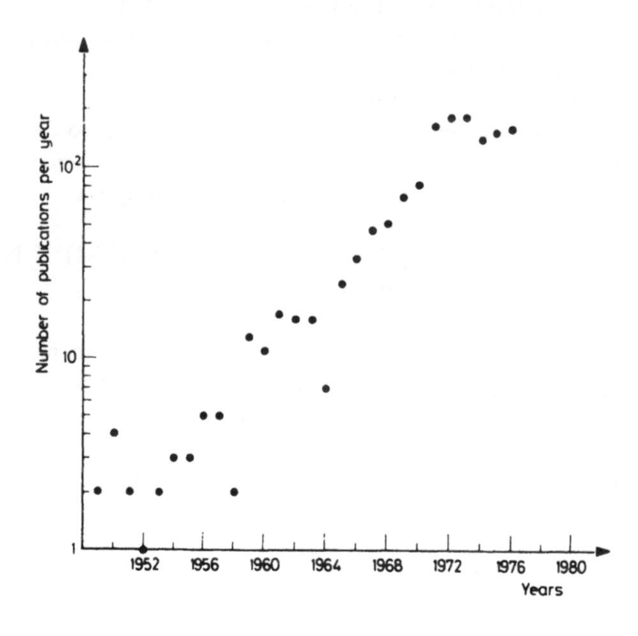

FIGURE 63. Publication output in prompt nuclear analysis (1949 to 1976). (From Bujdosó, E., Lyon, W. S., and Noszlopy, I., *J. Radioanal. Chem.*, 74, 197, 1982. With permission.)

stant, whereas in 28 years the number of papers published in conference proceedings has grown by a factor of five. Reports dropped by 50% (Table 56).

Review papers form an important component of the journal article category. They try to condense and order the acquired information. The first review on prompt nuclear analysis appeared in 1961, and since 1971 the reviews have formed a constant proportion (about 10%) of the ever-growing set of papers (Figure 64).

The focus of research intensity as reflected in the number of publications shifted from time to time. Figure 65 shows the yearly number of publications on the use of various nuclear reactions as shares of the total. As seen, in the early days photoneutron- and neutron-induced reactions dominated. In the second half of the 1950s charged particles, and from 1960 onwards, channeling, were added to the arsenal.

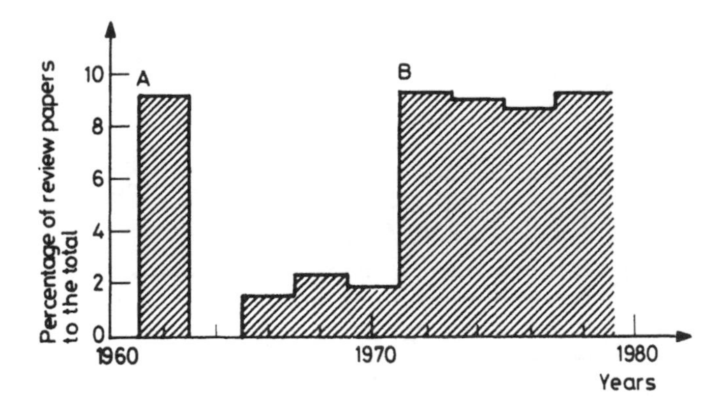

FIGURE 64. Percentage of review papers to the total in prompt nuclear analysis. The number of publications at points A and B were 80 and 600, respectively. (From Bujdosó, E., Lyon, W. S., and Noszlopy, I., *J. Radioanal. Chem.*, 74, 197, 1982. With permission.)

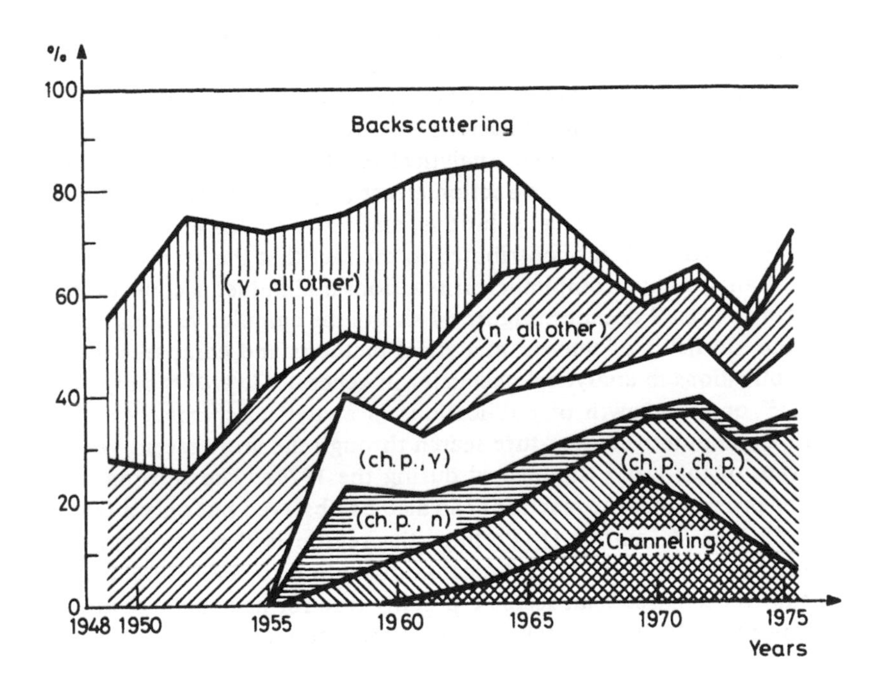

FIGURE 65. Percentage distribution of publications on various types of nuclear reactions (or methods) used in prompt nuclear analysis; ch. p., charged particles. (From Bujdosó, E., Lyon, W. S., and Noszlopy, I., *J. Radioanal. Chem.*, 74, 197, 1982. With permission.)

Since the middle of the 1960s backscattering has gained even more ground whereas photoreactions have faded into the background.

The sudden growth of a new subject field can occur when scientists working in other areas perceive the new one as a greener pasture. The exploration of the new area is thus usually set in motion by a process of demographic migration.[142] The growth of prompt nuclear analysis seems to be measurably influenced by the migration of some nuclear physicists who realized that their outdated equipment and their knowledge could be better used for competitive research in the fast advancing field of nuclear analytical methods.[143]

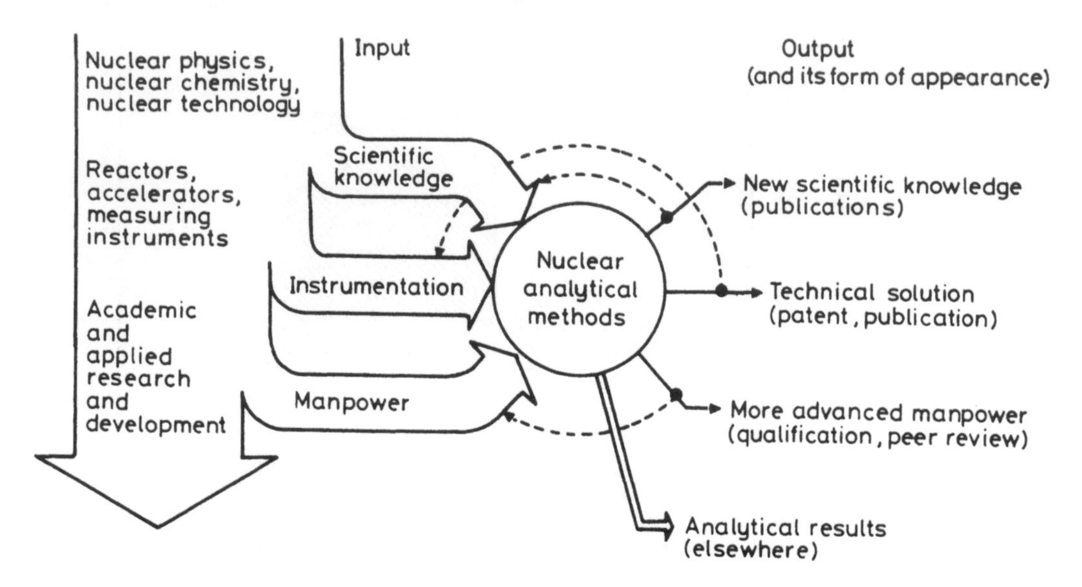

FIGURE 66. The R&D mechanism of the nuclear analytical methods.

On the input side of the R&D mechanism of prompt nuclear analysis, the equipment and knowledge originating from the basic nuclear research acted as a driving force. On the output side the demand for new analytical results applicable to various fields of research and development exerted a suction effect (Figure 66). This mechanism generated new knowledge, new technical solutions, and more skilled manpower which were fed back to the system. All this contributed to the fact that the literature of activation analysis and of analysis by prompt nuclear reactions maintained a constant exponential growth rate for the last 2 decades. Figure 67 shows the cumulative number of publications in various subfields of nuclear analytical methods and compares them with the growth of publications in analytical chemistry in general. Girardi[144] supplemented our previous data[20] on the growth of publications in analytical chemistry and activation analysis with a computerized literature search through the more recent years. As seen, the growth rate of the latter diminished during the 1970s. The slowing down of the growth rate of prompt nuclear reactions can also be observed but only a few years later (see Figures 68 and 76, curves 1). Perhaps the reason is that the opportunities for making a noticeable scientific or technical contribution to the subfield of activation analysis began to decrease. That seems especially true for neutron activation analysis. Figure 68 shows also the growth dynamics of published methods on the use of various nuclear reactions.

The overall doubling time of 3.2 years of the literature of prompt reactions includes the doubling times of less than 2 years. For example, the literature on analytical use of charged particle-charged particle reactions show doubling time T_d = 1.7 years; that of Rutherford backscattering, T_d = 1.7 years; and that of channeling, T_d = 1.3 years. On the other hand, the literature of photon reactions shows a doubling time of 9.7 years, reached after an initial steep increase, while that of charged particle-neutron reactions have had a steady doubling time of 5.8 years from their onset.

The relative distribution of literature on the analytical use of different nuclear reactions, grouped according to the bombarding and emitting particles, is shown in Table 57. The most frequently applied reactions were of the (n, γ) and (p, γ) types. In 38.4% of the cases the analyses published between 1949 and 1977 were done by (n, γ) reactions; in 15.9% of the cases by (p, γ) reactions; and in 15.9% of the cases by (p, γ) reactions.

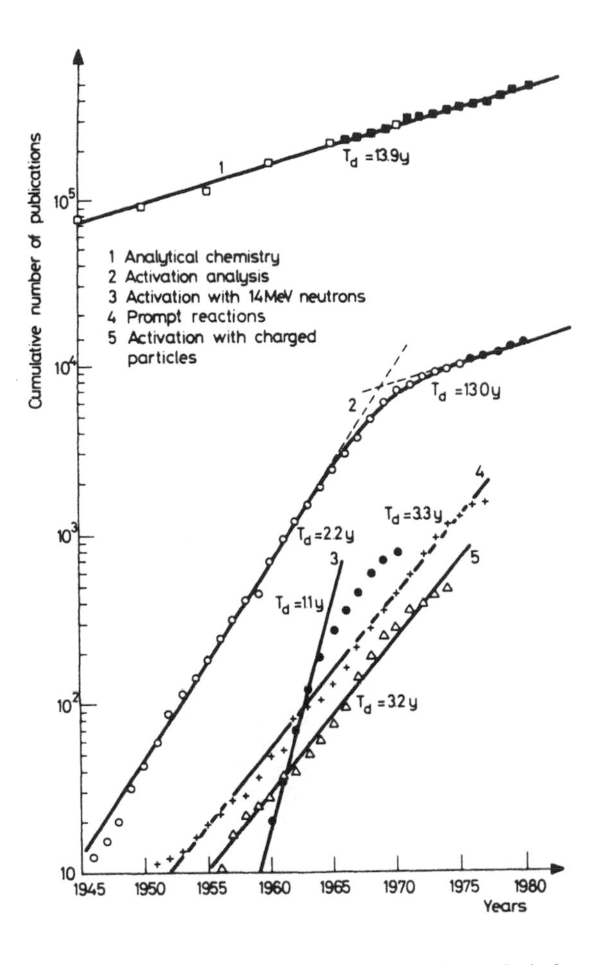

FIGURE 67. Growth of publications in analytical chemistry and some fields of nuclear analytical methods. The black dots and squares show the result of a computerized literature search by Girardi.[144] T_d, doubling time. (From Bujdosó, E., Lyon, W. S., and Noszlopy, I., *J. Radioanal. Chem.*, 74, 197, 1982. With permission.)

The growth of the number of determinations using nuclear reactions induced by various particles is shown in Figure 69. In most of the cases growth can be described by an exponential. The growth of the literature on heavy ion reactions (T_d = 0.8 years); (p, d) (T_d = 2.0 years); (p, γ), (T_d = 2.1 years); (d, α), (T_d = 2.3 years); (n, γ), (T_d = 2.4 years); and ^{3}He reactions (T_d = 2.8 years) is characterized by the shortest doubling times, i.e., by the fastest growth rate. Slowest growth is found for (n, n) (T_d = 21.6 years) and for (p, p) (T_d = 8.7 years) reactions. Figure 69 also shows the rapid initial growth and later slowdown of the use of deuteron, triton, and α reactions, among which (d, α), (α, α), and (α, γ) reactions are exceptions.

Table 58 collects the most frequently determined elements and the number of their analysis both by prompt nuclear reactions and backscattering published from 1949 to 1977. In the papers published between 1950 and 1977, 4240 analyses were counted. The 6 elements O, Si, C, Al, N, and B account for 38% of the analyses, among which 22% were performed by nuclear reactions and 16% by backscattering.

The growth in the number of determinations of elements by the use of backscattering as well as the sum of analyses by backscattering and by nuclear reactions are plotted

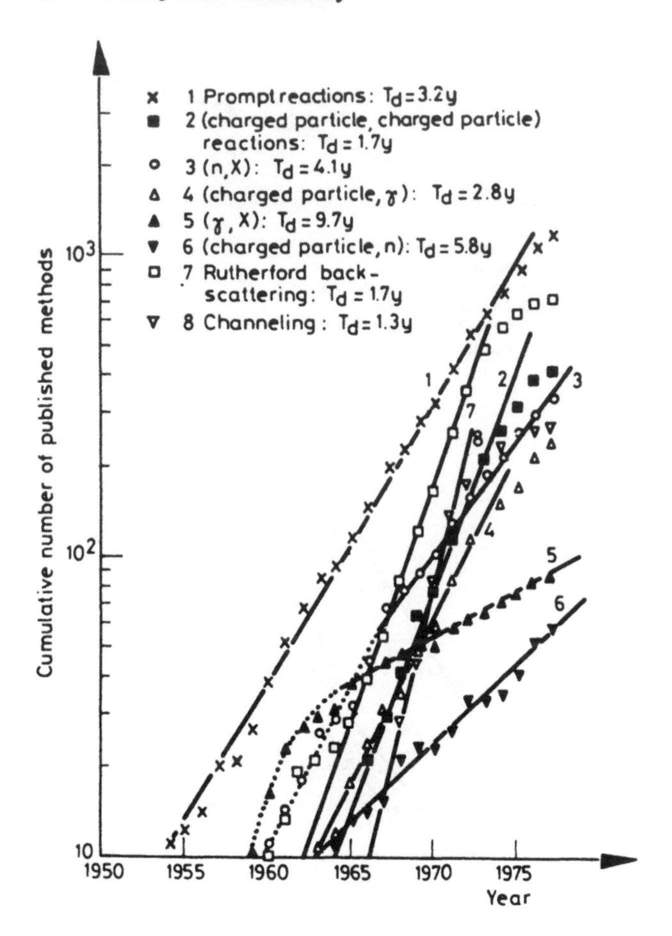

FIGURE 68. Growth of various nuclear reactions used in prompt nuclear analysis. (From Bujdosó, E., Lyon, W. S., and Noszlopy, I., *J. Radioanal. Chem.*, 74, 197, 1982. With permission.)

Table 57
THE PERCENT USAGE OF NUCLEAR REACTIONS OF VARIOUS KINDS ARRANGED ACCORDING TO BOMBARDING AND EMITTED PARTICLES IN PROMPT NUCLEAR REACTIONS, 1949 TO 1977

Bombarding particle	Emitted particle					
	n	p	α	γ	d	Other
n	1.16	0.26	2.47	38.37		
p	1.26	0.34	5.20	15.90		
d	1.13	10.50	2.78	1.60		
t	0.05	0.13		0.22		
α	2.03	2.51	0.80	1.60		
^{3}He		0.61	0.26	0.22		
γ	4.08			0.95		
Heavy ions	0.08	0.91	1.26	1.43	1.04	
Other						0.82

Bujdosó, E., Lyon, W. S., and Noszloply, I., *J. Radioanal. Chem.*, 74, 197, 1982. With permission.

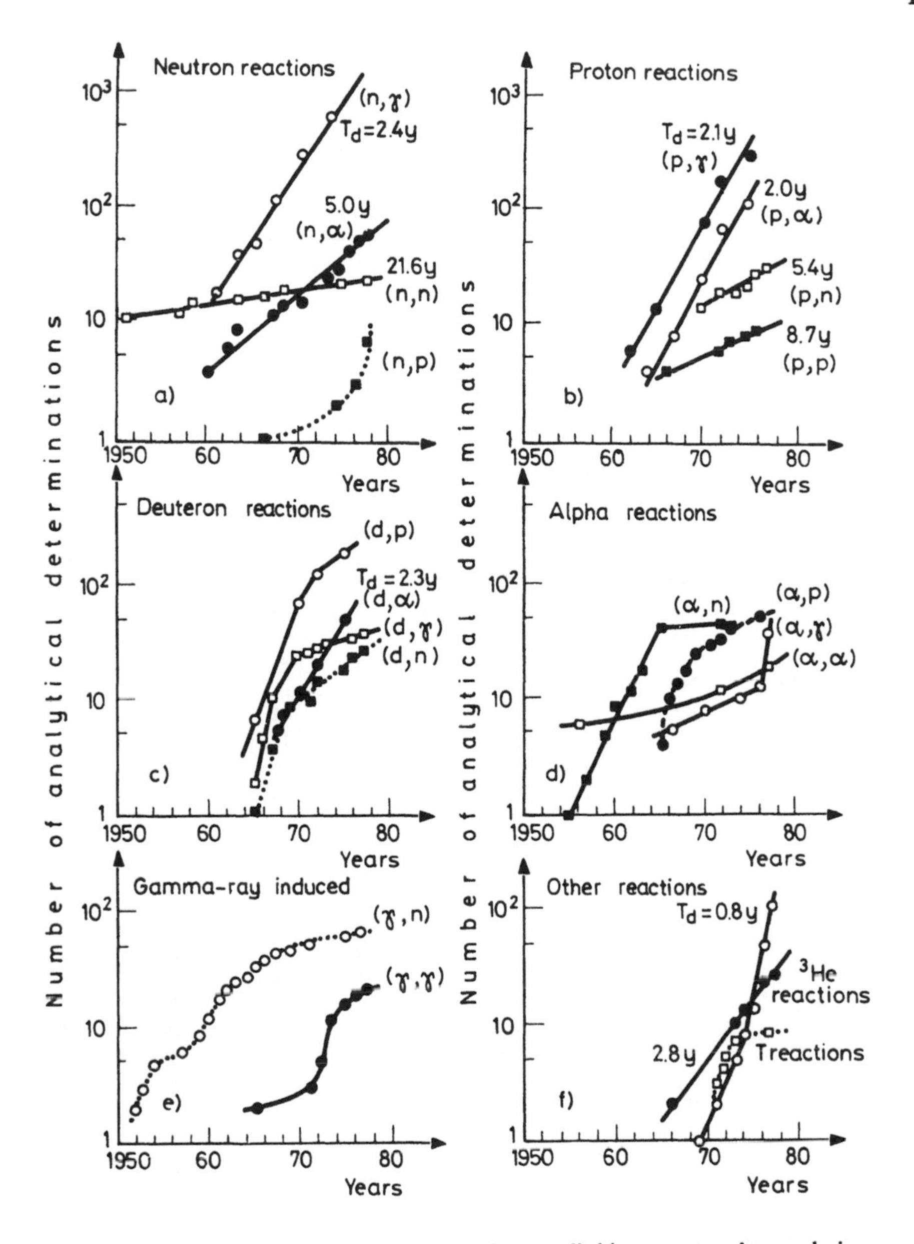

FIGURE 69. Growth of various nuclear reactions applied in prompt nuclear analysis as reflected in the number of analytical determinations accomplished by them. (From Bujdosó, E., Lyon, W. S., and Noszlopy, I., *J. Radioanal. Chem.*, 74, 197, 1982. With permission.)

in Figure 70. The growth of published analyses can be described for most of the elements by an exponential with doubling times between 1.8 and 3.6 years.

Figures 71 to 75 show in 5-year intervals the cumulative number of elements analyzed by Rutherford backscattering and by (n, γ) and (n, p) nuclear reactions, respectively. As seen in Figure 71 a, prior to 1961 only 8 elements were determined by backscattering. In the following 10 years 80% of the periodic table was covered, and after 5 more years only the lanthanide region remained unexploited. A similar situation prevails for (n, γ) reactions (Figures 73 and 74), while the use of (p, γ) reactions (Figure 75) has gradually spread from elements of low to those of high atomic number.

Table 58
ELEMENTS DETERMINED BY VARIOUS METHODS OF PROMPT NUCLEAR ANALYSIS AND THE NUMBER OF THEIR OCCURRENCES IN PUBLICATIONS, 1949 TO 1977

		Reactions induced by						
Rank	Element	n	p	d	α	Other	Back-scattering	Total
1	O	26	88	117	15	27	152	425
2	Si	59	16	20	12	—	262	369
3	C	36	55	67	4	21	87	270
4	Al	39	42	6	13	—	103	203
5	N	20	38	74	7	10	48	197
6	B	42	36	17	17	37	19	168
7	Fe	78	7	—	2	—	73	160
8	H	45	—	1	—	20	3	139
9	F	—	74	3	14	6	23	120
10	Cu	35	3	2	2	9	60	111
11	Au	18	3	—	2	—	75	98
12	Mg	24	22	6	10	—	31	93
13	Ni	39	3	3	1	7	36	89
14	Ca	39	6	3	1	—	40	89
15	S	27	13	7	1	—	39	87
16	Na	22	20	—	16	—	28	86
17	Be	7	7	2	10	42	15	83
18	Li	20	27	8	3	16	6	80
19—20	Ti	27	1	1	1	1	35	66
	Cl	28	9	—	2	—	27	66
21	P	8	10	—	1	—	44	63
22	As	7	—	—	1	1	51	60
23	Ga	2	2	—	—	—	52	56
24	Mn	37	3	—	1	—	13	54
25	Ge	4	2	—	—	—	39	45
26	Cr	19	2	—	4	—	18	43
27	Pb	6	1	—	1	1	32	41
28—31	Sb	2	—	—	—	—	38	40
	Cd	21	—	—	—	—	19	40
	Ag	13	3	—	1	—	23	40
	K	17	1	—	2	—	20	40
32	W	9	1	1	1	—	25	37

From Bujdosó, E., Lyon, W. S., and Noszlopy, I., *J. Radioanal. Chem.*, 74, 197, 1982. With permission.

III. THE EPIDEMICS OF PROMPT NUCLEAR ANALYSIS

As already mentioned in Chapter 7, the transmission of ideas within a population is similar to the transmission of infectious diseases. Some diseases may be transmitted if an infective subject and a susceptible subject come into close proximity. Other diseases may be transmitted by means of biological carriers. The model assumes that the role of carriers in a scientific community is played by publications. The exact mathematical formulation of the evolution of an epidemic presents enormous difficulties. Most of the real transmission problems are too difficult to be exactly described by such a treatment. However, simple models can be created which help pinpoint the parameters which determine whether or not an epidemic will develop and which parameters will influence its evolution.[145] There are two kinds of models for the mathematical treatment of epidemics: the deterministic and the stochastic ones. Deterministic models represent the process as a system of differential equations; stochastic models describe it as a finite Markov process.

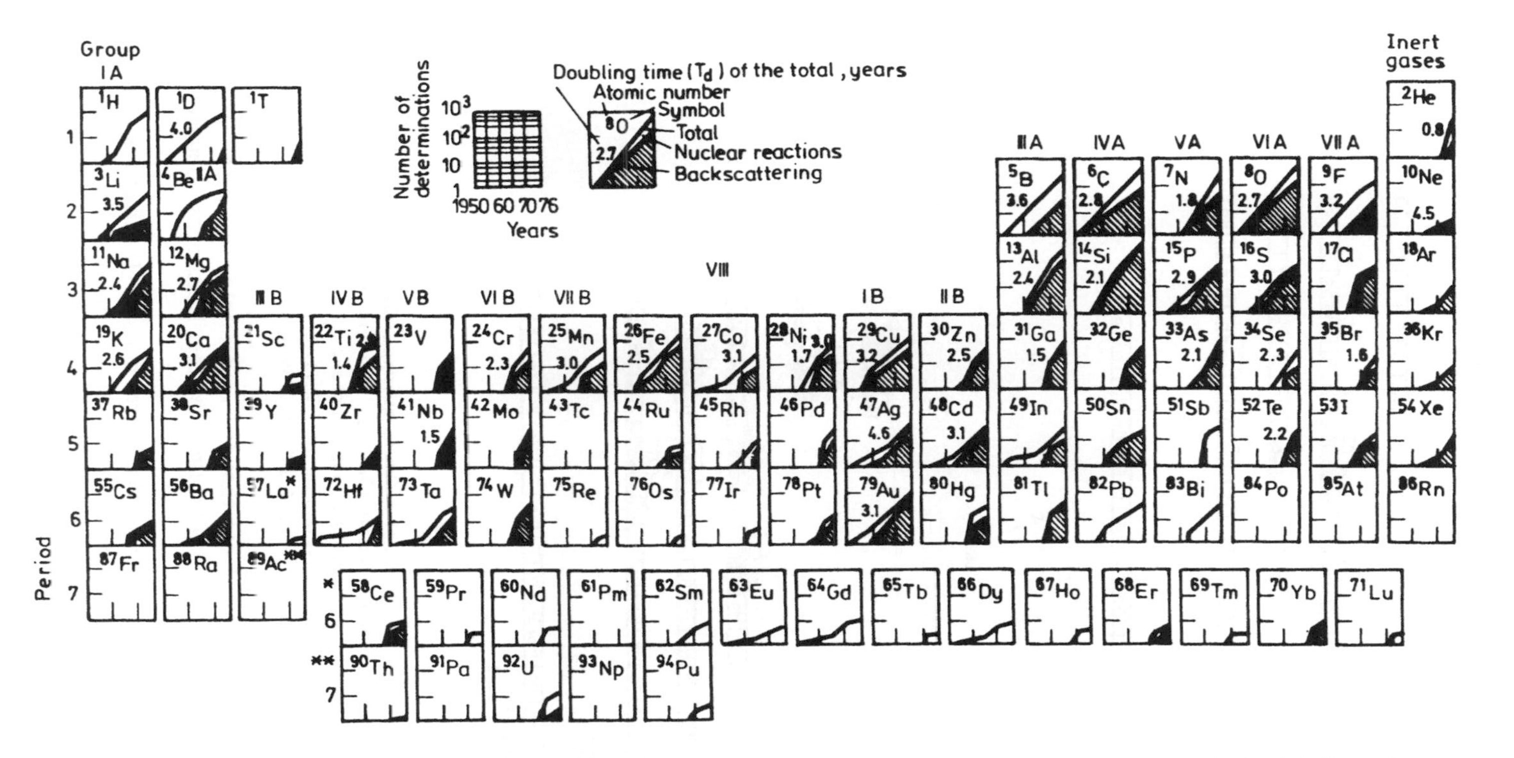

FIGURE 70. Growth of the number of analyses by prompt nuclear methods and their doubling times for the elements of the periodic table. (From Bujdoso, E., Lyon, W. S., and Ncszlopy, I., *J. Radioanal. Chem.*, 74, 197, 1982. With permission.)

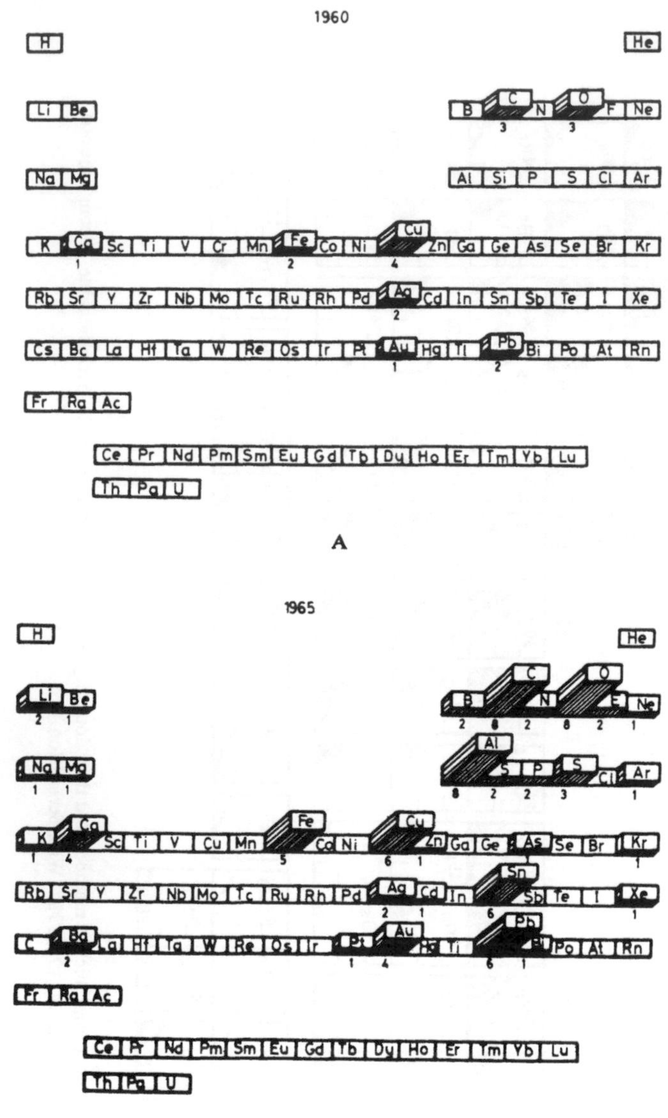

FIGURE 71. (A to C) Growth of the backscattering technique over the periodic table in 5-year intervals between 1960 and 1970. The height of the columns is proportional to the number of analyses published up to the year indicated. (From Bujdosó, E., Lyon, W. S., and Noszlopy, I., *J. Radioanal. Chem.*, 74, 197, 1982. With permission.)

Goffman and Newill[97] have pointed out that the epidemiological spread of ideas within the scientific community can be treated by the Reed-Frost deterministic model.[146]

The members of a population belong to one of three classes: (I) infectives, (S) susceptibles, or (R) removals. The identification of I and R in a subfield of science can be relatively easily made by determining the date of first and last publication on the subject. It is more difficult, sometimes impossible, to estimate the number of S. Whether or not an epidemic will develop depends mainly upon the existence of a sufficiently large population of S.

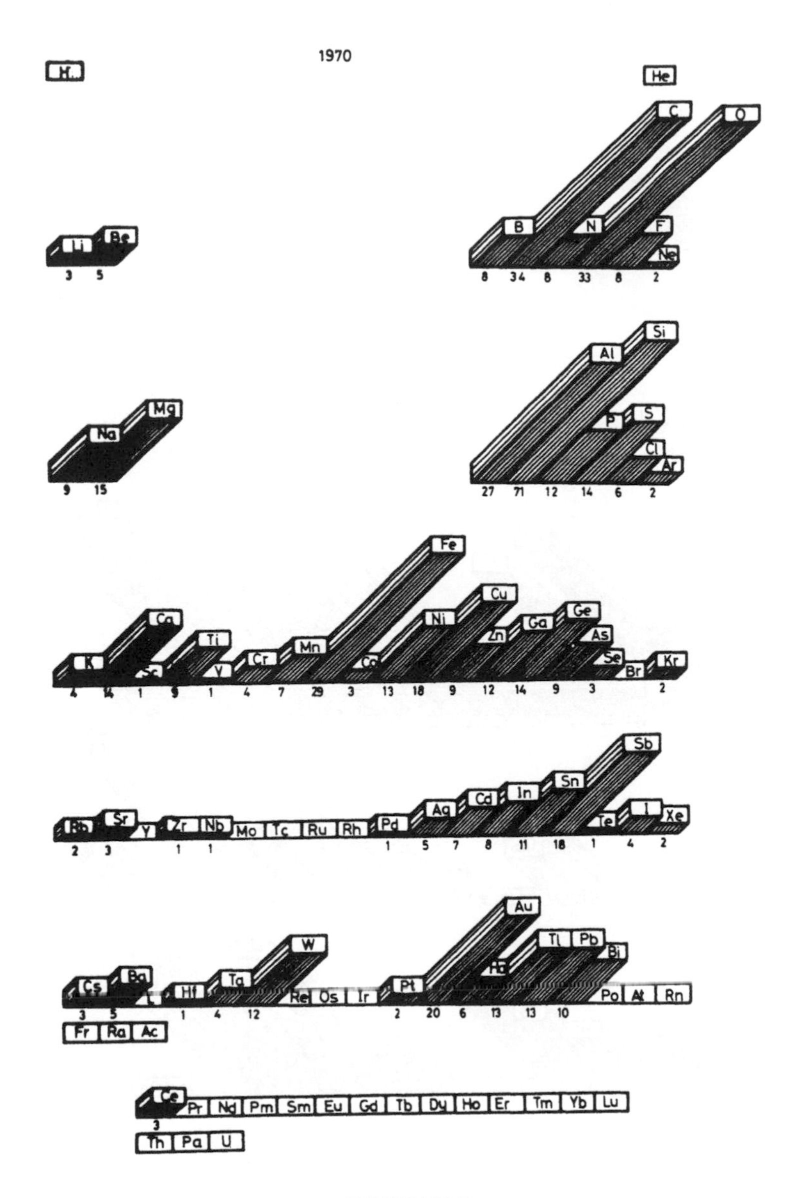

FIGURE 71C

According to the deterministic model the process enters an epidemic state, i.e., dI/dt becomes greater than zero, when the number of S reaches a threshold value. The process peaks as d^2I/dt^2 becomes zero and dR/dt approaches a constant value. The number of remaining S eventually falls below the threshold value, and the epidemic slows down.

In the stochastic model a scientific idea (D) may be treated as a complete, ordered, and finite set of permutable elements of information: $D = (a, b, c, \ldots, n)$. The elements of the set may be ordered sets of earlier scientific contributions. This set passes through various development stages. The transition is not continuous but occurs in discrete steps. Goffman and Harmon[147] distinguish four possible stages of the information set D.

Stage I — Insufficient and unordered information. The problem is to acquire information. Once more information has been acquired, ordering problems appear.

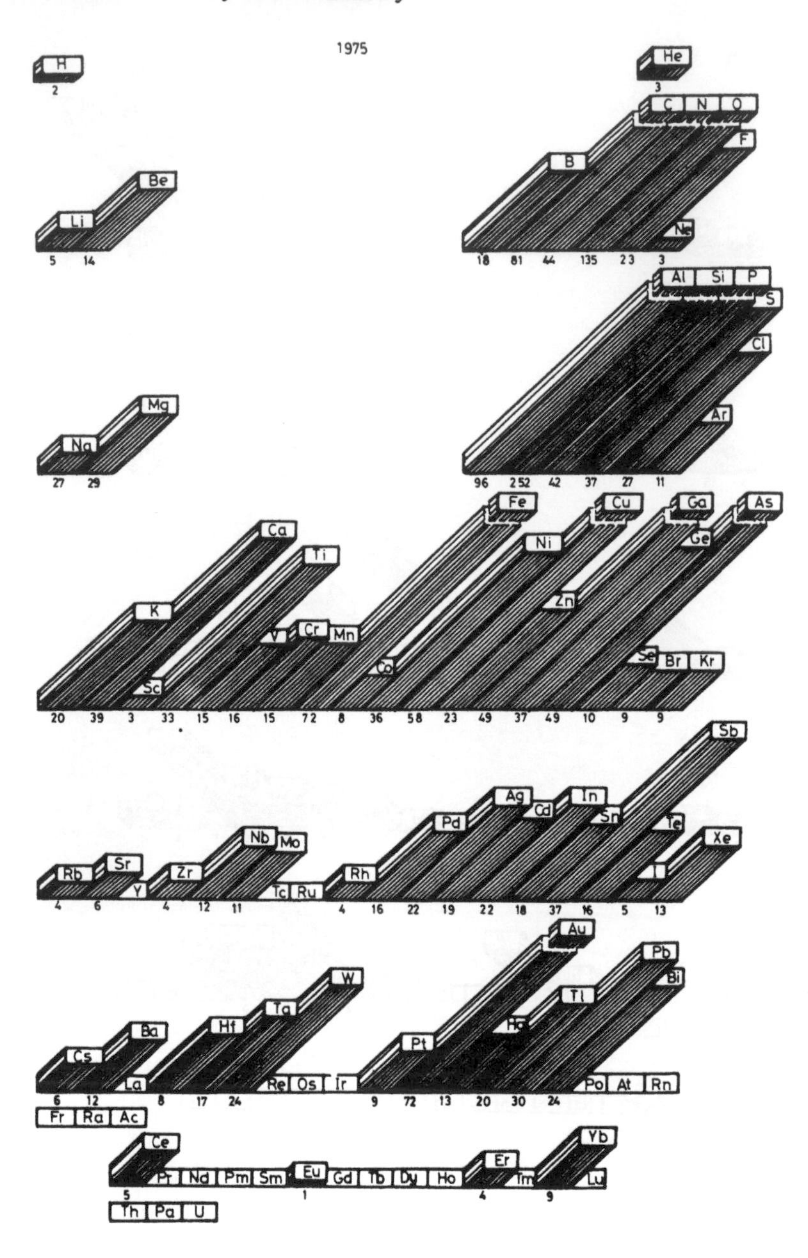

FIGURE 72. Growth of the backscattering technique over the periodic table between 1970 and 1975. The height of the columns is proportional to the number of analyses published up to the year indicated. (From Bujdosó, E., Lyon, W. S., and Noszlopy, I., *J. Radioanal. Chem.*, 74, 197, 1982. With permission.)

Stage II — Insufficient but ordered information. The task becomes primarily the increase of the information in magnitude.

Stage III — Sufficient but unordered information. As the acquisition of information continues, sufficient or even surplus information elements become available. The problem of ordering arises again.

Stage IV — Sufficient and ordered information. Redundant and nonrevelant information is excluded. With the acquisition of new information the set may be partially or completely decomposed. The cycle then returns to Stage II or III or in some cases to Stage I.

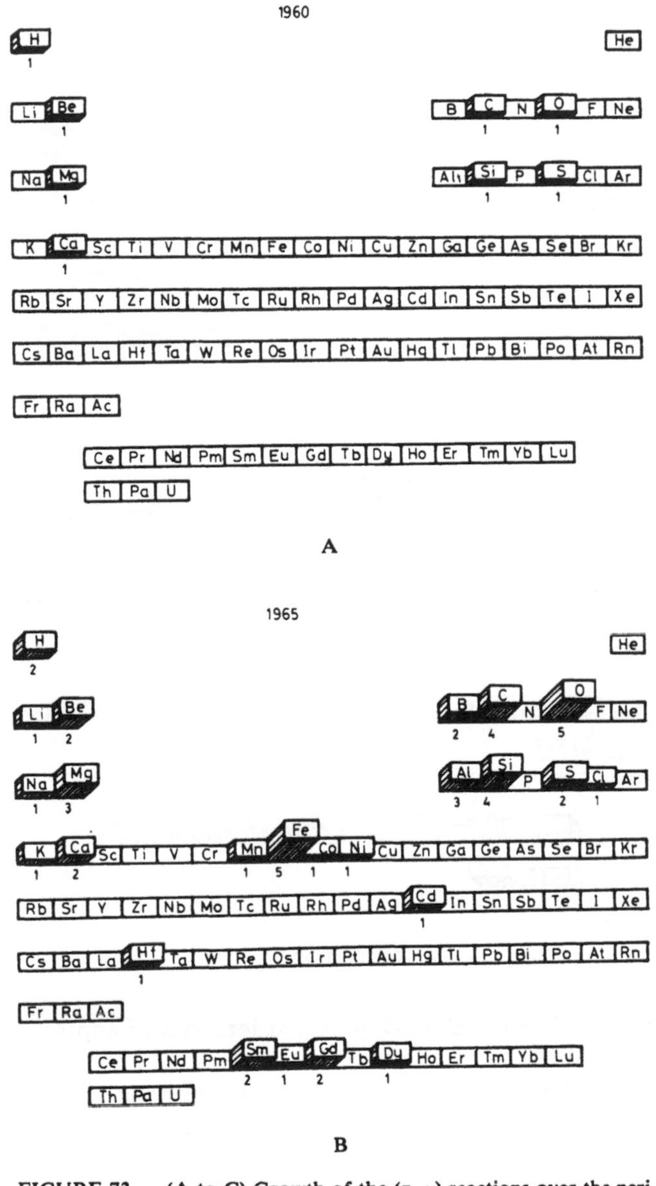

FIGURE 73. (A to C) Growth of the (n, γ) reactions over the periodic table in 5-year intervals between 1960 and 1970. The height of the columns is proportional to the number of analyses published up to the year indicated. (From Bujdosó, E., Lyon, W. S., and Noszlopy, I., *J. Radioanal. Chem.*, 74, 197, 1982. With permission.)

For prompt nuclear analysis, the cumulative growth of I and R is shown in Figure 76. The growth of both populations can be described by a constant rate exponential with a doubling time of 3.5 and 3.8 years, respectively. These values are very close to the doubling time of the growth of publications (T_d = 3.3 years, Figure 67, curve 4) and of methods (T_d = 3.2 years, Figure 68, curve 1). The curve characterizing the epidemic growth $\Delta(I - R)/\Delta t$ or the yearly changes of active contributors as a function of time, i.e., the number of I minus R in the same year, is plotted in Figure 77. There are three separate waves on the curve showing that prompt nuclear analysis has devel-

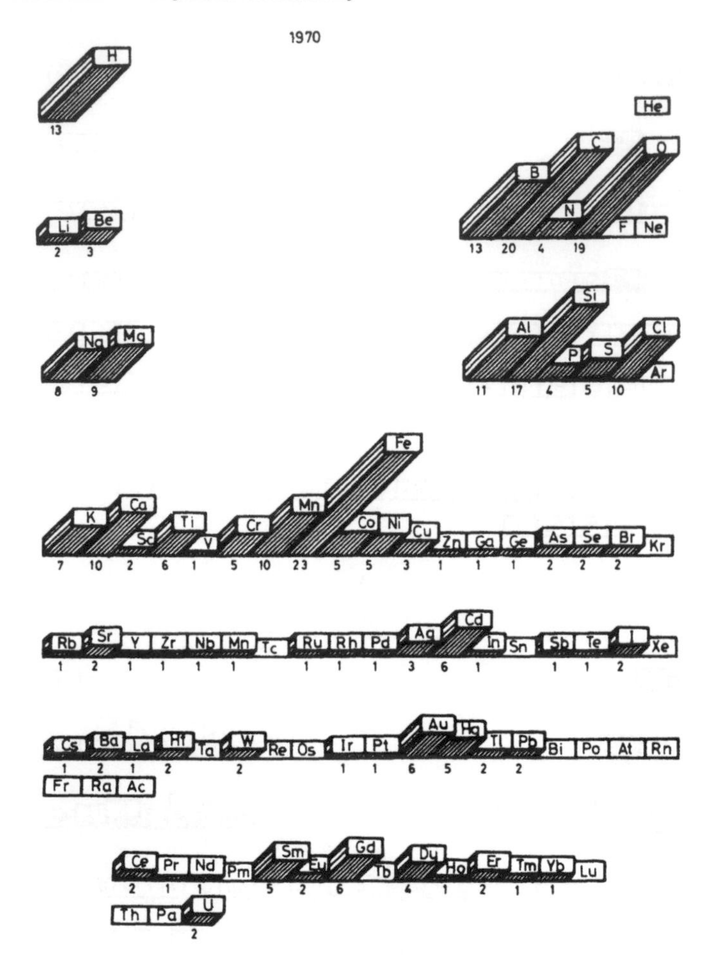

FIGURE 73C

oped as a recurring epidemic with peak points at intervals of approximately 11 years; i.e., in 1950, 1961, and 1972.

Having identified the epidemic waves of prompt nuclear analysis let us try to outline the events that might have given rise to them. Papers published between 1949 and 1952 were byproducts of nuclear physics research, neutron spectroscopy and transmission measurements, and γ-ray-induced reactions as applied in analyses.

The first explicitly emerging epidemic (beginning in the 1950s) was caused by the appearance and widespread use of NaI(T1) detectors. This process achieved the stage of ordered information by 1961 as characterized by the review papers published in 1961 and 1962.

The next and largest epidemic was triggered by the use of semiconductor detectors between 1965 and 1974. The highly cited papers trace and the growth of the field until about 1971 when the epidemic reached its climax with yearly I of about 100. Ordering of the acquired information again proceeded as shown by the many review papers published. At the end of the third epidemic wave, bibliographies appear.[138-140] It is tempting to predict the fourth epidemic wave of prompt nuclear analysis with a peak point around 1983.

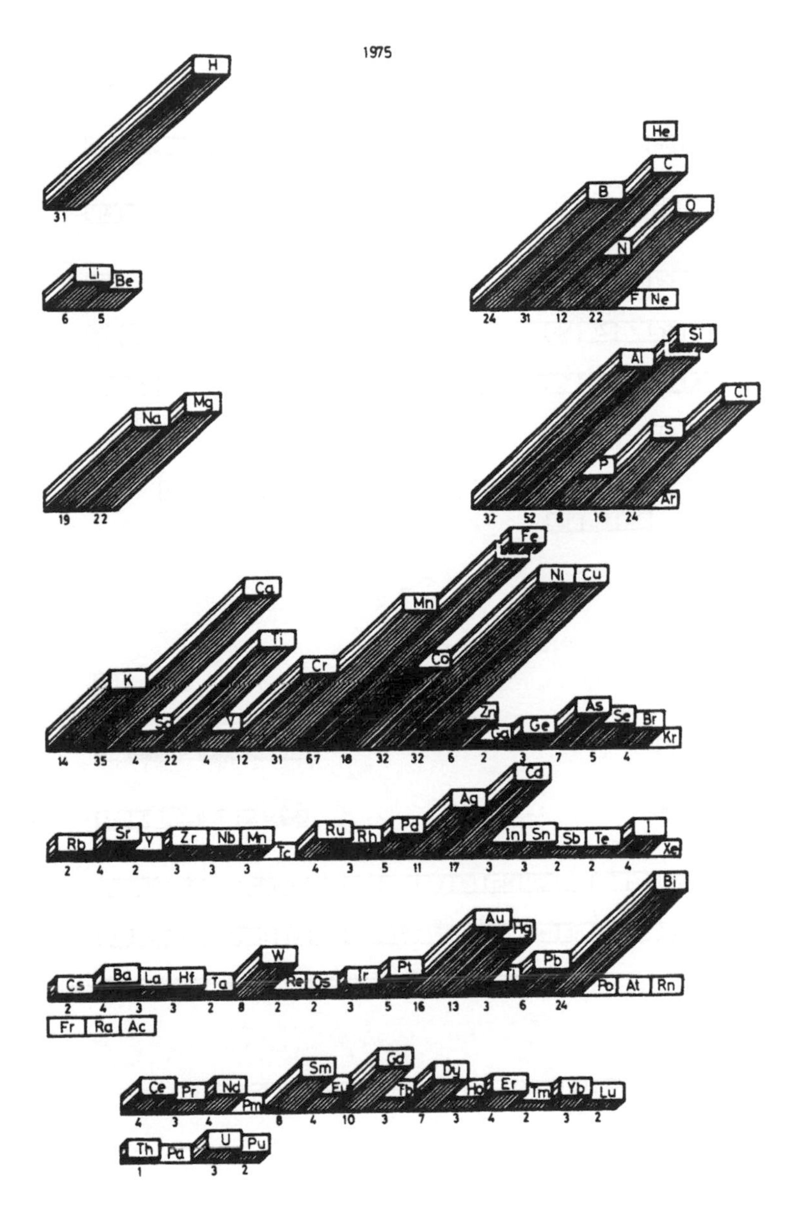

FIGURE 74. Growth of the (n, γ) reactions over the periodic table between 1970 and 1975. The height of the columns is proportional to the number of analyses published up to the year indicated. (From Bujdosó, E., Lyon, W. S., and Noszlopy, I., *J. Radioanal. Chem.*, 74, 197, 1982. With permission.)

IV. SCATTER AND NATIONAL DISTRIBUTION OF THE LITERATURE ON PROMPT NUCLEAR ANALYSIS

As seen in Table 56, 60 to 70% of the publications in the field of prompt nuclear analysis are journal articles. The scatter of information can be empirically represented by Bradford's law discussed in Chapter 6, Section I.

Figure 78 shows the Bradford distribution of the literature on prompt nuclear analysis. As seen, the 1030 journal papers appeared in 183 different journals. The subfield has about 10 core journals that carry nearly 60% of the total journal literature. As a comparison it may be mentioned that the distribution is the same in nuclear analytical

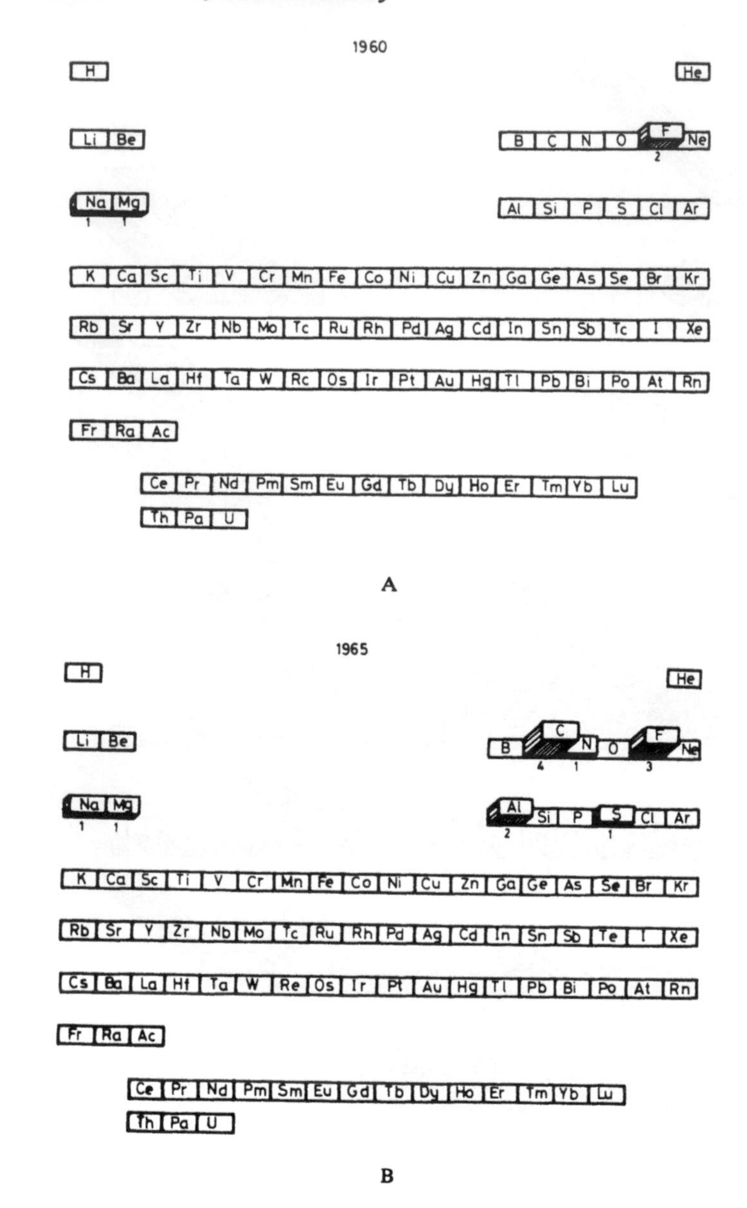

FIGURE 75. (A to D) Growth of the (p, γ) reactions over the periodic table in 5-year intervals. The height of the columns is proportional to the number of analyses published up to the year indicated. (From Bujdosó, E., Lyon, W. S., and Noszlopy, I., *J. Radioanal. Chem.*, 74, 197, 1982. With permission.)

chemistry,[20] whereas in analytical chemistry about 30 journals concentrate 60% of the journal papers (see Chapter 6, Section I). These two data show the difference in the dispersion or concentration of the literature of these fields. Three journals *(Analytical Chemistry, Journal of Radioanalytical Chemistry,* and *Radiochemical and Radioanalytical Letters)* are core journals of all three subject fields, i.e., of analytical chemistry, activation analysis, and prompt nuclear analysis.

The national distribution of papers on prompt nuclear analysis was determined according to the affiliation of the authors. Two thirds of the authors belong to three

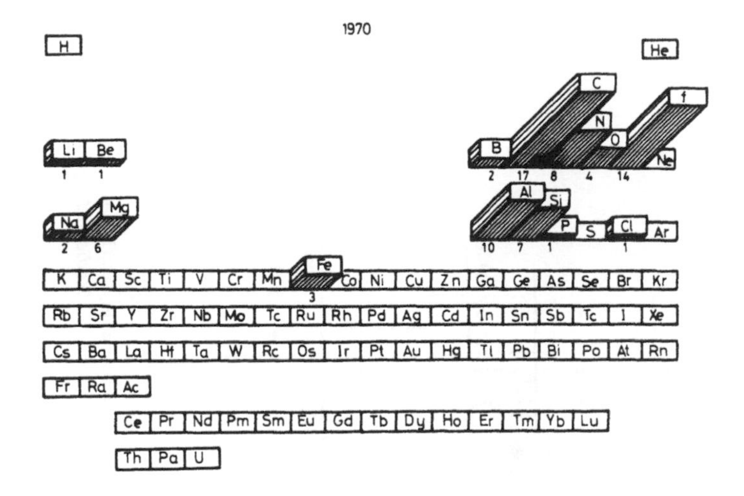

FIGURE 75C

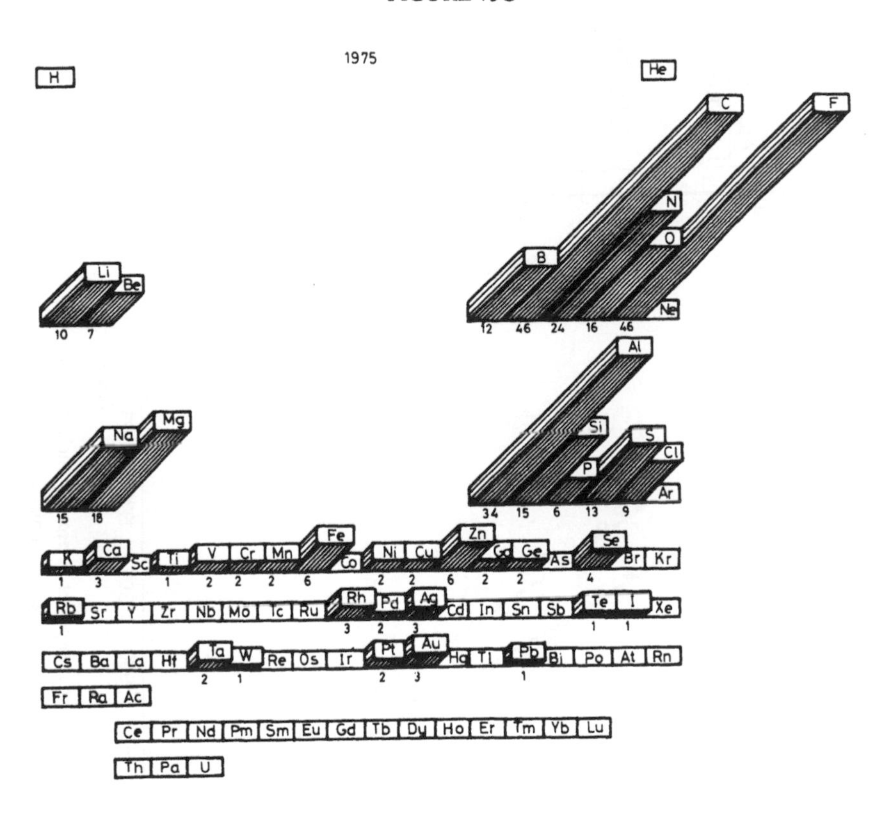

FIGURE 75D

countries only. This represents a concentration that is more intensive than that of the authors of analytical chemistry publications (Table 59).[12]

V. PRODUCTIVITY DISTRIBUTION OF PROMPT NUCLEAR ANALYSIS AUTHORS

In examining the productivity in the field of prompt nuclear analysis, the co-authors were taken into account with identical weight, and a unity was added to the number

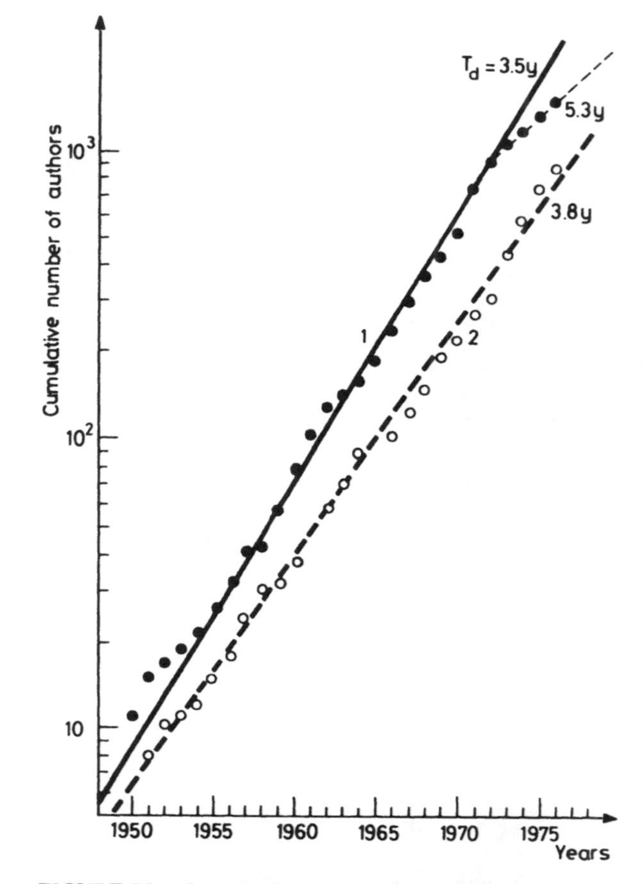

FIGURE 76. Growth of manpower in prompt nuclear analysis. Curve 1 — cumulative number of publishing authors (at least one paper in every other year), curve 2 — authors leaving the field. (From Bujdosó, E., Lyon, W. S., and Noszlopy, I., *J. Radioanal. Chem.*, 74, 197, 1982. With permission.)

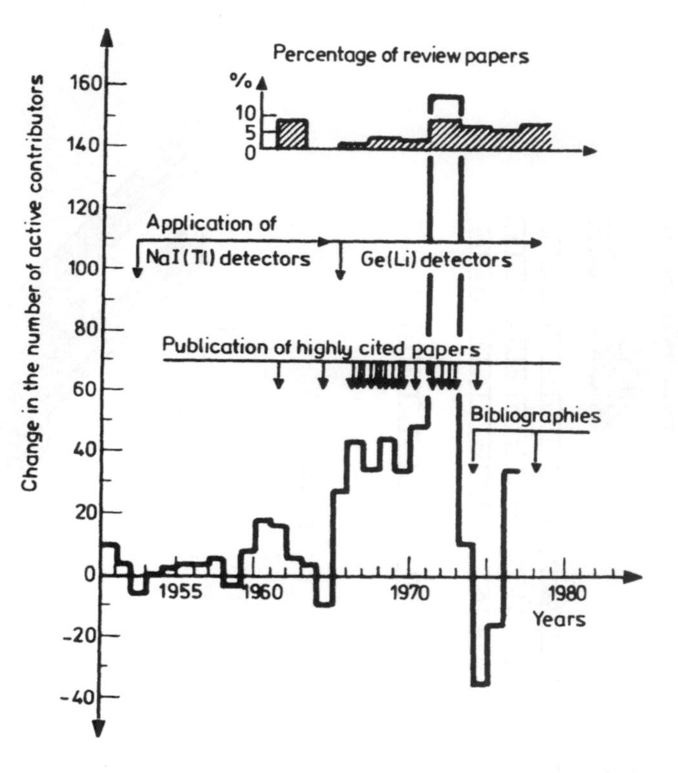

FIGURE 77. Epidemic curve of prompt nuclear analysis. (From Bujdosó, E., Lyon, W. S., and Noszlopy, I., *J. Radioanal. Chem.*, 74, 197, 1982. With permission.)

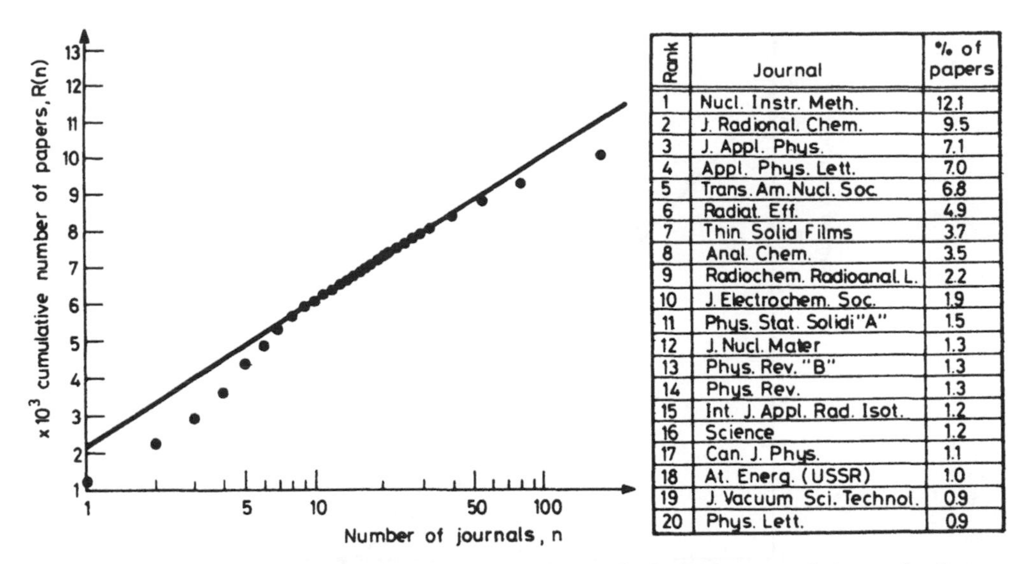

x 10^3 cumulative number of papers, R(n)
13
12
11
10
9
8
7
6
5
4
3
2
1
1
5
10
50
100
Number of journals, n
Rank
Journal
% of papers
1 Nucl. Instr. Meth. 12.1
2 J. Radional. Chem. 9.5
3 J. Appl. Phys. 7.1
4 Appl. Phys. Lett. 7.0
5 Trans. Am. Nucl. Soc. 6.8
6 Radiat. Eff. 4.9
7 Thin Solid Films 3.7
8 Anal. Chem. 3.5
9 Radiochem. Radioanal. L. 2.2
10 J. Electrochem. Soc. 1.9
11 Phys. Stat. Solidi "A" 1.5
12 J. Nucl. Mater 1.3
13 Phys. Rev. "B" 1.3
14 Phys. Rev. 1.3
15 Int. J. Appl. Rad. Isot. 1.2
16 Science 1.2
17 Can. J. Phys. 1.1
18 At. Energ. (USSR) 1.0
19 J. Vacuum Sci. Technol. 0.9
20 Phys. Lett. 0.9

FIGURE 78. Scatter of the literature on prompt nuclear methods. R(n) — cumulative totals of papers contributed by the journals ranked 1 to n. (From Bujdosó, E., Lyon, W. S., and Noszlopy, I., *J. Radioanal. Chem.*, 74, 197, 1982. With permission.)

Table 59
PERCENTAGE OF PUBLICATIONS ON PROMPT NUCLEAR ANALYSIS AND TOTAL ANALYTICAL CHEMISTRY CARRIED OUT IN VARIOUS COUNTRIES

	Percentage of papers on	
Country	Prompt nuclear analysis	Total analytical chemistry[a]
U.S.	41.7	18.0
France	13	3.0
U.K.	10	5.4
Germany	4.8	5.7[b]
U.S.S.R.	4.6	25.5
Canada	3.5	
Sweden	2.9	
South Africa	2.8	
Belgium	2.4	
Australia	2.3	
Japan	2.1	8.8
Czechoslovakia	1.6	4.9
Italy	1.4	1.7
Poland	1.3	2.4
Denmark	1.2	
Finland	0.6	
Hungary	0.6	
Netherlands	0.5	1.0
India	0.4	3.7
Venezuela	0.3	
New Zealand	0.3	
Rest of the world	2.0	19.9

[a] Average of the years 1960—1970.[12]
[b] Includes both East and West Germany.

From Bujdosó, E., Lyon, W. S., and Noszlopy, I., *J. Radioanal. Chem.*, 74, 197, 1982. With permission.

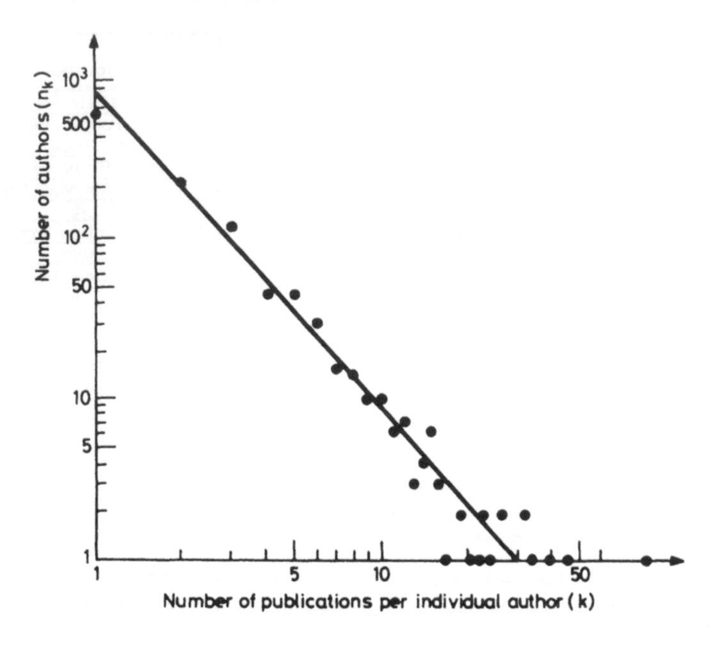

FIGURE 79. Distribution of productivity in prompt nuclear analysis. (From Bujdosó, E., Lyon, W. S., and Noszlopy, I., *J. Radioanal. Chem.*, 74, 197, 1982. With permission.)

of their publications instead of their fractional share. The distribution is shown in Figure 79. Upon determining the value of the exponent in Equation 19, using the maximum likelihood method and neglecting values where $n > 20$, the exponent was found to be 1.98. In other words, actual single-paper authors amount to 52%; authors of two papers, 19.5% and authors of three papers, to 10% of the total. For an exponent of 2, in the original Lotka distribution these values would be 60, 15, and 7%, respectively, as mentioned in Chapter 8, Section I. Table 60 shows the most prolific authors in the field of prompt nuclear analysis as well as the number of their co-authors. The close correlation ($r = 0.84$) between the number of publications and the number of co-authors points out that the collaboration and productivity are related to each other. In the range studied (i.e., authors of 13 or more papers) $n = 0.66\ c + 2.32$, where c is the number of co-authors and n is the number of papers.

The productivity of authors depends upon survival, i.e., how long they have been working in the field. In order to investigate the survival rate of authors in the field of prompt nuclear analysis we assumed that the publication of a paper, even for co-author, needs at least 1 year of active work. We have therefore taken the difference of publication dates of the first and the last papers of the authors and added 1 more year, i.e., the time needed for preparation of the first paper. There was a group of authors whose future behavior was uncertain, i.e., those who entered the field between 1975 and 1977. They were treated separately. It was assumed that the survival distribution of this population will be equal to that for which our data were certain.

The results are shown in Figure 80. Curve 2 shows that after 3 years in the field, 65% of the scientists left prompt nuclear analysis; after 10 years, 86% left; and after 20 years, only 12% of the original population can be found in the field. The number of those (58%) who worked only for 2 years or less is very large. Of these, 90% are single-paper authors, 8% are authors of two papers, while 2% are authors of three or more papers. The survival of publishing authors in the fields of experimental particle physics[149] and oscillating chemical reaction[150] is also shown for comparison. The difference in behavior of authors in the three fields is quite remarkable. About 40% of

Table 60
MOST PRODUCTIVE AUTHORS IN THE SUBJECT FIELD OF PROMPT NUCLEAR ANALYSIS, 1949 TO 1976

Rank	Name	Number of publications	Number of co-authors
1	Mayer, J.W.	87	68
2	Amsel, G.	46	31
3	Picraux, S.T.	39	24
4	Pierce, T.B.	34	11
5—6	Peisach, M.	32	13
	Ziegler, J.F.	32	22
7—8	Davies, J.A.	26	21
	Meyer, O.	26	16
9	Dearnaley, G.	24	19
10—11	Senftle, F.E.	23	21
	Poate, J.M.	23	28
12	Chu, W.K.	22	21
13	Duffey, D.	21	10
14—15	Turkevich, A.L.	19	8
	Eisen, F.H.	19	16
16	Nicolet, M.A.	18	19
17	Mitchell, I.W.	17	24
18—20	Ligeon, E.	16	18
	Starchik, L.P.	16	10
	Wiggins, P.F.	16	8
21—26	Behrisch, R.	15	10
	David, D.	15	16
	Cohen, C.	15	26
	Croset, M.	15	12
	Mackintosh, W.D.	15	7
	Patterson, J.H.	15	7
27—30	Beranger	14	12
	Feldman, L.C.	14	11
	Franzgrote, E.J.	14	7
	Rimini, E.	14	13
31—33	Eriksson, L.	13	16
	Sigurd, D.	13	18
	Turos, A.	13	7

From Bujdosó, E., Lyon, W. S., Noszlopy, I., *J. Radioanal. Chem.*, 74, 197, 1982. With permission.

the authors of prompt nuclear analysis and in oscillating chemical reactions leave after about 2 years of work. The same percentage of particle physics authors leaving the area is reached only after 30 years. The short lifetime of many prompt nuclear analysis and oscillating reactions authors may be due to graduate and postdoctoral students who participate in a 2-year work and then leave for other areas.

VI. CO-AUTHORSHIP IN THE FIELD OF PROMPT NUCLEAR ANALYSIS

Co-authorship reflects an active mutual influence, collaboration, and is inextricably related to the research technique of the field in question. As the complexity of research methodology increases requiring the formation of teams to build equipment and obtain data, the number of authors per paper has to increase parallely.[151]

The distribution of the number of authors of papers in prompt nuclear analysis is

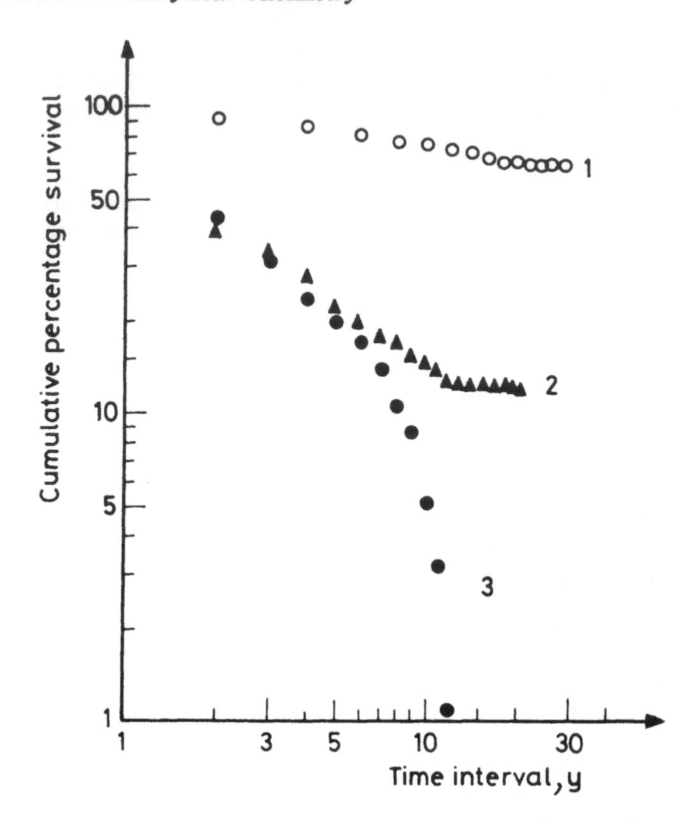

FIGURE 80. Survival of publishing authors in the field of prompt nuclear reactions (▲) of particle physics research (O), and of oscillating chemical reactions (•). (From Bujdosó, E., Lyon, W. S., and Noszlopy, I., *J. Radioanal. Chem.*, 74, 197, 1982. With permission.)

shown in Figure 81. The mean number of co-authors is 2.7, and 3 authors are more frequent than 1. For radioanalytical chemistry this value is 2.6.[20]

As shown in the former section, collaboration and productivity are related to each other. The more productive a scientist, the more co-authors he had. The scientists around whom groups were formed have become communication centers for the informal dissemination of information. The highest level of information exchange occurred between co-authors, but communication was established between centers when co-authors were associated with different centers. It can be assumed that channels of communication remain open once established. The pattern of the information flow can be revealed by co-authorship multigraphs.[150,152] On these, points and lines represent authors and channels of information, respectively.

Figure 82 shows the co-authorship pattern between the 33 most productive authors in the field of prompt nuclear analysis in Table 60 during the 27 years investigated. There are four co-authorship groups identifiable on the graph. Three of them are small involving only eight highly productive authors as compared to the working group centered around the Mayer-Amsel-Picraux-Ziegler group. This is the largest co-authorship group in prompt nuclear analysis and identifies the most "visible" authors.

The above-mentioned could be taken as an empirical proof of the statement by de Beaver et al.[153] that collaboration is associated with higher productivity and higher quality of research and also tends to increase visibility.

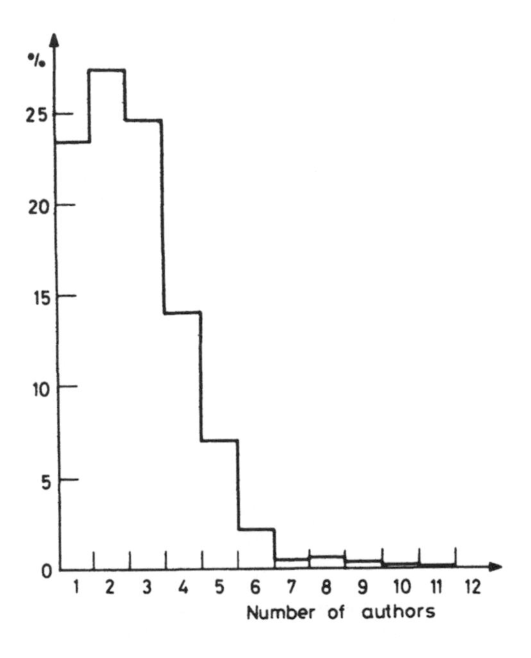

FIGURE 81. Distribution of multiple authorship in the field of prompt nuclear analysis. (From Bujdosó, E., Lyon, W. S., and Noszlopy, I., *J. Radioanal. Chem.*, 74, 197, 1982. With permission.)

VII. CITATION ANALYSIS OF PAPERS ON PROMPT NUCLEAR ANALYSIS

The number of citations in the research literature can be viewed as a form of recognition awarded to a scientist or to a particular paper. Citation counts have also come to be used as indicators of quality or worth of contributions to science. It was shown that there is a strong correlation between the number of citations and recognition by any other measure. The more recognition a scientist receives, the more productive he is subsequently likely to become.[142] Accordingly the most-cited authors are to be found primarily among those of high productivity. We have, therefore, counted the citations of the authors listed in Table 60 in the 1965 to 1974 volumes of the *SCI*. Because the *SCI* shows citations only to first authors, this may lead to deviations from the actual values. The data are given in Table 61 according to the number of citations received in the intervals 1965 to 1969, 1970 to 1974, and 1965 to 1974.

The citations for the decade 1965 to 1974 show a significant correlation with the number of published papers ($r = 0.728$). For the interval 1965 to 1969 the correlation coefficient is $r = 0.416$. The interval 1970 to 1974 reveals the greatest correlation coefficient ($r = 0.918$). The explanation is, perhaps, that authors whose work in the years 1965 to 1969 made a significant impact upon the development of prompt nuclear analysis received 30% more citations in the next 5 years, thus the correlation between citations and productivity became stronger.

Following the idea of Mulkay[142] the most-cited papers were sought among the papers of the most productive authors. The papers most frequently cited in the years between 1965 and 1974 are given in Table 62. The ranking follows the citation rate, i.e., the number of citations per year.

Examination of Table 62 reveals that the most-cited papers are those concerned with ion implantation in semiconductors, a very hot field from the middle 1960s on. The

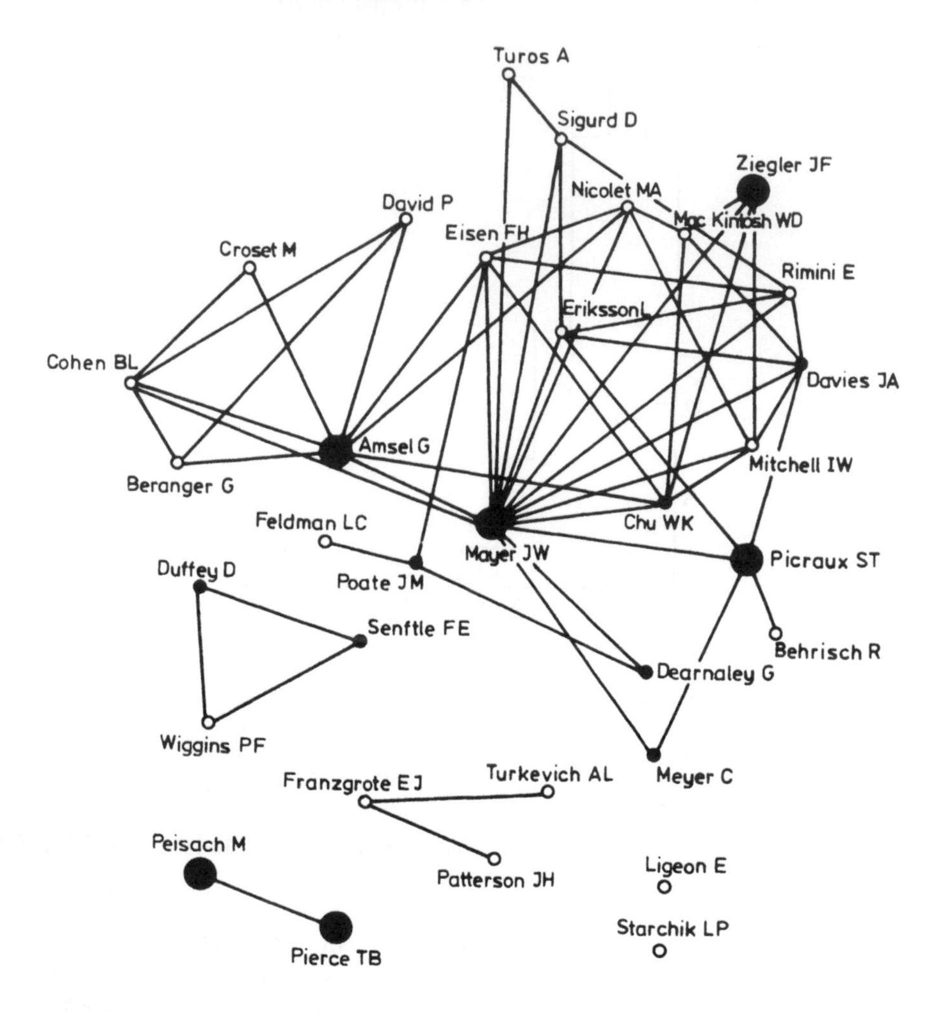

FIGURE 82. Co-authorship links among the 33 most productive authors in prompt nuclear analysis (see Table 60). Points show the approximate number of publications of the authors between 1949 and 1976: ● more than 30, • 20 to 30, and ○ 13 to 20. (From Bujdosó, E., Lyon, W. S., and Noszlopy, I., *J. Radioanal. Chem.*, 74, 197, 1982. With permission.)

papers by Mayer and co-workers account for 192, 116, 71, 59, and 61 citations. These papers, though pertaining primarily to the field of physics, do imply an analytical application. The number of citations they received were, however, far greater than those of some of the most popular purely analytical methods. Thus one sees that physics-based analytical techniques continue to be important to physicists. This is discussed in more detail in the following section.

It is useful to attempt to assess the actual scientific importance of the papers published on a specific area. Jean Jacques Rousseau argued in his *Contrat Social* that the size of the so-called "elite" of a population varies as the square root of the population. Rousseau's rule was tested for people in general by Galton and for scientific papers by De Solla Price.[1] The number of genuinely significant papers in a research area is roughly equivalent to the square root of the total number of papers. This idea was extended by Rescher[36] to include various quality levels of scientific papers (as already mentioned in Chapter 3). Using these quality levels[36] we tried to find out the significance level of prompt nuclear analysis papers.

The total number of publications in prompt nuclear analysis is 1250 which can be taken as "at least routine". Then, according to Rescher, there should be 210 "at least

Table 61
HIGHLY CITED FIRST AUTHORS IN THE SUBJECT FIELD OF PROMPT NUCLEAR ANALYSIS

Rank	Name	Number of citations 1965—1969	1970—1974	1965—1974	Total number of the authors's publications (1949—1976)
1	Davies, J.A.	489	392	881	26
2	Mayer, J.W.	216	572	788	87
3	Dearnaley, G.	396	252	648	24
4	Turkevich, A.	231	183	414	19
5	Amsel, G.	172	224	396	46
6	Pierce, T.B.	206	152	358	34
7	Vook, F.L.	179	117	296	12
8	Eriksson, L.	129	157	286	13
9	Meyer, O.	78	179	257	26
10	Eisen, F.H.	120	108	228	19
11	Ziegler, J.F.	25	192	217	32
12	Hart, R.R.	43	131	174	10
13	Behrisch, R.	48	106	154	15
14	Chu, W.K.	1	130	131	22
15	Feldman, L.C.	9	118	127	14
16	Meek, R.L.	1	108	109	11
	Total	2343	3121	5464	410

From Bujdosó, E., Lyon, W. S., and Noszlopy, I., *J. Radioanal. Chem.*, 74, 197, 1982. With permission.

significant", 35 "at least important", and 6 "very important" papers. These classes should be growing with doubling times of 4.3, 6.4, and 12.8 years, respectively.

Table 62 lists the 21 most-cited papers of the subject field of prompt nuclear analysis. So it is very probable that the list includes most of the six "very important" papers.

VIII. DIFFUSION OF THE RESULTS OF RESEARCH IN PROMPT NUCLEAR ANALYSIS INTO OTHER SCIENCE FIELDS

The number of papers on prompt nuclear analysis published in a set of journals characterizes the extent of participation of the topic in the field or subfield of that set.

The cumulated number of prompt nuclear analysis papers published in journals of various science fields, their percentage distribution as a function of time, the doubling times of the diffusion into the given field, and the dominant subfield are shown in Table 63. The growth rate is the largest in the field of physics, followed by chemistry. Engineering and technology are in third place. Biomedical sciences represent the most restricted area of applications of prompt nuclear analysis.

Citations can also be used for the study of diffusion of prompt nuclear analysis. The August 18, 1980 issue of *Current Contents*, "Physical, Chemical and Earth Sciences" section, reported that the paper by Amsel et al.[154] published in the journal *Nuclear Instruments and Methods* in 1971 had become a "citation classic".* The paper had been cited a total of 165 times up to January 1980. From a 10-year distance in time

* "Citation classic" is a highly cited publication as identified by the *SCI* or by the *Social Sciences Citation Index*. Citation rates differ for each discipline. However, one can appreciate the relative impact of each classic by considering that the average article published in a journal covered by the *SCI* in 1973 received 5.7 citations between 1973 and 1976.

Table 62
HIGHLY CITED PUBLICATIONS (BETWEEN 1965 AND 1974) OF THE MOST PRODUCTIVE AUTHORS IN THE SUBJECT FIELD OF PROMPT NUCLEAR ANALYSIS

Rank	Item	Total number of citations (1965—1974)	Citations per year
1	Mayer, J. W., Eriksson, L., and Davis, J. A., The Channelling effect technique, in *Ion Implantation in Semiconductors, Si and Ge,* Academic Press, New York, 1970.	192	38.4
2	Davies, J. A., Denhartog, J., Eriksson, L., and Mayer, J. W., Ion implantation of silicon. I. Atom location and lattice disorders by means of 1.0-MeV helium ion scattering, *Can. J. Phys.*, 45, 4054, 1967.	116	14.5
3	Amsel, G., Nadai, J. P., D'Artemare, E., David, D., Girard, E., and Moulin, J., Microanalysis by direct observation of nuclear reactions using a 2-MeV Van de Graaf, *Nucl. Instrum. Methods,* 92(4), 481, 1971.	50	12.5
4	Nicolet, M. A., Mayer, J. W., and Mitchell, I. V., Microanalysis of materials by backscattering spectrometry, *Science,* 177(4052), 841, 1972.	37	12.3
5	Bogh, E., Defect studies in crystals by means of channelling, *Can. J. Phys.*, 46, 653, 1968.	85	12.1
6	Mayer, J. W., Eriksson, L., Davies, J. A., and Picraux, S. T., Ion implantation of silicon and germanium at room-temperature. Analysis by means of 10-MeV helium ion scattering, *Can. J. Phys.*, 46, 663, 1968.	71	10.1
7	Eriksson, L., Davies, J. A., Johansson, N. G. E., and Mayer, J. W., Implantation and annealing behavior of group III and V dopants in silicon as studied by channeling technique, *J. Appl. Phys.*, 40, 842, 1969.	59	9.8
8	Davies, J. A., Denhartog, J., and Whitton, J. L., Channelling of MeV projectiles in tungsten and silicon, *Phys. Rev.*, 165, 345, 1968.	61	8.7
9	Chu, W. K., Mayer, J. W., Nicolet, M. A., Buck, T. M., Amsel, G., and Eisen, F., Principles and applications of ion beam techniques for the analysis of solids and thin films, *Thin Solid Films,* 17, 1, 1973.	17	8.5
10	Chu, W. K. and Powers, D., Alpha-particle stopping cross-section in solids from 400 keV to 2 MeV, *Phys. Rev.*, 187(2), 478, 1969.	48	8
11	Ziegler, J. F. and Baglin, J. E. E., Determination of lattice disorder profiles in crystals by nuclear backscattering, *J. Appl. Phys.*, 43(7), 2973, 1972.	23	7.7
12	Amsel, G. and Samuel, D., Microanalysis of stage isotopes of oxygen by means of nuclear reactions, *Anal. Chem.*, 39, 1689, 1967.	57	7.1
13	Picraux, S. T., Davies, J. A., Eriksson, L., Johansson, G., and Mayer, J. W., Channelling studies in diamond-type lattices, *Phys. Rev.*, 180, 873, 1969.	43	7.1
14	Turkevich, A. L., Franzgrote, E. J., and Patterson, J. H., Chemical analysis of moon at Surveyor V landing site, *Science,* 158, 635, 1967.	57	7.1
15	Eisen, F. H., Clark, G. J., Bottiger, J., and Poate, J. M., Stopping power of energetic He ions transmitted through thin Si crystals in channelling and random directions, *Radiat. Eff.*, 13, 13, 1972.	21	7
16	Turkevich, A. L., Franzgrote, E. J., and Patterson, J. H., Chemical composition of lunar surface in Mare Tranquillitatis, *Science,* 165, 277, 1969.	36	6

Table 62 (continued)
HIGHLY CITED PUBLICATIONS (BETWEEN 1965 AND 1974) OF THE MOST PRODUCTIVE AUTHORS IN THE SUBJECT FIELD OF PROMPT NUCLEAR ANALYSIS

Rank	Item	Total number of citations (1965—1974)	Citations per year
17	Patterson, J. H., Turkevich, A. L., and Franzgrote, E. J., Chemical analysis of surfaces using alpha particles, *J. Geophys. Res.*, 70, 1311, 1965.	56	5.6
18	Feldman, L. C. and Rodgers, J. W., Depth profiles of lattice disorder resulting from ion-bombardment of silicon single-crystals, *J. Appl. Phys.*, 41(9), 3776, 1970.	27	5.4
19	Turkevich, A. L., Franzgrote, E. J., and Patterson, J. H., Chemical analysis of moon and Surveyor 7 landing site. Preliminary results, *Science*, 162, 117, 1968.	38	5.4
20	Amsel, G. and Samuel, D., Mechanism of anodic oxidation, *J. Phys. Chem. Solids*, 23, 1707, 1962.	50	5
21	Picraux, S. T. and Vook, F. L., Multilayer thin-film analysis by ion-backscattering, *Appl. Phys. Lett.*, 18, 191, 1971.	20	5
22	Picraux, S. T., Ion-channelling studies of epitaxial layers, *Appl. Phys. Lett.*, 20, 91, 1972.	15	5
23	Ziegler, J. F. and Brodsky, M. H., Specific energy loss of the 4 MeV ions in Si (amorphous, polycrystalling and single crystals), *J. Appl. Phys.*, 44, 188, 1973.	10	5

From Bujdosó, E., Lyon, W. S., and Noszlopy, I., *J. Radioanal. Chem.*, 74, 197, 1982. With permission.

Amsel wrote: "The ideas for this paper came in the sixties when, as a young nuclear physicist, I had the luck to work on semiconductor detectors soon after their discovery. . . . It suddenly occurred to me after the discovery of a very narrow resonance in the $^{18}O(p, \alpha)^{15}N$ reaction that the position of this resonance may be an indicator where the ^{18}O nuclei are. . . . The simplicity of the experiments struck me and I soon realized the great analytical potentialities of such an technique."[155]

We used the Amsel et al. paper to study citations in prompt nuclear analysis literature. From the *SCI*, papers citing the Amsel at al. paper were selected. The yearly distribution of citations is shown in Figure 83. Data were smoothed by a moving average of 3 years. As seen, after an initial increase the curve reached saturation at 18 citations per year, and this level has been constant over the last 7 years.

The scatter of citations in the various journals follows a power law similar to that of the previously discussed Bradford distribution.[65] The 165 citations of the Amsel et al. paper are coming from papers published in 64 journals. The ranking is led by *Nuclear Instruments and Methods* (17% of all the citations); then the *Journal of Radioanalytical Chemistry* (10%), *Journal of Applied Physics* (7%), *Thin Solid Films* (6%), and the *Journal of the Electrochemical Society* (6%) follow in that order. Of the total citations, 60% are carried by 10 journals, a finding that is in good agreement with the distribution of the papers discussed in the section of the scatter of publications.

The citations were classified into science fields (Table 64) on the basis of the journal in which they appeared.[156] The distribution of citations is practically identical to that of publications (see the data calculated up to 1975 in Table 63). This shows the close relationship between the production and use of scientific information.

Figure 84 illustrates the dynamics of the diffusion of information from the Amsel et al. paper. It shows a linearly increasing diffusion with a relative velocity of 2.4, 1.4, and 1 into the fields of physics, chemistry, engineering and technology, respectively.

Table 63
CUMULATIVE NUMBER AND DISTRIBUTION OF JOURNAL ARTICLES ON PROMPT NUCLEAR ANALYSIS IN VARIOUS SCIENCE FIELDS

	Journal papers									
	—1960		—1965		—1970		—1975			
Field	Number	%	Number	%	Number	%	Number	%	Doubling time (years)	Dominant subfield (% of the field in 1975)
Physics	9	31.0	27	39.1	117	48.8	337	51.1	2.7	Applied physics (62)
Chemistry	6	20.7	15	21.7	56	23.3	225	30.5	2.8	Analytical chemistry (50), physical chemistry (40)
Engineering and technology	9	31.0	15	21.8	41	17.1	83	11.2	3.9	Nuclear technology (67), metals and metallurgy (13)
Earth and space sciences	1	3.5	4	5.8	10	4.2	21	2.8	3.4	Earth and planetary science (62), geology (24)
Biomedical research	3	10.3	6	8.7	11	4.5	19	2.6	5.6	General biomedical research (89)
Clinical medicine	1	3.5	2	2.9	5	2.1	13	1.8	4.0	Radiology and nuclear medicine (77)
Total	29	100	69	100	240	100	738	100	3.1	

From Bujdosó, E., Lyon, W. S., Noszlopy, I., *J. Radioanal. Chem.*, 74, 197, 1982. With permission.

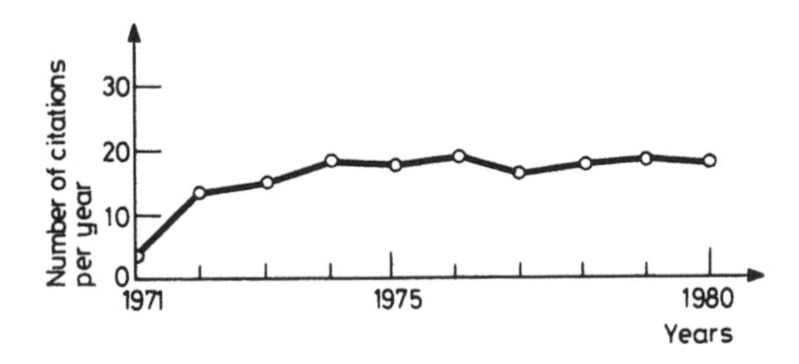

FIGURE 83. Number of papers citing the paper by Amsel et al.[154] (From Bujdosó, E., Lyon, W. S., and Noszlopy, I., *J. Radioanal. Chem.*, 74, 197, 1982. With permission.)

Table 64
FIELD DISTRIBUTIONS OF CITATIONS TO THE PAPER BY AMSEL ET AL.[154]

Field	Percentage of citations (1971—1981 January)	Dominant subfield (% of the field)
Physics	55.7	Applied physics (82)
Chemistry	29.6	Analytical chemistry (45), physical chemistry (38)
Engineering and technology	8.4	Metals and metallurgy (50), nuclear technology (33)
Other	6.3	

From Bujdosó, E., Lyon, W. S., and Noszlopy, I., *J. Radioanal. Chem.*, 74, 197, 1982. With permission.

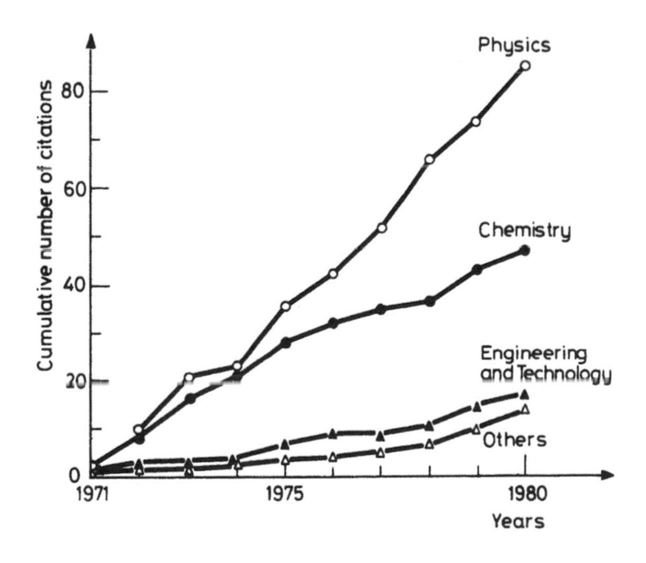

FIGURE 84. Diffusion of prompt nuclear analysis. Cumulative number of citations from various science fields to the paper by Amsel et al.[154] (From Bugdosó, E., Lyon, W. S., and Noszlopy, I., *J. Radioanal. Chem.*, 74, 197, 1982. With permission.)

IX. A CUMULATIVE ADVANTAGE APPROACH TO THE LITERATURE OF PROMPT NUCLEAR ANALYSIS

In the previous sections of this chapter both manpower growth (as reflected by the number of authors) and distribution of publication productivity of authors in prompt nuclear analysis have been alluded to already. However, as usual in such kinds of analyses, both were treated as unrelated features. Surprisingly enough, no reference has been made so far in the vast literature of scientometrics to an apparently obvious relationship between literature growth and productivity distribution. Namely, in a rapidly expanding field the distribution of authors is bound to be shifted towards the low productivity region, while slowly growing fields are unavoidably dominated by more productive authors. A quantitative theoretical model accounting for this relationship has recently been proposed.[157,158]

The model actually combines two simple and familiar postulates: that of *self-reproduction* and *cumulative advantage*. The first one asserts that newcomers in a given field join the population of authors at a rate proportional to the actual number of authors (obviously, newcomers are supposed to belong at first to the class of single-paper authors). The second postulate establishes a linear relation between the number of papers already published by a given author and his chances to produce a subsequent paper in the given field.

In mathematical terms, the model can be formulated as a system of differential equations:

$$\begin{aligned} dn_1/dt &= \sigma n - \alpha n_1 \\ dn_{i+1}/dt &= [\alpha + \beta(i - 1)]n_i - (\alpha + \beta i)n_{i+1} \qquad (i = 1, 2, \ldots) \end{aligned} \tag{20}$$

where t denotes time, n_i is the number of authors having exactly i publications, n is the total number of authors, and σ, α, and β are positive constants. Solution of the differential equations leads to the following conclusions:

1. Under constant circumstances, the system evolves toward a "steady-growth" regime in which the total population of authors as well as the number of authors having 1, 2, 3, . . . publications grows exponentially with a common time coefficient σ (doubling time = $\ln 2/\sigma$).
2. In the "steady-growth" regime, the fraction of authors having exactly i publications will follow the Waring distribution:

$$f_i = \frac{ak^{[i]}}{(a + k)^{[i+1]}} \qquad (i = 1, 2, \ldots) \tag{21}$$

with parameters

$$k = \alpha/\beta \tag{22}$$

($k^{[i]}$ denotes the i-th ascending factorial product of k, i.e., $k^{[i]} = (k+i)!/k!$). Since the Waring distribution asymptotically conforms with Lotka's law ($\lim_{i\to\infty} f_i \sim i^{-a}$), the second equation of Equation 22 suggests a strong relation between the classical models of manpower growth and productivity distribution.

Prompt nuclear analysis data offer an unique possibility for demonstrating the model. Namely, as indicated by Figure 76, growth rate of authors markedly decreased by the early 1970s: a change in doubling time from 3.5 to 5.3 years can be observed.

The question is now whether a parallel change in productivity distribution can be observed. The answer is given in Table 56, where estimated parameters $\hat{k}$ and $\hat{a}$ of productivity distribution are presented for a 10-year period (1967 to 1976).

Apparently, parameters $\hat{k}$ and $\hat{a}$ exhibit the pattern predicted by the model: $\hat{k}$ varies quite randomly, while $\hat{a}$ decreases abruptly by 1970. Calculating average values (by a proper averaging method), one may obtain

$$\tilde{k} = 1.15 \quad \text{(10-year average, 1967 to 1976)}$$

$$\tilde{a}_1 = 2.50 \quad \text{(5-year average, 1967 to 1971)}$$

$$\tilde{a}_2 = 1.78 \quad \text{(5-year average, 1972 to 1976)}$$

Moreover, we have

$$\sigma_1 = \ln 2/3.5 \text{ years} = 0.198/\text{year}$$

$$\sigma_2 = \ln 2/5.3 \text{ years} = 0.131 \text{ year}$$

Finally, using Equation 22, "intrinsic" parameters, α and β, of the model can be calculated for both half-periods:

$$\beta_1 = \sigma_1/\tilde{a}_1 = 0.078/\text{year}$$

$$\beta_2 = \sigma_2/\tilde{a}_2 = 0.074/\text{year}$$

$$\alpha_1 = \tilde{k}\beta_1 = 0.090/\text{year}$$

$$\alpha_2 = \tilde{k}\beta_2 \quad 0.086/\text{year}$$

It is impressive that the intrinsic parameters remain practically constant all the time; thus, changes in the productivity distribution can be attributed solely to the decrease in population growth rate. It is worth mentioning that parameter β, measuring the enhancement affected by previously published papers to later publication activity, may be considered the ***cumulative advantage coefficient***. Indirectly, it reflects the reward of publication in the given scientific community.

For comparison, β values 0.035/year (research on discrete distributions) and 0.051/year (research on scientific performance) were found in our earlier investigations.[157] Publication in prompt nuclear analysis seems thus to have higher reward than the mathematical or scientometric research fields mentioned.

Chapter 10

SCIENTOMETRIC INDICATORS FOR CROSS-NATIONAL COMPARISON OF PUBLICATION PRODUCTIVITY AND CITATION IMPACT IN ANALYTICAL CHEMISTRY (1978 TO 1980)

I. FROM "BROWNIE POINTS" TO INDICATORS

In some or another form, number magic has always been practiced, from the Pythagoreans to the present days. Recently, ordinal numbers have gained a particular popularity. Having their prototypes in such competitive enterprises as sports and business, different kinds of rankings are widespread in all areas of human activity.

The objects ranked may extend from individuals to whole countries or nations. In certain cases, position on lists of this latest kind may have decisive effects on the life of a nation (e.g., the rank on liquidity lists determines the chances for gaining loans); in others it has only a value of curiosity.

Science, by its very nature, is considered one of the most competitive human activities. Therefore, it is not surprising to see as early as in the first quarter of this century cross-national comparisons of scientific activities. As already mentioned Cole and Eales[3] as well as Hulme[159] used the number of published papers to compare the scientific productivity of countries. Due attention was given to these pioneering results only around the middle of this century when the need for the evaluation of effectiveness of scientific research became imperative.

It has to be, however, clearly seen that any ranking may exceed the level of a "ranking game" only if the quantities serving as a base of the ranking are sufficiently well defined and reliable. Numerical differences between ranked objects must exceed a reasonable "critical limit of comparability", i.e., differences must prove to be statistically significant at a properly chosen confidence level.

The need for evaluative assessment called for the systematic development of *statistical indicators.* These are selected or constructed from empirical statistical data in a way

1. To form a coherent system based explicitly or implicitly on some theoretical model of the phenomena under study
2. To permit aggregation and segregation as required
3. To render possible the uncovering of trends in the form of time series

Statistical indicators have been used first to compare the economy of nations. Social indicators, though their *ad hoc* use goes back 1 century, have become generally and systematically used only during the last few decades. The pioneering event in using statistical indicators for characterizing the scientific enterprise was the compilation of the *Science Indicators 1972* volume. It was published in 1973 by the National Science Board and has been followed since then by similar volumes biannually.[160] These volumes have been designed to provide both the U.S. government and the general public with data reflecting the state of health of scientific research in the U.S. and offering an objective picture by assessment of research in an international setting.

In each subsequent volume of *Science Indicators* a continually increasing weight has been given to indicators based on the scientific literature, i.e., to the so-called *scientometric indicators.* This is not because other kinds of data would not be of great interest and importance, but largely since "most of the S[cience] & T[echnology] data suffer from problems of reliability. . . . Manpower, funding and education data are often collected by inexperienced individuals. Serious definitional problems associated with the data further complicate their collection and may render them inadequate for making precise international comparisons. The only S&T indicators which appear to be reliable are counts of scientific papers, since these counts are straightforward and unambiguous and can be made without depending on the data collection efforts of untrained personnel."[161]

In the *Science Indicators* volumes as well as in some related compilations[56] cross-national comparisons are restricted to comparing the U.S. with a few major countries and/or geopolitical regions. A volume of scientometric indicators for comparative analysis of scientific research in 32 "small" countries was published recently.[162]

Having reliable data sources represents a crucial condition for building unobtrusive science indicators. Dozens of statistics on science are published on a national level year by year all over the world; international organizations, especially the UNESCO, Washington, D.C., have long taken great efforts to collect, standardize, and compile them.[163-166] However, in spite of all endeavors, notions and rules are interpreted ambiguously even now; therefore, these compilations can scarcely be used as sources of regular and reliable comparisons. For evaluative purposes it is never advantageous to depend on data supplied by the countries, individuals, or institutions under survey; the use of independent data sources warrants more objective and more reliable comparisons. Moreover, few if any of the statistical compilations contain detailed enough data for building indicators in the depth of science subfields, i.e., analytical chemistry.

For reasons previously mentioned and in concordance with the spirit of this volume, only indicators based on the scientific subject literature, i.e., scientometric indicators, will be considered in the following.

Since the volume of the literature of analytical chemistry is, as outlined in Chapter 2, by far too large to permit the collection of the original papers, we had to resort to one or another of the bibliographic data bases available. Several reasons turned our choice to the *Science Citation Index (SCI)* of the Institute for Scientific Information, Philadelphia. The use of this one instead of the more specialized chemistry data bases, i.e., *Chemical Abstracts* or *Analytical Abstracts,* was motivated by following:

1. Being a multidisciplinary data base, the *SCI* offers the possibility of a unified procedure for parallel analyses of different scientific subject fields as well as for estimating the relative weight of analytical chemistry in the global scientific publication performance and citation impact of a country.
2. The *SCI* provides a unique means for assessing the impact of publications as it processes the citations to all previously published items in more than 4500 science journals.
3. *SCI* data are available on magnetic tapes in a format particularly suited to statistical data processing.

The adequacy of the *SCI* data base in building scientometric indicators has been previously subjected to serious scrutiny.[167-169] The most frequently claimed objections against the *SCI* are its apparent biases in favor of biomedical sciences and chemistry as well as of journals published in English. Since the present analysis is restricted to a single subject field, analytical chemistry, the only question is whether the relative contribution of different countries to the literature of this field is properly reflected in the *SCI* data base. As it will be demonstrated later, a positive answer can be given to this question.

The period under study is 1978 to 1980. Analytical chemistry papers published in 1978 to 1979 (according to cover date) were counted as source publications, and citations to them in the 1980 *Citation Index* volume of the *SCI* were considered. It might be objected that this period is too short to assess the real impact of papers. There were, however, several reasons motivating our choice.

1. This choice of citation/source period length conforms with that used in the *Journal Citation Reports (JCR)* volumes of the *SCI*[53] in calculating impact factors and also proved to be suitable in developing a statistical test for assessing the reliability of comparisons based on citation indicators.[170] Although impact factor values were recalculated for the purposes of the present study, *JCR* experience is of utmost importance in interpreting citation data. Citations in the 1-year period considered represent about 20% of the lifetime citation rate of the source papers. There does not seem to be any evidence that within analytical chemistry or even within any other given subject field any systematic bias is present in favor of or against any country.
2. Using only a rather short citation period may render possible the assessment of the impact of relatively recent research.
3. Increase in time periods (especially that of the cited period) would cause an enormous increase in the required computer time.

Nevertheless, it seems advisable or even necessary to repeat the same type of evaluations from time to time, since the trends so obtained would surely give much more information than a single snapshot.

In the *SCI* no subject field classification of papers is given. One way to overcome this is to prepare a subject field classification of journals and to identify the literature of a given field with the set of papers published in the corresponding journals. This method is to be used only if the literature of the subject field in question is duly concentrated in a well-defined set of core journals. In analytical chemistry, as it was mentioned earlier in this volume, such a set of core journals really exists; this validates the use of the above-mentioned classification method.

After thorough consideration, 22 analytical chemistry core journals were selected (from a total of 3160 source journals of the *SCI* between 1978 and 1979) as representing the international analytical chemistry literature. In this analysis only papers labeled as articles, notes, letters, and reviews were taken into account. Papers within these categories will be termed in the following *relevant publications.* Other types of papers, i.e., editorials, obituaries, meeting abstracts, book reviews, etc., were considered irrelevant and were discarded from the analysis. Full and abbreviated titles (11- character abbreviation, as used in the *SCI*) of the 22 journals as well as the number of relevant publications they published in 1978 to 1979 and the corrected impact factors are presented in Table 65.* Corrected impact factor[170] is defined as the number of citations recorded in the 1980 *Citation Index* volume of the *SCI* to relevant publications published in the journal in 1978 to 1979 (according to cover date) divided by the number of these publications. In other words, it is the expected number of citations in 1980 to relevant publications published in the journal in 1978 to 1979.

The total number of papers considered is 12,027. This represents 68.8% of the papers included in the *Analytical Abstracts* (*AA*) data base in the same period. The difference is mainly due to nonjournal items (books, conference proceedings, reports, etc.) within the *AA*. The *AA* mentions also papers published in nonspecialized chemistry and multi- and interdisciplinary journals.

In spite of the difference in number, there is a close similarity in the national distribution of publications within the *SCI* and *AA* data bases. For the purposes of this cross-national study, the first 36 countries of the productivity ranking were looked at,

* Tables and figures for this chapter are presented at the end of the text.

and these were the same whichever of the 2 data bases were used. In Table 66 the percentage share of the 36 countries in world analytical chemistry literature is recorded both according to the *SCI* and the *AA* data base together with their respective ranks. The Spearman rank correlation coefficient is 0.9739, indicating a rather close similarity between the 2 rankings. It can thus be seen that using the *SCI* data base instead of the more extensive and analytically oriented *AA*, practically no country suffers from any bias concerning its relative share in world analytical chemistry literature.

For comparison, papers in chemistry and in total science were also considered. Similar to analytical chemistry, chemistry papers have been identified on the basis of their source journals; to this end a set of 247 journal titles representing chemistry (including the 22 analytical chemistry journals) has been selected (see Appendix). Total science incorporates all relevant papers in a total of 3160 source journals of the 1978 to 1979 *SCI* data base.

Papers were assigned to countries by the nationality of their first author according to the mail address on the byline of the paper (recorded in the *Corporate Index* of the *SCI*). In this way, each paper was counted only once and this allowed the summing or averaging of indicators of different countries as required. On the other hand, each country loses in this way a part of its papers produced in international cooperation. This may be a substantial loss especially for small- or medium-size countries, since it may include the results of cooperation with the leading institutions of major countries on work having a first author from one of those major countries. To assess the degree of this bias, a sample of 18 small- and medium-size (with respect to the size of their scientific endeavor) countries was considered for which both first-author- and all-author-based country assignment could be done. The number of first-authored papers as a percentage of all papers (i.e., those having at least one — not necessarily the first — author from the given country), as well as the number of citations to first-authored papers as a percentage of citations to all papers, were determined. The results presented in Table 67 support the above-mentioned conjecture: the percentage of citations to first-authored papers is always lower than the percentage of the papers themselves both in chemistry and in total science (most likely, similar results could be gained for analytical chemistry). Consequently, using first-author-based country assignment, a slight underestimation of citation impact indicators of small- and medium-size countries must be reckoned as compared to all-author-based assignment. Nevertheless, the deficiency scarcely amounts to more than 5%.

II. DEFINITION AND INTERPRETATION OF SCIENTOMETRIC INDICATORS

A. Publication Counts and Citation Rates

Fundamental building blocks of scientometric indicators are the number of publications and citations. As previously mentioned, relevant publications (articles, notes, letters, and reviews) published in the 1978 to 1979 issues (according to cover date) of *SCI* source journals and citations to them recorded in the 1980 *Citation Index* volume of the *SCI* served here as a basis for building the indicators. Chemistry and analytical chemistry papers were identified on the basis of their source journals. First-author-based country assignment was used.

B. Activity and Attractivity Indexes

Some global data of the *SCI* data base for source years 1978 to 1979 and citation year 1980 are presented in Table 68. The most direct way of comparing national data to these global ones is to calculate the relative share of the countries in the world total of publications or citations. This has already been done for analytical chemistry publications in Table 66, comparing national composition of the *SCI* and *AA* data bases. Due to the multidisciplinary character of the *SCI* data base, consistent estimations for national shares in publications in chemistry and in total science can also be given.

Let us now define the *activity index* of a given country (as proposed by Bujdosó and Braun in their analysis of physics subfields[171]) in analytical chemistry by the ratio

$$\text{activity index} = \frac{\text{share of the country in world analytical chemistry publications}}{\text{share of the country in world total science publications}}$$

or equivalently

$$\text{activity index} = \frac{\text{analytical chemistry publications of the country as a proportion of publications in total science}}{\text{world analytical chemistry publications as a proportion of publications in total science}}$$

The activity index measures the relative research effort as reflected in publication productivity that the country in question devotes to analytical chemistry. Its value is unity if, in a country, analytical chemistry has the same weight as it has in world total science. Its value exceeds unity for countries where the weight of analytical chemistry is higher than the world average; for countries devoting less-than-average effort to analytical chemistry, the activity index has a value between zero and one.

It has to be stressed that relative efforts in one science field or subfield can be increased only at the cost of decreasing the efforts in other(s). The activity index, indeed, is a measure of priorities allocated to analytical chemistry research (as reflected in publications) by the national science policy of a country.

It has been demonstrated[172] that certain groups of countries are typically oriented to biomedical research, others to physics, chemistry, and so on. It is, therefore, reasonable to ask whether a high (resp. low) activity index in analytical chemistry simply reflects the degree of priority given to chemistry in a certain country as a whole or whether it indicates some specific feature of analytical chemistry research there. An answer can be offered by using the *activity index within chemistry,* defined as

$$\text{activity index within chemistry} = \frac{\text{share of the country in world analytical chemistry publications}}{\text{share of the country in world chemistry publications}}$$

or equivalently

$$\text{activity index within chemistry} = \frac{\text{analytical chemistry publications of the country as a proportion of publications in chemistry}}{\text{world analytical chemistry publications as a proportion of publications in chemistry}}$$

This index measures the relative weight of efforts, reflected in publication productivity, devoted to analytical chemistry within chemistry. Its interpretation is completely analogous to that of the activity index in total science.

Global data in Table 68 show that the percentage of citations to analytical chemistry publications is slightly higher than the percentage of publications, both in total science and in chemistry. This might — at least partly — be attributed to the wide interaction of analytical chemistry particularly with the biomedical fields. Anyway, it is no surprise that the proportions follow a similar pattern in each country as a rule. There are, however, marked quantitative differences in what fraction of the citations to the publications of a country in total science or in chemistry is attracted by analytical chemistry publications. This invokes the building of some additional indexes. The analytical chemistry *attractivity index* in total science is defined as

$$\text{attractivity index} = \frac{\text{share of the country in citations to world analytical chemistry publications}}{\text{share of the country in citations to world total science publications}}$$

or equivalently

$$\text{attractivity index} = \frac{\text{citations to analytical chemistry publications of the country as a proportion of citations to total science publications}}{\text{citations to world analytical chemistry publications as a proportion of citations to total science publications}}$$

Similarly to the activity index, it is reasonable to build an *attractivity index within chemistry,* defined as

$$\text{attractivity index within chemistry} = \frac{\text{share of the country in citations to world analytical chemistry publications}}{\text{share of the country in citations to world chemistry publications}}$$

or equivalently

$$\text{attractivity index within chemistry} = \frac{\text{citations to analytical chemistry publications of the country as a proportion of citations to chemistry publications}}{\text{citations to world analytical chemistry publications as a proportion of citations to chemistry publications}}$$

Attractivity indexes measure the relative attention (manifested in the form of citations) attracted by analytical chemistry publications both in total science and within chemistry. Thereby, they assess — at least partially — the returns of the relative publication efforts measured by activity indexes.

C. Indicators of Citation Impact

Keeping in mind the reservations mentioned earlier in this book, citation data nevertheless provide a valid source of information about the scientific impact of publications. Papers receiving no citations at all in the first several years after their publication certainly do not contribute to the current paradigm system of their field. Hence, the *fraction of cited publications* helps to assess the potentially influential portion of the total publication output of a certain country.

An obvious indicator for measuring citation impact of the analytical chemistry publications of a country is their *mean citation rate,* i.e.,

$$\text{mean citation rate} = \frac{\text{number of citations in 1980 to analytical chemistry papers published in 1978 to 1979}}{\text{number of analytical chemistry papers published in 1978 to 1979}}$$

Two facts influencing the mean citation rate of any given set of papers must clearly be distinguished.

1. Papers published in different journals may receive a different number of citations due to the differences in language, publicity, diffusion of the journals, etc., which together with the different average professional quality of the papers is reflected in the impact factor of the journals.
2. Papers published in the same journal may differ in citation rate because of differences in their inherent merits: novelty, importance, clarity, etc. These qualities can be evaluated by comparing the actual citation rate to the expected one, i.e., to the average citation rate of the journal, i.e., the impact factor.

In order to reveal this distinction, the *expected citation rate* and *relative citation rate* of analytical chemistry publications are also reported here for each country. Expected citation rate is the sum of the expected number of citations for every single paper, i.e., the impact factor of the journals where they were published. Dividing the expected citation rate by the number of papers in question gives the *mean impact factor* of the publications. Mean impact factor characterizes the average quality of the journals analytical chemists of the given country use to publish their results. It is worth noting that the average impact factor of the 22 analytical chemistry journals concerned is 1.467. It is of undeniable interest to compare the mean impact factor of the single countries to this global average. However, one should remember that availability of publication channels for scientists of a country is determined by several factors besides the professional level of their work: language, requirements of science policy, and even politics.

Relative citation rate is defined as the actual number of citations received divided by the expected citation rate or, equivalently, as the mean citation rate divided by the mean impact factor. Being the ratio of observed to expected citation rate, it assesses the relative contribution of scientists from a given country to the citation impact of the journals in which they happen to publish. Relative citation rate has a value of unity if the publications in question are cited just on the average; values between 0 and 1 indicate a lower-than-average citation rate; and values greater than 1 indicate that the publications of the given country have a citation impact beyond the average of the respective journal.

To put plainly once more, while the mean impact factor assesses the quality level of the publication channels, the relative citation rate measures the relative contribution of the countries within the standards of the publication channels they use; the mean citation rate reflects the result of both effects.

D. Error Estimation of Scientometric Indicators

As already mentioned, a necessary prerequisite of indicators to be used for comparative purposes is the possibility of complementing them with reliable error estimates. Unfortunately, the literature of science indicators painfully lacks a systematic treatment of questions of error estimation and reliability tests, although valuable contributions have been made to some particular cases.[124,170,173] An attempt is made here to apply some unsophisticated but mathematically well-established methods of error estimation and reliability tests to the indicators used.

Several indicators mentioned are simple proportions (proportion of publications or citations and fraction of cited publications). In their case the relative frequency, f, of a certain class of objects is in a finite sample of size n and affected by random error. We need, then, the error bounds within which f can be considered the estimator of the probability, p, that a randomly drawn object will fall into the class in question. The answer is given by the error formula of the binomial distribution

$$\sigma_p^2 = \frac{f(1 - f)}{n} \tag{23}$$

As an example, we consider the case when 10 analytical chemistry publications are found in a total of 100 chemistry publications. The relative frequency of analytical chemistry papers with an error of $\sigma_p = (0.1 \times 0.9/100)^{1/2} = 0.03$. If, in another case, 100 analytical chemistry papers from 1000 chemistry publications are found, the estimated probability is, of course, likewise $p = 0.1$; however, the standard error is significantly lower: $\sigma_p = (0.1 \times 0.9/1000)^{1/2} = 0.00095$.

Mean citation rate represents another type of indicator, i.e., that of sample means. As it is well known, sample mean, $\bar{x}$, is an unbiased, efficient estimator of the theoretical expected value, m, with a standard error of $\sigma_m = \sigma/n^{1/2}$ where σ is the standard deviation of the distribution in question (in the present case the distribution of citations over publications) and n is the sample size. In order to estimate σ, both theoretical and practical considerations suggested the use of a method based on assuming a negative binomial distribution of citations.[170] Accordingly, σ can be calculated from the mean citation rate, $\bar{x}$, and the fraction of uncited publications, f_0, by the equation $\sigma^2 = \bar{x}Q_0$ where Q_0 is the solution of the equation

$$\frac{\log Q}{Q - 1} = -\frac{\log f_0}{\bar{x}} \tag{24}$$

Taking as a typical example the values mean citation rate, $\bar{x} = 1.0$; fraction of cited publications, $f_0 = 0.5$; and sample size, $n = 10$, 100, and 1000, the standard error of mean will be 0.447, 0.141, and 0.045, respectively.

E. Reliability Tests

The main aim of estimating errors of the indicators is to enable one to test the reliability of comparisons between indicator values. Two typical cases of comparisons may be considered:

a. Comparison of indicators of the same type for different countries
b. Comparison of observed indicator values with given fixed values

1. Case a

Let y be a certain indicator (proportion of publications or citations or mean citation rate), let y_1, y_2 denote its value for two countries, and let σ_1, σ_2 denote their estimated standard errors. If the size of the populations which the indicators are related to (number of publications in analytical chemistry, chemistry, or total science) is large enough (in practice, if population size is greater than 30), then the difference $(y_1 - y_2)$ is an approximately normally distributed random variable with an estimated error of $\sigma_1^2 + \sigma_2^2$. Consequently, the test statistic

$$t_{12} = \frac{y_1 - y_2}{\sigma_1^2 + \sigma_2^2} \tag{25}$$

is a random variable of the student's t-distribution, which in view of the previously assumed large sample sizes (which means a large degree of freedom for the t-distribution) can be approximated by a standard normal distribution.

Thus, if $t_{12} = \tau$ (τ is a suitably chosen positive real number), then y_1 and y_2 do not differ significantly at a significance level of $2\phi(\tau) - 1$, where $\phi(\cdot)$ is the cumulative distribution function of the standard normal distribution. With the particular choice $\tau = 2$, deviations can be tested at a significance level of 0.95, which is the most commonly used value in similar kind of studies. The milder condition $\tau = 1$ and the sharper condition $\tau = 3$ correspond to significance levels of about 0.7 and 0.995, respectively.

Example — Let us consider the fraction of cited analytical chemistry publications in Australia, Austria, and Belgium (the first three countries of the alphabetical country-by-country indicator lists of Section III). Indicator values and their standard errors are

Australia	y_1	= 0.6839
	σ_1	= 0.0335
Austria	y_2	= 0.5063
	σ_2	= 0.0562
Belgium	y_3	= 0.7090
	σ_3	= 0.0330

Test statistics of pair comparisons are

$$t_{12} = \frac{0.6839 - 0.5063}{0.0335^2 + 0.0562^2} = 2.7145$$

$$t_{13} = \frac{0.6839 - 0.7090}{0.0335^2 + 0.0330^2} = -0.5338$$

$$t_{23} = \frac{0.5063 - 0.7090}{0.0562^2 + 0.0330^2} = -3.1102$$

It can thus be seen that indicator values of Australia and Austria differ significantly at the 0.95 level but not at the 0.995 level, Australia and Belgium do not differ significantly even at the 0.7 level, and Austria and Belgium differ significantly at the 0.995 level.

2. Case b

Let y be a certain indicator value, let σ denote its standard error, and let *a* be a given fixed value to which the indicator value in question is to be compared (in the case of proportions of publications or citations this may be the respective world average; in the case of mean citation rate, *a* may represent the mean impact factor). The test statistic defined by

$$t = \frac{y - a}{\sigma} \tag{26}$$

has now exactly the same properties as t_{12} in the previous case; that is, if t is less than 1, 2, or 3, then y does not differ significantly from *a* at a 0.7, 0.95, or 0.995 significance level, respectively.

Activity and attractivity indexes as well as relative citation rates are complemented in the tables of Section III (Tables 69 to 104) by this type of test statistic. While the indicators mentioned express the *ratio* of certain observed values to average or expected values, a test statistic helps to decide whether their *difference* is significant at a given level. The sign of the test statistic is positive or negative according to whether the ratio is above or below unity.

Example — In Section III, the following data can be found for Australia: activity index in chemistry: 1.029, test statistic: 0.419; activity index in total science: 0.749, test statistics: −4.680; and relative citation rate: 0.843, test statistic: −2.381. These results can be interpreted as follows: the proportion of Australian analytical chemistry publications within chemistry is slightly above the world average but the difference is not significant even at the 0.7 significance level. The proportion of Australian analytical chemistry publications in total science is significantly below the world average at the 0.995 level. The observed citation rate of the Australian analytical chemistry publications is below the expected value — the deviation is significant at the 0.95 level but not at the 0.995 level.

III. COUNTRY-BY-COUNTRY INDICATOR VALUES

In this section, indicator values, their standard errors, and test statistics are presented in alphabetical order of the countries. Numerical values of the data are listed and illustrated for each country on five charts designated as Figures XXX-A, XXX-B, XXX-C, XXX-D, and XXX-E with XXX denoting the triliteral code of the country in question (Figures 85 to 264).

The pie charts in Figures XXX-A illustrate the proportion of chemistry papers in total science as well as that of analytical chemistry papers within chemistry both for the given country and for the whole world. The segment delineated represents the share of chemistry in the whole pie of total science papers while the shaded area represents the share of analytical chemistry papers. The chart reveals the contribution of chemistry papers to total science papers as well as that of analytical chemistry papers within chemistry relative to the world average.

The analogous pie charts in Figures XXX-B illustrate citations to analytical chemistry papers as a proportion of citations to total science papers as well as a proportion of citations to all chemistry papers. Again, citations to chemistry are shown by the segment of the whole pie of citations to total science, the shaded area corresponding to citations to analytical chemistry papers.

Figures XXX-C exhibit the number of analytical chemistry papers and their cited part (shaded area) on a logarithmic scale. Due to the elementary properties of logarithms, the difference between the logarithms of the number of cited and total papers (unshaded area) is just the logarithm of the fraction of cited ("citedness") papers, as illustrated by the black triangle at the lower part of the figure.

Figures XXX-D provide a parallel chart of expected and observed citation rates, again on a logarithmic scale. The difference of their logarithm represents now the logarithm of their ratio — the relative citation rate as indicated by an arrow. In case of relative citation rates greater than unity (observed citation rate exceeds its expected value), the arrow points upwards; in the opposite case, it points downwards.

The statistical significance of difference between observed and expected citation rates is illustrated in Figures XXX-E. The shaded area ranges twice the standard error around the circle representing the mean citation rate. The circle is full or void according to whether the full square designating the expected mean citation rate (mean impact factor) is outside or within the double standard error range. Thus, a full circle indicates that the difference between the observed and expected mean citation rates is significant at a 0.95 confidence level. A void circle stands if that is not the case. If, moreover, the expected value is within the single standard error range marked by the bars attached to the circles, there is no significant difference even at a 0.7 confidence level.

IV. CROSS-NATIONAL COMPARISONS

The rest of this chapter presents charts to facilitate cross-national comparisons. The charts are more or less self-explanatory; only some of the most conspicuous remarks will be explicitly mentioned.

On Figure 265, countries are ranked by the number of publications, the fraction of cited publications being also indicated. The second position of Japan in the productivity ranking is striking; the scientific "super-powers" (U.S., Japan, U.S.S.R., West Germany, England, and France) are followed by India and, rather surprisingly, by Poland and Czechoslovakia, both preceding Canada and the major west European countries. Differences in the fraction of cited publications appear much more striking here than on the logarithmic charts in the country-by-country section.

A ranked "citedness" chart is presented in Figure 266. While pointing to the value of "citedness", the length of the black triangles represents the *uncited* fraction of the papers, i.e., in this respect the shorter the better. Amazingly, both Wales and Scotland far outrank England; West Germany appears to be outstripped not only by Austria and East Germany, but also by Hungary and Czechoslovakia.

At this point we have to return to the question of reliability of comparisons. As stressed before, any ranking based on indicator values makes sense only if differences between the ranked objects are statistically significant at a suitably chosen confidence level. To put it simply, if the difference between two values is not sufficiently large as compared to their standard errors, they will be rated as ties. The greater the reliability required (the higher confidence level is given), the greater the difference needed to break a tie. As mentioned in Section II, t-statistics are suited for assessing the significance of differences, i.e., if t is greater than 1, 2, or 3, the difference is significant at a confidence level of about 0.7, 0.95, or 0.999, respectively. Figure 267 provides a tool for statistically reliable ranking of countries by "citedness". The symbol in each field specifies the degree of significance of the difference between the indicator values of the given pair of countries. After the required confidence level is chosen, proceeding from the highest at the top to the lowest at the bottom, groups of countries with indicator values not differing from each other significantly can be separated; e.g., at a 0.7 confidence level, the ranking by "citedness" is as follows:

1—7	Wales, Sweden, Denmark, Scotland, U.S., Republic of South Africa, Belgium
8—13	Canada, New Zealand, Australia, Ireland, Norway, Greece
14—15	England, Switzerland
16—19	Netherlands, Finland, Italy, France
20—27	Spain, Japan, Pakistan, Austria, Yugoslavia, East Germany, Israel, Brazil
28—30	Hungary, Czechoslovakia, West Germany
31—34	Poland, Egypt, India, Bulgaria
35	Romania
36	U.S.S.R.

It can thus be seen that eventually the apparent difference in "citedness" of analytical chemistry papers of Hungary, Czechoslovakia, and West Germany proved to be not significant at a 0.7 confidence level.

Figures 268 through 271 are of the same type as Figure 267 and help in the formation of statistically reliable rankings by the proportion of analytical chemistry papers in total science and within chemistry, as well as by the proportion of citations to analytical chemistry papers in citations to total science papers and in citations to chemistry papers, respectively. Most conspicuous is the top position of Pakistan in all of these charts; it has to be emphasized here that activity and attractivity indexes are not by any means characterizing the quality of analytical chemistry research of the given country; what they do assess is the relative effort devoted to it, particularly its relative effect as compared with other science fields. Thus, the exceedingly high Pakistani index values reflect a certain internal disproportion of Pakistani contribution to the international scientific communication activity rather than some special eminence of Pakistani analytical chemistry research.

Figures 272 and 273 offer a global view on activity and attractivity indexes in total science and within chemistry, respectively. Efforts and effects appear to be closely correlated.

A statistically reliable ranking of countries by the mean citation rate of analytical chemistry papers can be performed with the aid of Figure 274 (it works analogously to Figure 267 through 271). For example, at a 0.7 confidence level, the following ranking can be made:

1—3	New Zealand, Denmark, Sweden
4—6	U.S., Ireland, Wales
7—10	Switzerland, Canada, Netherlands, Scotland
11—14	England, Australia, Norway, Belgium
15—16	France, Brazil
17—18	Republic of South Africa, Japan
19—25	West Germany, Finland, Italy, Greece, Spain, Yugoslavia, Austria
26—27	Czechoslovakia, Hungary
28—30	Israel, East Germany, Pakistan
31—33	Poland, Egypt, India
34	Bulgaria
35—36	Romania, U.S.S.R.

Figure 275 displays observed and expected number of citations per paper, mean citation rate, and mean impact factor. Standard error of mean citation rate is also indicated on the chart (standard error of mean impact factor is usually negligibly small). Full or void circles show whether observed and expected values differ significantly at a 0.95 confidence level. Points above the diagonal represent countries with relative citation rates greater than unity.

A ranking of countries by relative citation rate is presented in Figure 276. The comparison of Figures 275 and 276 reveals that countries with the highest relative citation rates (Denmark, New Zealand, Sweden, Wales, and Ireland) also publish in journals with rather high impact factors. This feature indicates that the analytical chemistry research publication activity and impact of these countries are fairly remarkable.

Table 65
PUBLICATION AND CITATION DATA OF ANALYTICAL CHEMISTRY CORE JOURNALS

Journal title	Number of publications (1978 and 1979)	Corrected impact factor (1980)
Analusis	157	0.701
Analyst	327	1.446
Anal. Chim. Acta	756	1.714
Anal. Biochem.	1294	2.128
Anal. Chem.	1282	2.821
Anal. Lett.	230	1.250
Bunseki Kagaku	408	0.319
Chem. Anal.	261	0.261
Chromatographia	245	1.406
Crit. Rev. Anal. Chem.	11	2.727
Fres. Z. Anal. Chem.	581	0.189
J. Anal. Chem. USSR	814	0.021
J. Chromatogr. Sci.	218	2.321
J. Chromatogr.	2271	2.132
J. Electroanal. Chem. Interfacial Electrochem.	733	1.843
J. Radioanal. Chem.	508	0.669
J. Thermal Anal.	229	0.445
Microchem. J.	136	0.757
Radiochem. Radioanal. Lett.	469	0.473
Sep. Sci. Technol.	140	1.029
Talanta	410	1.034
Thermochim. Acta	547	0.757

Table 66
DISTRIBUTION OF ANALYTICAL CHEMISTRY PAPERS BY COUNTRIES

Country	Code	*AA* data base		*SCI* data base	
		%	Rank	%	Rank
Australia	AUS	1.76	13	1.60	14
Austria	AUT	0.74	24	0.66	24
Belgium	BEL	1.22	19	1.54	15
Brazil	BRA	0.32	32	0.37	31
Bulgaria	BGR	0.82	21	0.72	22—23
Canada	CAN	3.49	7	3.08	10
Czechoslovakia	CSK	2.85	9	3.30	9
Denmark	DNK	0.58	28	0.47	27
Egypt	EGY	0.84	20	0.91	19
England	GBR	6.12	5	5.61	5
West Germany	DEU	7.62	3	6.36	4
Finland	FIN	0.51	29	0.44	28
France	FRA	2.47	10	5.25	6
East Germany	DDR	1.69	14	0.93	18
Greece	GRC	0.25	34	0.27	33—34
Hungary	HUN	1.37	17	1.43	16
India	IND	4.69	6	4.42	7
Ireland	IRL	0.11	36	0.10	36
Israel	ISR	0.44	30	0.38	30
Italy	ITA	2.05	12	2.22	12
Japan	JPN	7.60	4	9.30	2
Netherlands	NLD	2.10	11	2.52	11
New Zealand	NZL	0.36	31	0.27	33—34
Norway	NOR	0.62	27	0.62	25
Pakistan	PAK	0.13	35	0.24	35
Poland	POL	3.04	8	4.18	8
Romania	ROM	0.77	23	0.35	32
Scotland	SCO	0.81	22	0.76	21
South African Republic	ZAF	0.73	25	0.49	26
Spain	ESP	1.28	18	0.78	20
Sweden	SWE	1.56	15	1.72	13
Switzerland	CHE	1.48	16	1.09	17
U.S.	USA	26.22	1	26.65	1
U.S.S.R.	SUN	11.09	2	8.18	3
Wales	WAL	0.29	33	0.41	29
Yugoslavia	YUG	0.65	26	0.72	22—23

Table 67
COMPARISON OF PUBLICATION AND CITATION DATA USING FIRST-AUTHOR- AND ALL-AUTHOR-BASED COUNTRY ASSIGNMENT

	First authored papers/all papers			
	Chemistry		Total science	
Country	Percent of publications	Percent of citations	Percent of publications	Percent of citations
Australia	94.1	93.0	94.4	89.9
Austria	93.8	91.0	92.1	83.7
Belgium	92.6	92.4	90.3	85.0
Canada	94.1	93.4	93.0	90.4
Czechoslovakia	96.2	94.1	94.5	90.6
Denmark	88.6	87.9	89.1	82.7
Egypt	95.5	90.6	91.7	81.7
Finland	96.1	93.9	93.3	88.8
East Germany	95.5	86.4	94.9	90.4
Hungary	92.7	86.1	92.8	88.8
India	97.8	93.6	97.9	94.7
Italy	95.3	92.6	92.8	85.7
Netherlands	94.8	92.5	91.4	87.3
New Zealand	90.8	88.8	94.4	89.7
Norway	88.8	88.6	90.6	85.5
Poland	94.3	89.2	94.5	85.7
Sweden	92.7	92.7	91.3	87.6
Switzerland	90.6	88.6	86.8	80.4

Table 68
GLOBAL PUBLICATION AND CITATION DATA FROM THE *SCI* DATA BASE

Number of publications/cover date: 1978—1979	
In analytical chemistry	12,027
In chemistry	99,398
In total science	696,314
Number of citations recorded in the 1980 *Citation Index* volume of the *SCI* to 1978—1979 papers	
To analytical chemistry	17,640
To chemistry	129,329
To total science	973,041
Percentage of publications	
Chemistry/total science	14.27
Analytical chemistry/chemistry	12.10
Analytical chemistry/total science	1.73
Percentage of citations	
Chemistry/total science	13.29
Analytical chemistry/chemistry	13.64
Analytical chemistry/total science	1.81

Table 69
SCIENTOMETRIC INDICATOR VALUES 1978—1980

Australia

Publication counts	
Analytical chemistry	193
Chemistry	1,550
Total science	14,914
Analytical chemistry publications as a proportion of publications	
In chemistry	0.1245 ± 0.0084
Activity index	1.029
Test statistics	0.419
In total science	0.0129 ± 0.0009
Activity index	0.749
Test statistics	−4.680
Citation rates	
Analytical chemistry	315
Chemistry	2,364
Total science	17,735
Citations to analytical chemistry publications as a proportion of citations	
To chemistry	0.1332 ± 0.0070
Attractivity index	0.977
Test statistics	−0.450
To total science	0.0178 ± 0.0010
Attractivity index	0.980
Test statistics	−0.370
Fraction of cited publications	0.6839 ± 0.0335
Expected citation rate	373.8
Mean impact factor	1.937
Mean citation rate	1.632 ± 0.128
Relative citation rate	0.843
Test statistics	−2.381

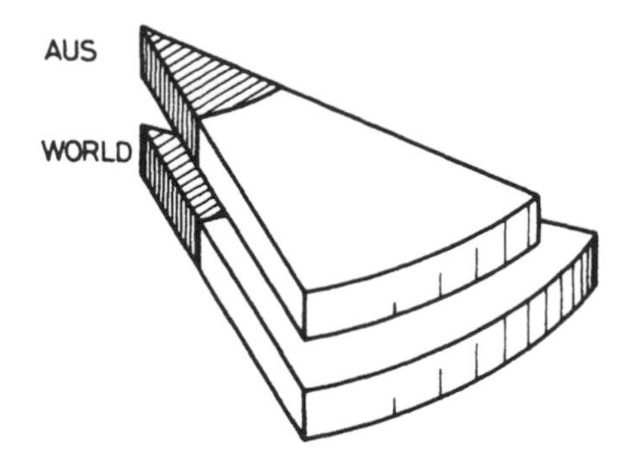

FIGURE 85. AUS-A. Proportion of chemistry and analytical chemistry papers in Australian science publications.

FIGURE 86. AUS-B. Citations to chemistry and analytical chemistry papers as a proportion of citations to Australian science publications.

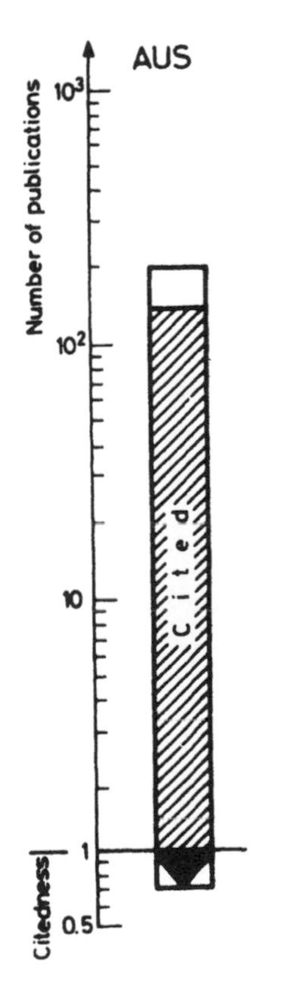

FIGURE 87. AUS-C. Number of Australian analytical chemistry papers and the cited fraction.

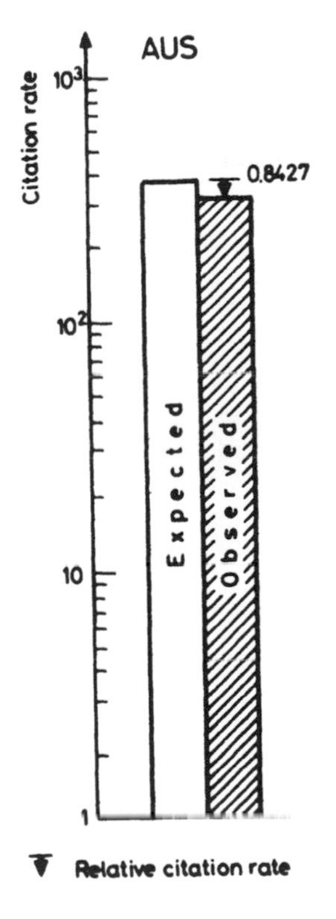

Relative citation rate

FIGURE 88. AUS-D. Expected and observed citation rates of Australian analytical chemistry papers.

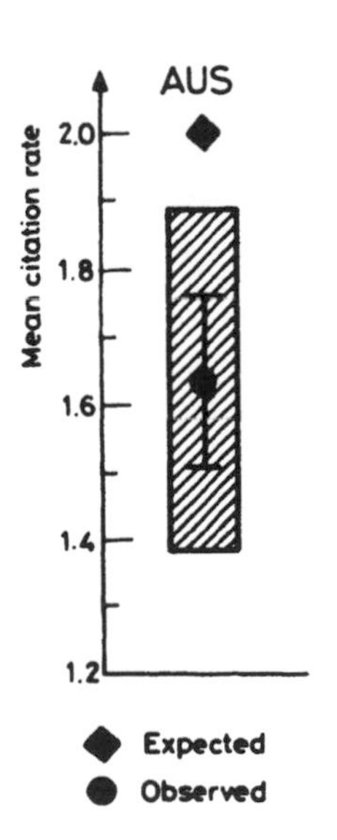

FIGURE 89. AUS-E. Expected and observed values of mean citation rate of Australian analytical chemistry papers.

Table 70
SCIENTOMETRIC INDICATOR VALUES 1978—1980

Austria

Publication counts	
Analytical chemistry	79
Chemistry	456
Total science	3952
Analytical chemistry publications as a proportion of publications	
In chemistry	0.1732 ± 0.0177
Activity index	1.432
Test statistics	2.948
In total science	0.0200 ± 0.0022
Activity index	1.157
Test statistics	1.221
Citation rates	
Analytical chemistry	79
Chemistry	471
Total science	3374
Citations to analytical chemistry publications as a proportion of citations	
To chemistry	0.1677 ± 0.0172
Attractivity index	1.230
Test statistics	1.820
To total science	0.0234 ± 0.0026
Attractivity index	1.292
Test statistics	2.030
Fraction of cited publications	0.5063 ± 0.0562
Expected citation rate	93.6
Mean impact factor	1.185
Mean citation rate	1.000 ± 0.157
Relative citation rate	0.844
Test statistics	−1.181

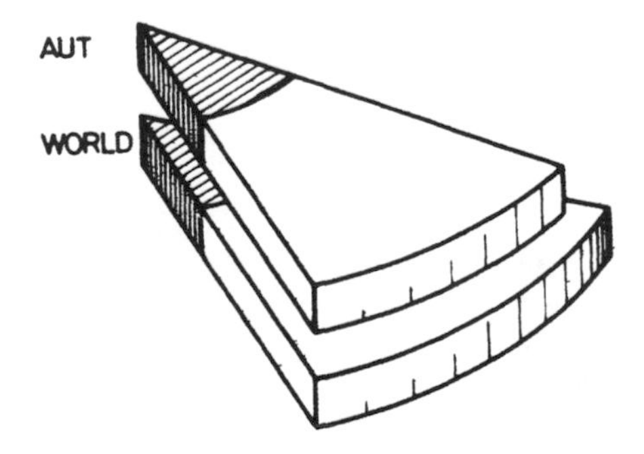

FIGURE 90. AUT-A. Proportion of chemistry and analytical chemistry papers in Austrian science publications.

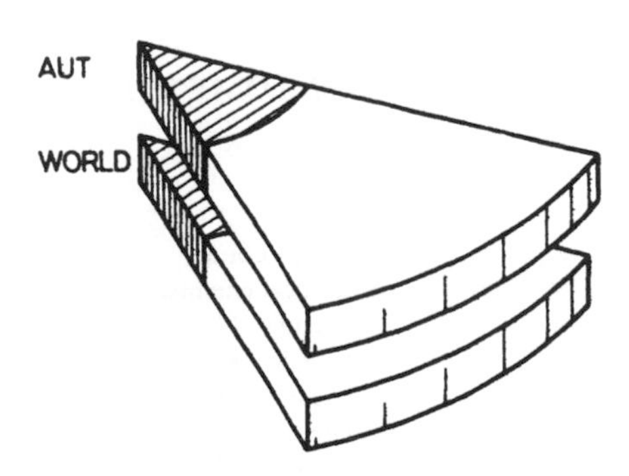

FIGURE 91. AUT-B. Citation to chemistry and analytical chemistry papers as a proportion of citations to Austrian science publications.

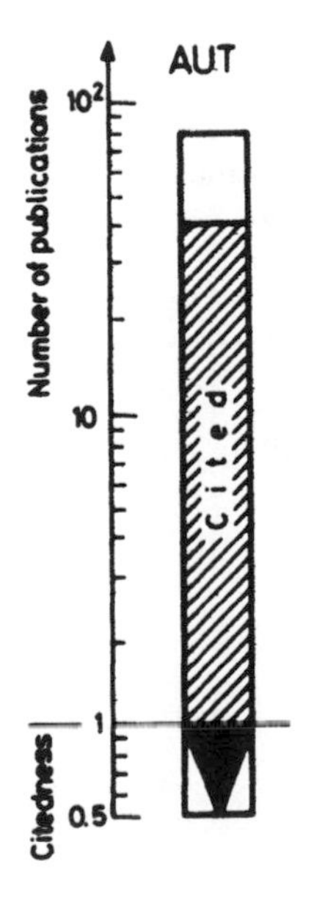

FIGURE 92. AUT-C. Number of Austrian analytical chemistry papers and the cited fraction.

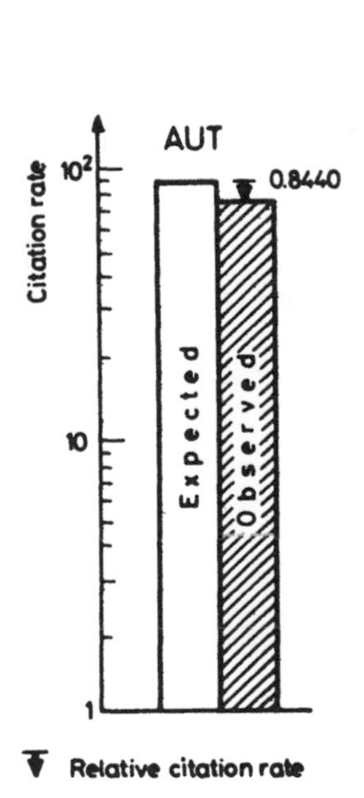

FIGURE 93. AUT-D. Expected and observed citation rates of Austrian analytical chemistry papers.

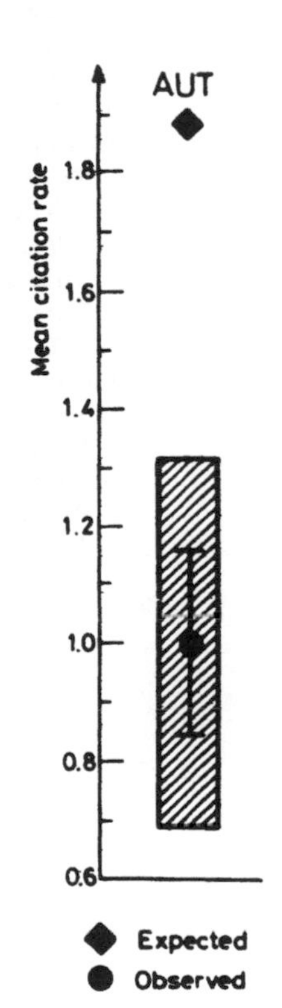

FIGURE 94. AUT-E. Expected and observed values of mean citation rate of Austrian analytical chemistry papers.

Table 71
SCIENTOMETRIC INDICATOR VALUES 1978—1980

Belgium

Publication counts	
Analytical chemistry	189
Chemistry	915
Total science	5929
Analytical chemistry publications as a proportion of publications	
In chemistry	0.2066 ± 0.0134
Activity index	1.707
Test statistics	6.393
In total science	0.0319 ± 0.0023
Activity index	1.846
Test statistics	6.402
Citation rates	
Analytical chemistry	297
Chemistry	1312
Total science	8193
Citations to analytical chemistry publications as a proportion of citations	
To chemistry	0.2264 ± 0.0116
Attractivity index	1.660
Test statistics	7.788
To total science	0.0363 ± 0.0021
Attractivity index	2.000
Test statistics	8.776
Fraction of cited publications	0.7090 ± 0.0330
Expected citation rate	291.1
Mean impact factor	1.540
Mean citation rate	1.571 ± 0.115
Relative citation rate	1.020
Test statistics	0.271

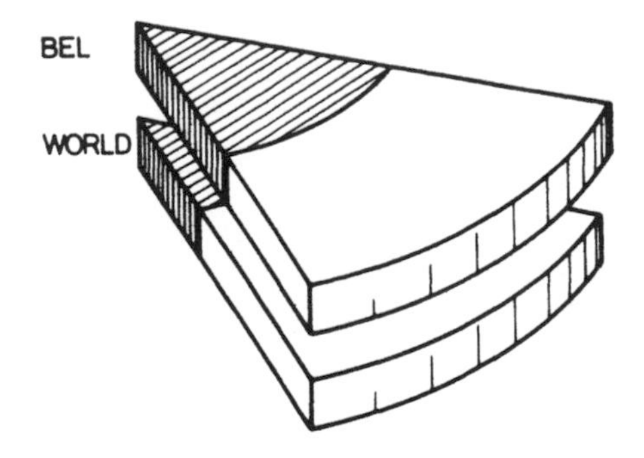

FIGURE 95. BEL-A. Proportion of chemistry and analytical chemistry papers in Belgian science publications.

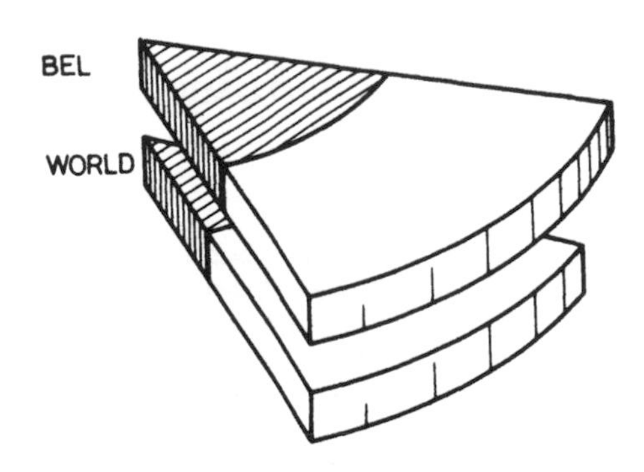

FIGURE 96. BEL-B. Citations to chemistry and analytical chemistry papers as a proportion of citations to Belgian science publications.

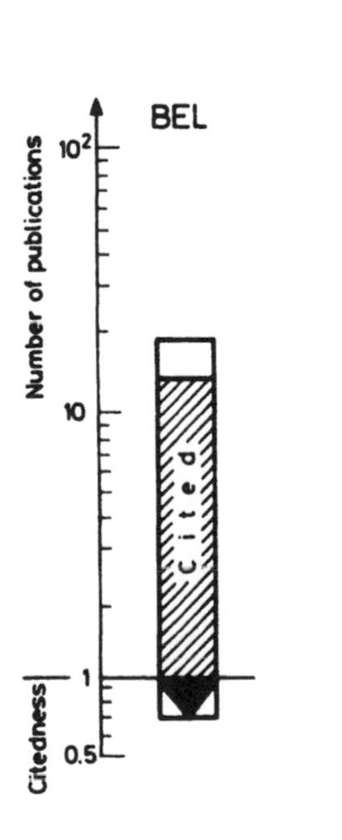

FIGURE 97. BEL-C. Number of Belgian analytical chemistry papers and the cited fraction.

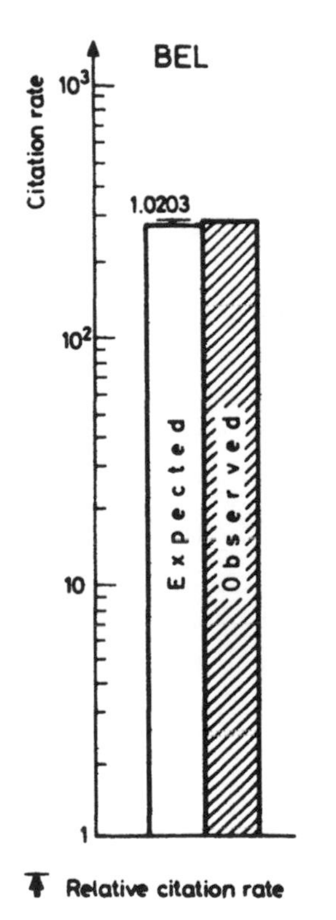

FIGURE 98. BEL-D. Expected and observed citation rates of Belgian analytical chemistry papers.

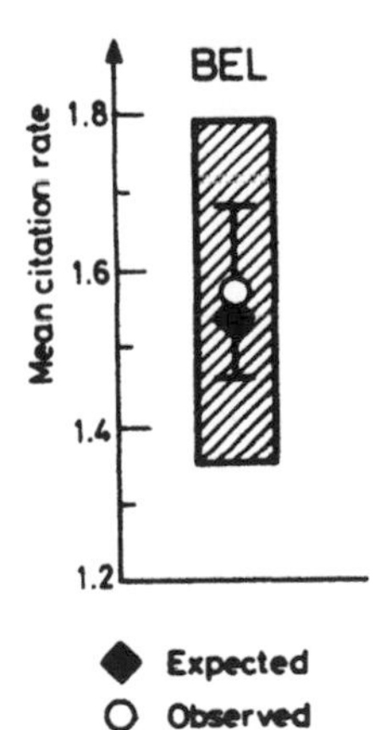

FIGURE 99. BEL-E. Expected and observed values of mean citation rate of Belgian analytical chemistry papers.

Table 72
SCIENTOMETRIC INDICATOR VALUES 1978—1980

Brazil

Publication counts	
Analytical chemistry	44
Chemistry	202
Total science	2296
Analytical chemistry publications as a proportion of publications	
In chemistry	0.2178 ± 0.0290
Activity index	1.800
Test statistics	3.334
In total science	0.0192 ± 0.0029
Activity index	1.110
Test statistics	0.661
Citation rates	
Analytical chemistry	64
Chemistry	218
Total science	1689
Citations to analytical chemistry publications as a proportion of citations	
To chemistry	0.2936 ± 0.0308
Attractivity index	2.152
Test statistics	5.096
To total science	0.0379 ± 0.0046
Attractivity	2.090
Test statistics	4.254
Fraction of cited publications	0.4773 ± 0.0753
Expected citation rate	54.6
Mean impact factor	1.241
Mean citation rate	1.455 ± 0.374
Relative citation rate	1.172
Test statistics	0.571

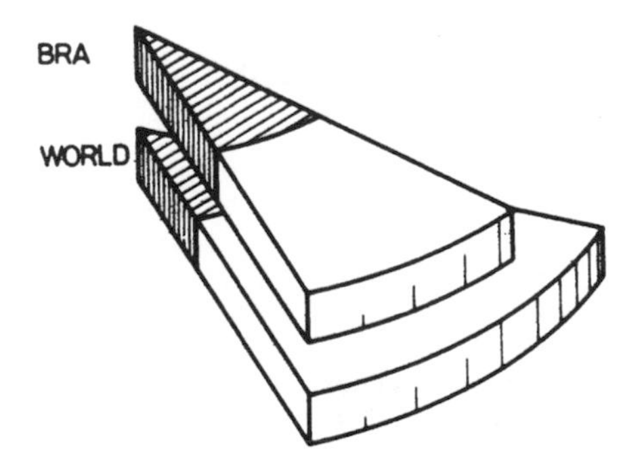

FIGURE 100. BRA-A. Proportion of chemistry and analytical chemistry papers in Brazilian science publications.

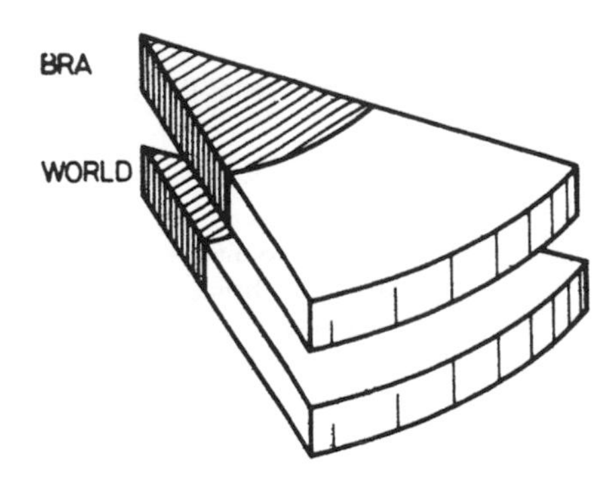

FIGURE 101. BRA-B. Citation to chemistry and analytical chemistry papers as a proportion of citations to Brazilian science publications.

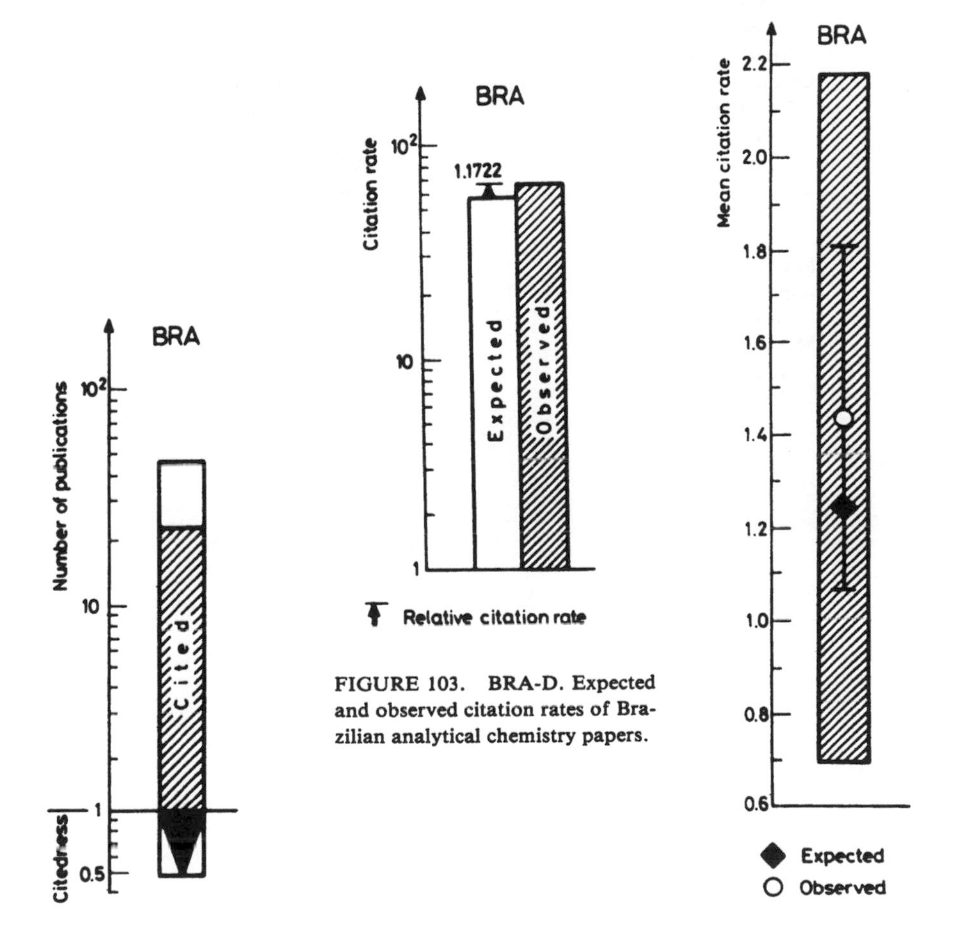

FIGURE 103. BRA-D. Expected and observed citation rates of Brazilian analytical chemistry papers.

FIGURE 102. BRA-C. Number of Brazilian analytical chemistry papers and the cited fraction.

FIGURE 104. BRA-E. Expected and observed values of mean citation rate of Brazilian analytical chemistry papers.

Table 73
SCIENTOMETRIC INDICATOR VALUES 1978—1980

Bulgaria

Publication counts	
Analytical chemistry	86
Chemistry	326
Total science	1862
Analytical chemistry publications as a proportion of publications	
In chemistry	0.2638 ± 0.0244
Activity index	2.180
Test statistics	5.851
In total science	0.0462 ± 0.0049
Activity index	2.674
Test statistics	5.944
Citation rates	
Analytical chemistry	43
Chemistry	157
Total science	543
Citations to analytical chemistry publications as a proportion of citations	
To chemistry	0.2739 ± 0.0356
Attractivity index	2.008
Test statistics	3.863
To total science	0.0792 ± 0.0116
Attractivity index	4.368
Test statistics	5.269
Fraction of cited publications	0.3372 ± 0.0510
Expected citation rate	91.8
Mean impact factor	1.067
Mean citation rate	0.500 ± 0.092
Relative citation rate	0.468
Test statistics	−6.157

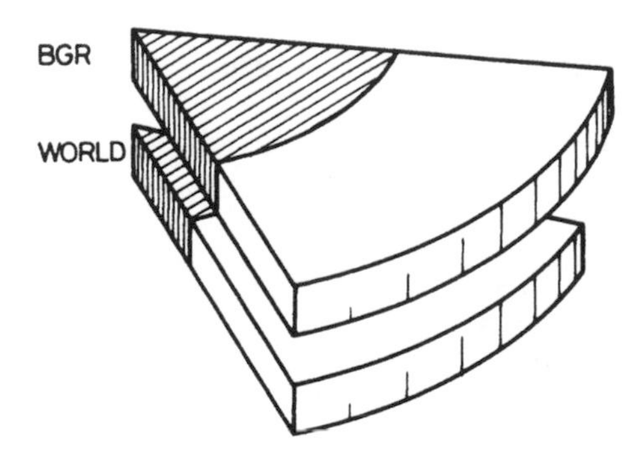

FIGURE 105. BGR-A. Proportion of chemistry and analytical chemistry papers in Bulgarian science publications.

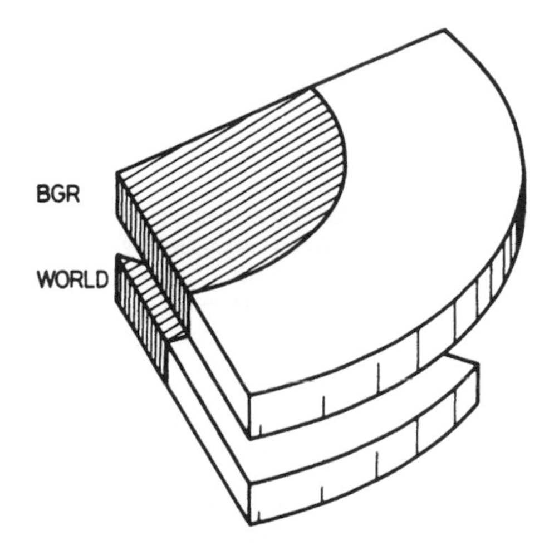

FIGURE 106. BGR-B. Citations to chemistry and analytical chemistry papers as a proportion of citations to Bulgarian science publications.

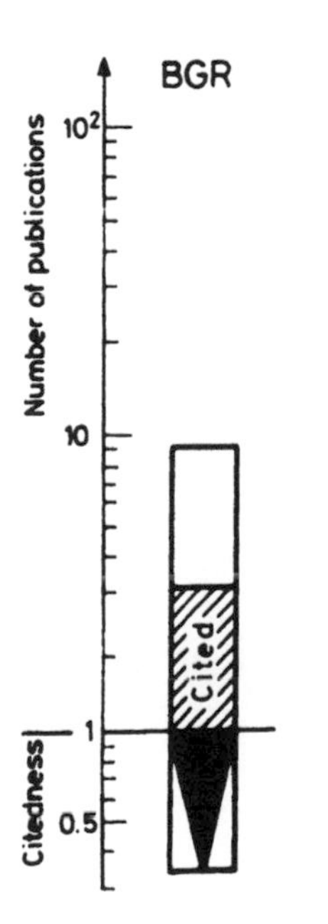

FIGURE 107. BGR-C. Number of Bulgarian analytical chemistry papers and the cited fraction.

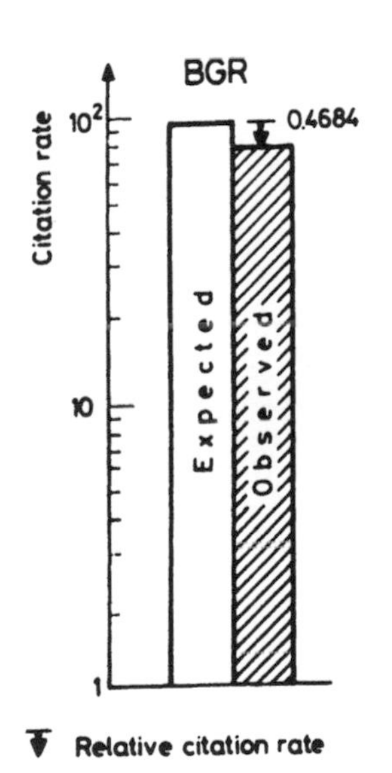

FIGURE 108. BGR-D. Expected and observed citation rates of Bulgarian analytical chemistry papers.

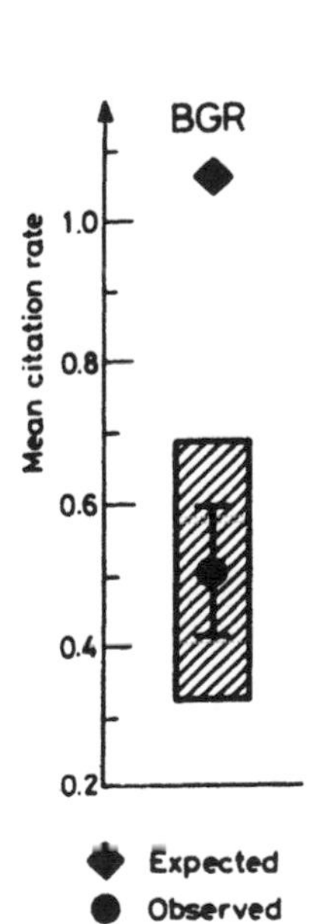

FIGURE 109. BGR-E. Expected and observed values of mean citation rate of Bulgarian analytical chemistry papers.

Table 74
SCIENTOMETRIC INDICATOR VALUES 1978—1980

Canada

Publication counts	
Analytical chemistry	371
Chemistry	3,274
Total science	29,265
Analytical chemistry publications as a proportion of publications	
In chemistry	0.1133 ± 0.0055
Activity index	0.937
Test statistics	−1.387
In total science	0.0127 ± 0.0007
Activity index	0.734
Test statistics	−7.026
Citation rates	
Analytical chemistry	686
Chemistry	5,982
Total science	40,774
Citations to analytical chemistry publications as a proportion of citations	
To chemistry	0.1147 ± 0.0041
Attractivity index	0.841
Test statistics	−5.272
To total science	0.0168 ± 0.0006
Attractivity index	0.928
Test statistics	−2.048
Fraction of cited publications	0.6927 ± 0.0240
Expected citation rate	719.7
Mean impact factor	1.940
Mean citation rate	1.849 ± 0.107
Relative citation rate	0.953
Test statistics	−0.845

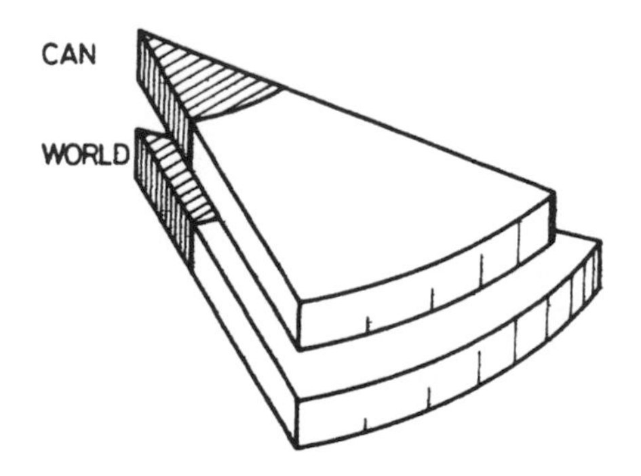

FIGURE 110. CAN-A. Proportion of chemistry and analytical chemistry papers in Canadian science publications.

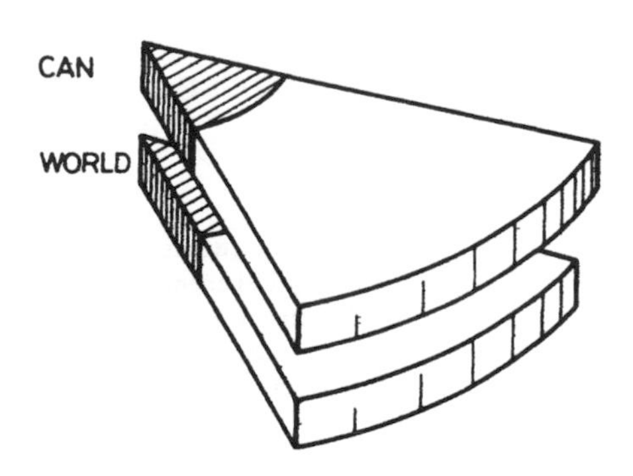

FIGURE 111. CAN-B. Citation to chemistry and analytical chemistry papers as a proportion of citations to Canadian science publications.

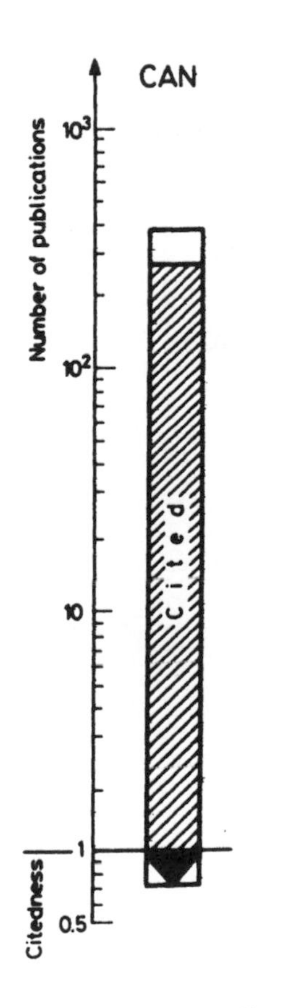

FIGURE 112. CAN-C. Number of Canadian analytical chemistry papers and the cited fraction.

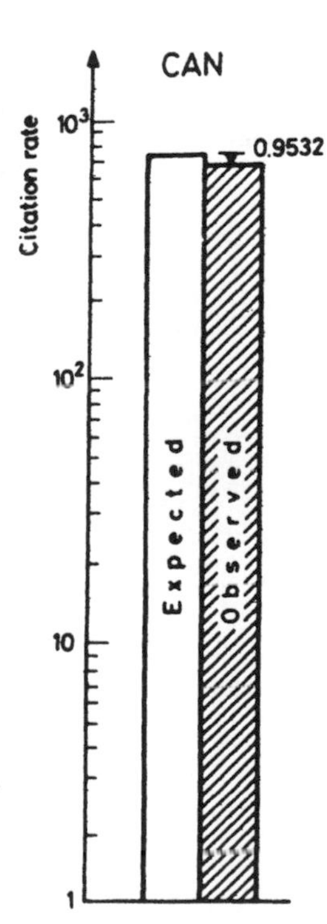

FIGURE 113. CAN-D. Expected and observed citation rates of Canadian analytical chemistry papers.

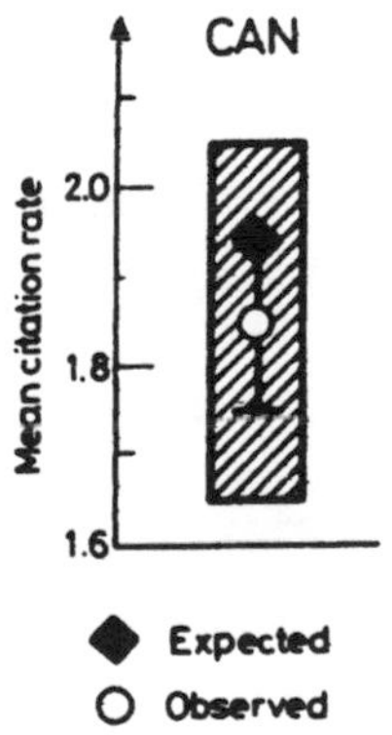

FIGURE 114. CAN-E. Expected and observed values of mean citation rate of Canadian analytical chemistry papers.

Table 75
SCIENTOMETRIC INDICATOR VALUES 1978—1980

Czechoslovakia

Publication counts	
Analytical chemistry	397
Chemistry	2074
Total science	6217
Analytical chemistry publications as a proportion of publications	
In chemistry	0.1914 ± 0.0086
Activity index	1.582
Test statistics	8.152
In total science	0.0639 ± 0.0031
Activity index	3.697
Test statistics	15.023
Citation rates	
Analytical chemistry	392
Chemistry	1435
Total science	3394
Citations to analytical chemistry publications as a proportion of citations	
To chemistry	0.2732 ± 0.0118
Attractivity index	2.003
Test statistics	11.628
To total science	0.1155 ± 0.0055
Attractivity index	6.371
Test statistics	17.748
Fraction of cited publications	0.4534 ± 0.0250
Expected citation rate	493.9
Mean impact factor	1.244
Mean citation rate	0.987 ± 0.079
Relative citation rate	0.794
Test statistics	−3.259

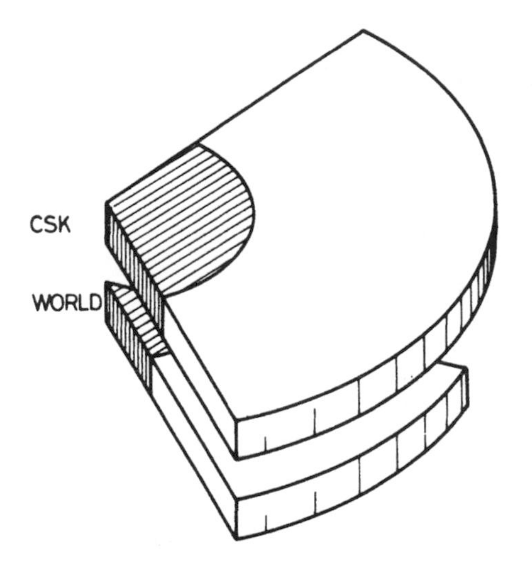

FIGURE 115. CSK-A. Proportion of chemistry and analytical chemistry papers in Czechoslovakian science publications.

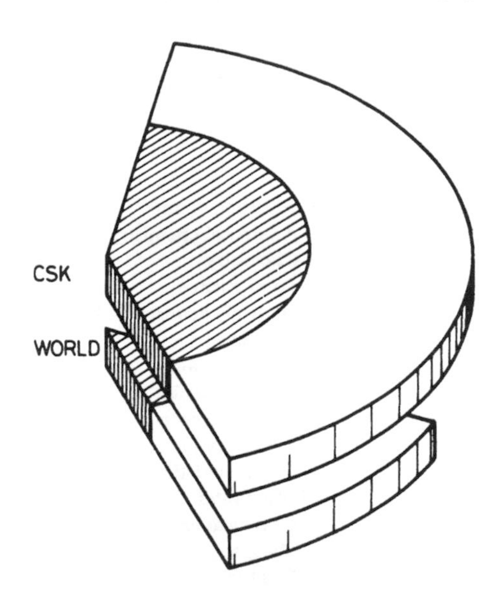

FIGURE 116. CSK-B. Citations to chemistry and analytical chemistry papers as a proportion of citations to Czechoslovakian science publications.

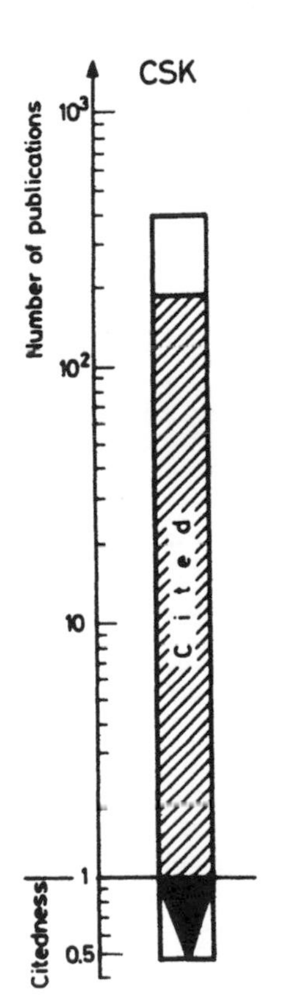

FIGURE 117. CSK-C. Number of Czechoslovakian analytical chemistry papers and the cited fraction.

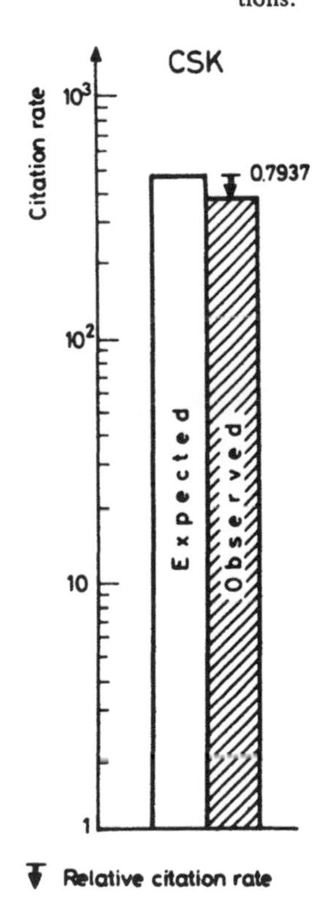

FIGURE 118. CSK-D. Expected and observed citation rates of Czechoslovakian analytical chemistry papers.

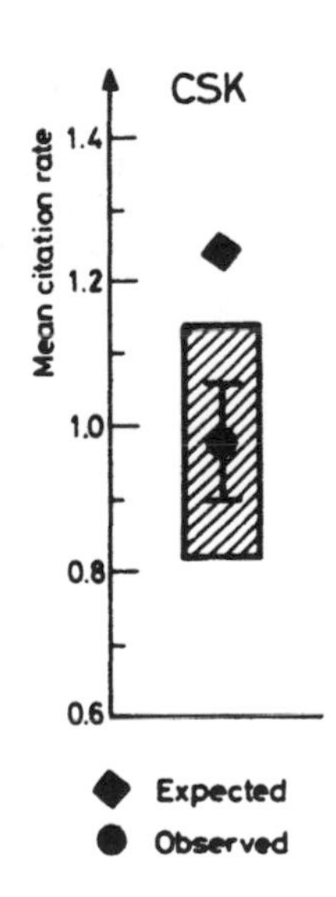

FIGURE 119. CSK-E. Expected and observed values of mean citation rate of Czechoslovakian analytical chemistry papers.

Table 76
SCIENTOMETRIC INDICATOR VALUES 1978—1980

Denmark

Publication counts	
Analytical chemistry	56
Chemistry	442
Total science	5744
Analytical chemistry publications as a proportion of publications	
In chemistry	0.1267 ± 0.0158
Activity index	1.047
Test statistics	0.360
In total science	0.0097 ± 0.0013
Activity index	0.564
Test statistics	−5.803
Citation rates	
Analytical chemistry	155
Chemistry	659
Total science	9553
Citations to analytical chemistry publications as a proportion of citations	
To chemistry	0.2352 ± 0.0165
Attractivity index	1.724
Test statistics	5.981
To total science	0.0162 ± 0.0013
Attractivity index	0.895
Test statistics	−1.473
Fraction of cited publications	0.7321 ± 0.0592
Expected citation rate	94.4
Mean impact factor	1.686
Mean citation rate	2.768 ± 0.434
Relative citation rate	1.642
Test statistics	2.493

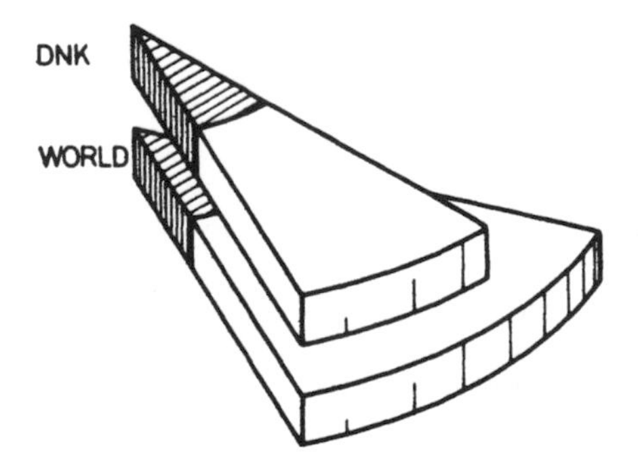

FIGURE 120. DNK-A. Proportion of chemistry and analytical chemistry papers in Danish science publications.

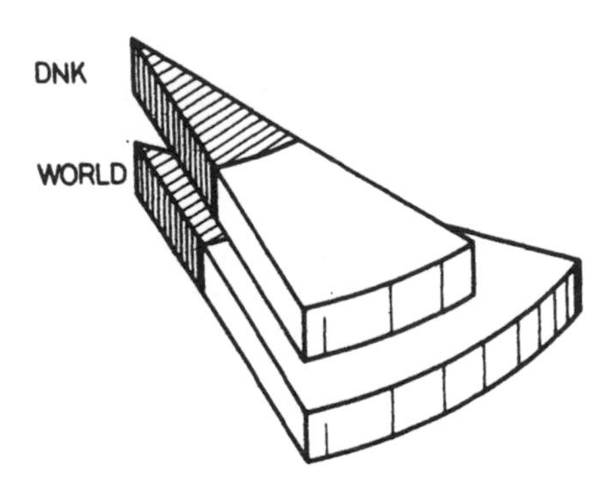

FIGURE 121. DNK-B. Citation to chemistry and analytical chemistry papers as a proportion of citations to Danish science publications.

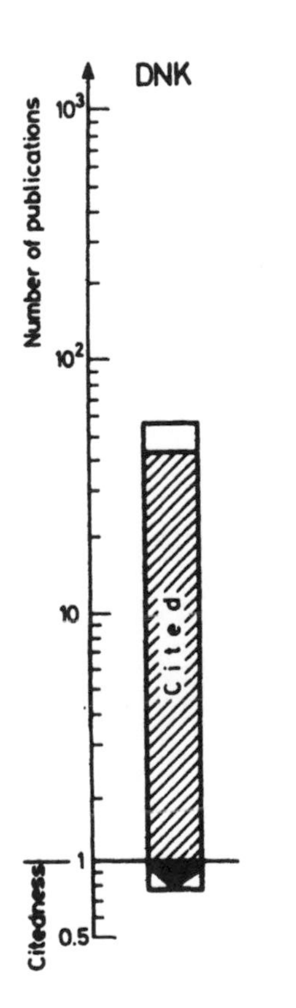

FIGURE 122. DNK-C. Number of Danish analytical chemistry papers and the cited fraction.

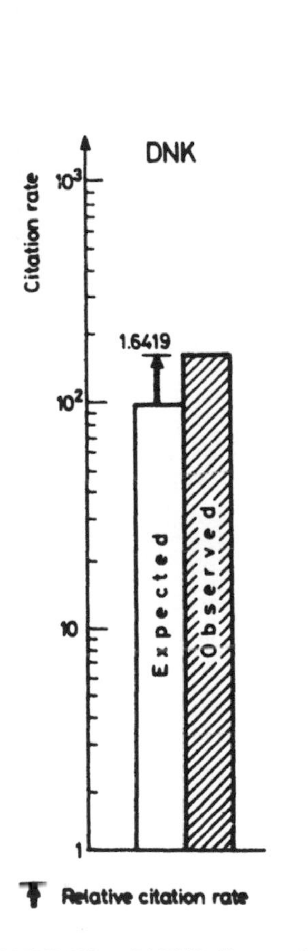

FIGURE 123. DNK-D. Expected and observed citation rates of Danish analytical chemistry papers.

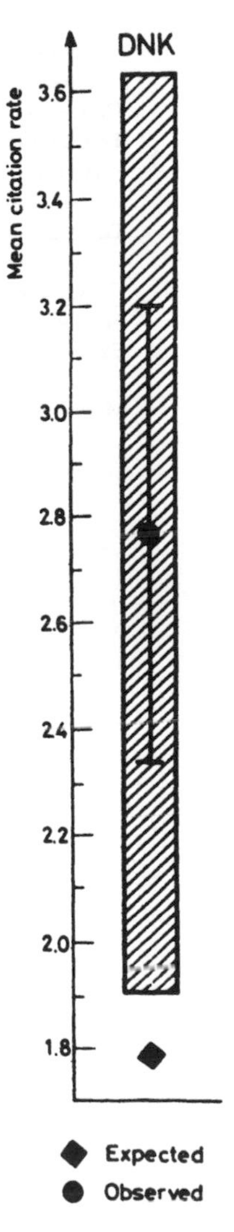

FIGURE 124. DNK-E. Expected and observed values of mean citation rate of Danish analytical chemistry papers.

Table 77
SCIENTOMETRIC INDICATOR VALUES 1978—1980

Egypt

Publication counts	
Analytical chemistry	109
Chemistry	576
Total science	1726
Anatlytical chemistry publications as a proportion of publications	
In chemistry	0.1892 ± 0.0163
Activity index	1.564
Test statistics	4.181
In total science	0.0632 ± 0.0059
Activity index	3.656
Test statistics	7.836
Citation rates	
Analytical chemistry	65
Chemistry	233
Total science	536
Citations to analytical chemistry publications as a proportion of citations	
To chemistry	0.2790 ± 0.0294
Attractivity index	2.045
Test statistics	4.852
To total science	0.1213 ± 0.0141
Attractivity index	6.689
Test statistics	7.315
Fraction of cited publications	0.3853 ± 0.0466
Expected citation rate	113.0
Mean impact factor	1.037
Mean citation rate	0.596 ± 0.090
Relative citation rate	0.575
Test statistics	−4.889

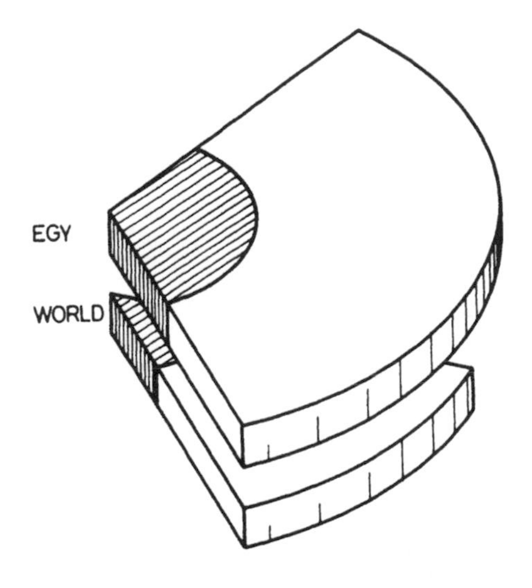

FIGURE 125. EGY-A. Proportion of chemistry and analytical chemistry papers in Egyptian science publications.

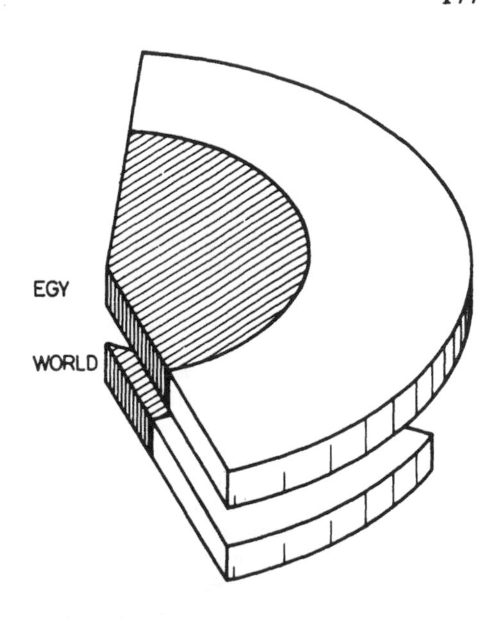

FIGURE 126. EGY-B. Citations to chemistry and analytical chemistry papers as a proportion of citations to Egyptian science publications.

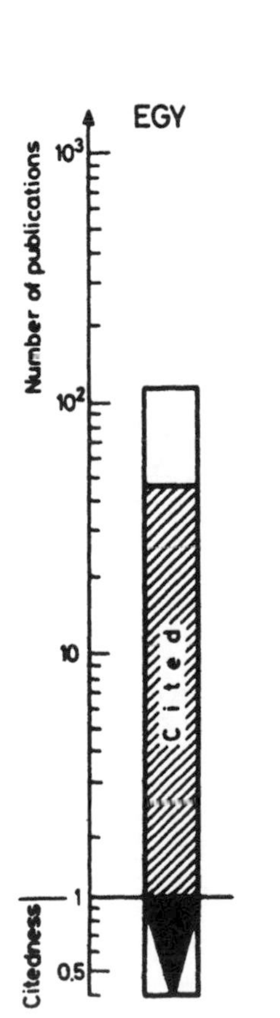

FIGURE 127. EGY-C. Number of Egyptian analytical chemistry papers and its cited fraction.

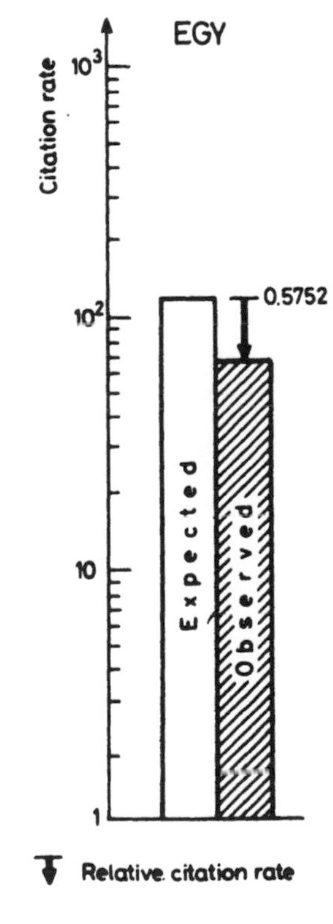

FIGURE 128. EGY-D. Expected and observed citation rates of Egyptian analytical chemistry papers.

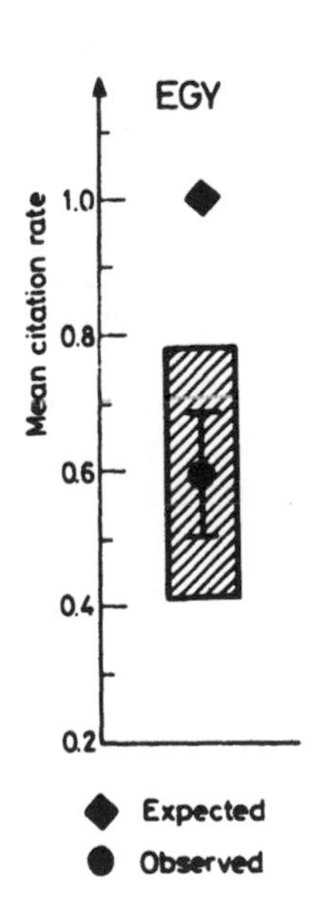

FIGURE 129. EGY-E. Expected and observed values of mean citation rate of Egyptian analytical chemistry papers.

Table 78
SCIENTOMETRIC INDICATOR VALUES 1978—1980

England

Publication counts	
Analytical chemistry	675
Chemistry	5,958
Total science	52,837
Analytical chemistry publications as a proportion of publications	
In chemistry	0.1133 ± 0.0041
Activity index	0.936
Test statistics	−1.877
In total science	0.0128 ± 0.0005
Activity index	0.740
Test statistics	−9.205
Citation rates	
Analytical chemistry	1,144
Chemistry	10,588
Total science	82,384
Citations to analytical chemistry publications as a proportion of citations	
To chemistry	0.1080 ± 0.0030
Attractivity index	0.792
Test statistics	−9.397
To total science	0.0139 ± 0.0004
Attractivity index	0.766
Test statistics	−10.406
Fraction of cited publications	0.6459 ± 0.0184
Expected citation rate	1,141.4
Mean impact factor	1.691
Mean citation rate	1.695 ± 0.079
Relative citation rate	1.002
Test statistics	0.049

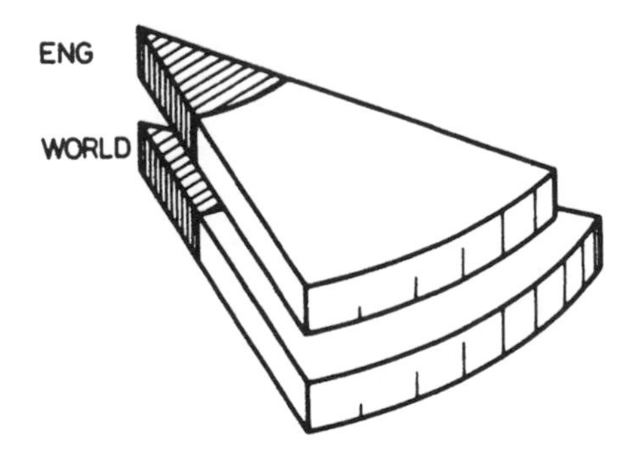

FIGURE 130. ENG-A. Proportion of chemistry and analytical chemistry papers in English science publications.

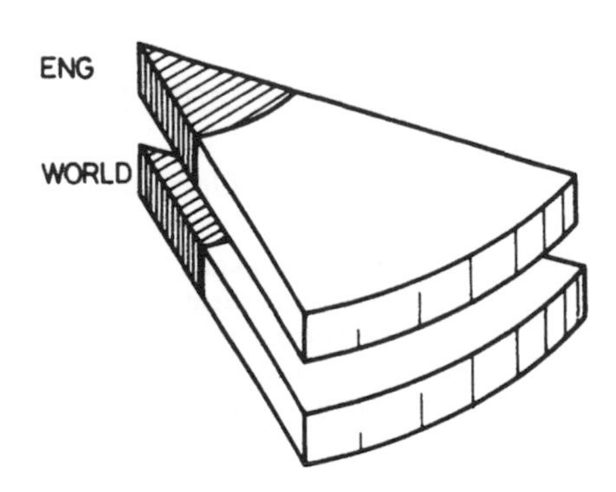

FIGURE 131. ENG-B. Citation to chemistry and analytical chemistry papers as a proportion of citations to English science publications.

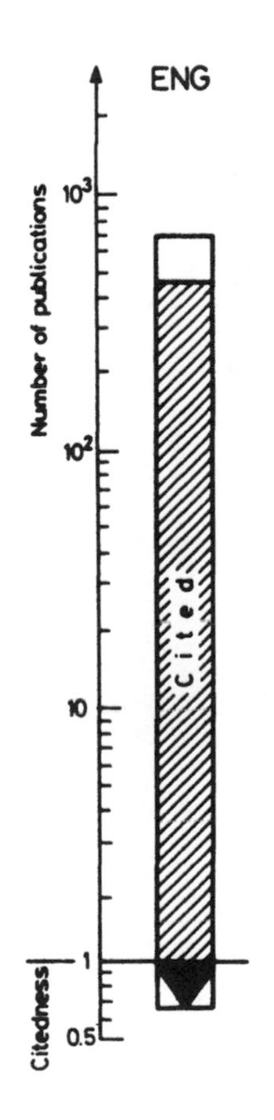

FIGURE 132. ENG-C. Number of English analytical chemistry papers and the cited fraction.

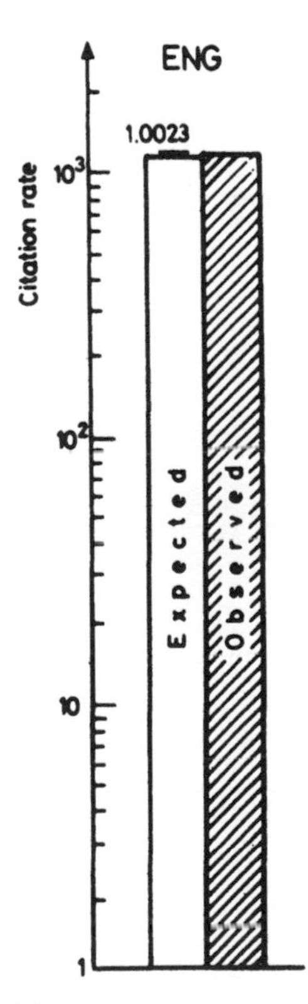

FIGURE 133. ENG-D. Expected and observed citation rates of English analytical chemistry papers.

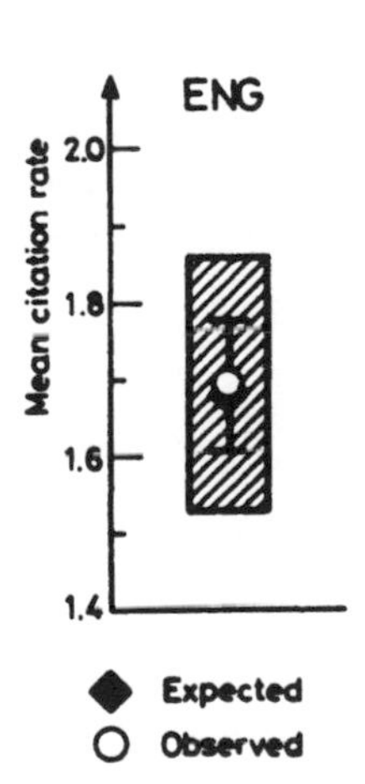

FIGURE 134. ENG-E. Expected and observed values of mean citation rate of English analytical chemistry papers.

Table 79
SCIENTOMETRIC INDICATOR VALUES 1978—1980

West Germany (Federal Republic of Germany)

Publication counts	
Analytical chemistry	765
Chemistry	6,470
Total science	43,697
Analytical chemistry publications as a proportion of publications	
In chemistry	0.1182 ± 0.0040
Activity index	0.977
Test statistics	−0.688
In total science	0.0175 ± 0.0006
Activity index	1.014
Test statistics	0.374
Citation rates	
Analytical chemistry	859
Chemistry	11,201
Total science	58,499
Citations to analytical chemistry publications as a proportion of citations	
To chemistry	0.0767 ± 0.0025
Attractivity index	0.562
Test statistics	−23.747
To total science	0.0147 ± 0.0005
Attractivity index	0.810
Test statistics	−6.927
Fraction of cited publications	0.4327 ± 0.0179
Expected citation rate	842.3
Mean impact factor	1.101
Mean citation rate	1.123 ± 0.071
Relative citation rate	1.020
Test statistics	0.306

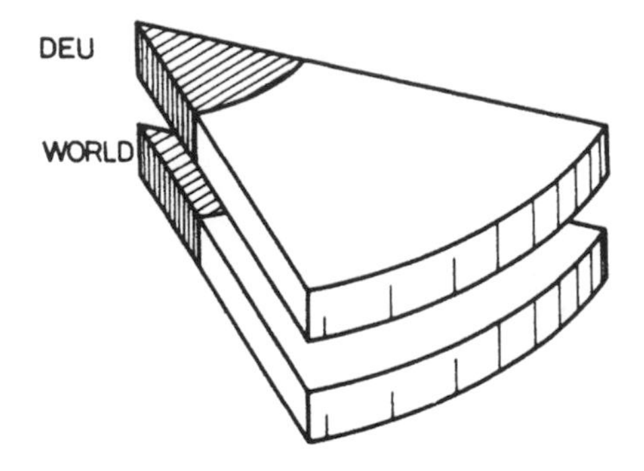

FIGURE 135. DEU-A. Proportion of chemistry and analytical chemistry papers in West German science publications.

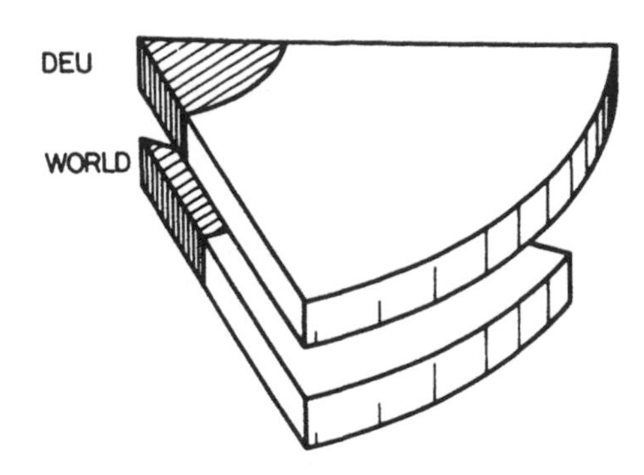

FIGURE 136. DEU-B. Citations to chemistry and analytical chemistry papers as a proportion of citations to West German science publications.

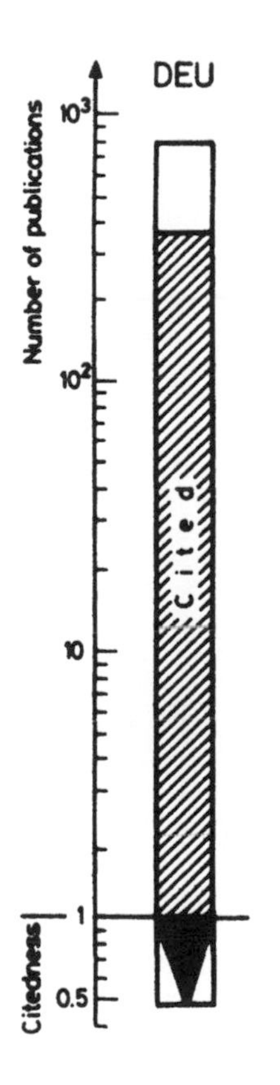

FIGURE 137. DEU-C. Number of West German analytical chemistry papers and the cited fraction.

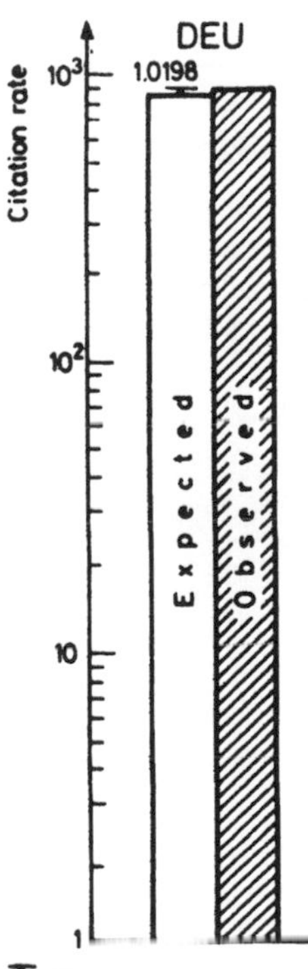

FIGURE 138. DEU-D. Expected and observed citation rates of West German analytical chemistry papers.

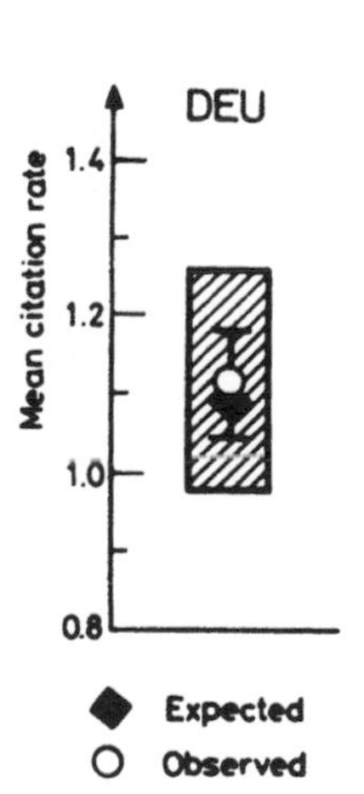

FIGURE 139. DEU-E. Expected and observed values of mean citation rate of West German analytical chemistry papers.

Table 80
SCIENTOMETRIC INDICATOR VALUES 1978—1980

Finland

Publication counts	
Analytical chemistry	53
Chemistry	379
Total science	4022
Analytical chemistry publications as a proportion of publications	
In chemistry	0.1398 ± 0.0178
Activity index	1.156
Test statistics	1.058
In total science	0.0132 ± 0.0018
Activity index	0.763
Test statistics	−2.277
Citation rates	
Analytical chemistry	59
Chemistry	319
Total science	5782
Citations to analytical chemistry publications as a proportion of citations	
To chemistry	0.1850 ± 0.0217
Attractivity index	1.356
Test statistics	2.234
In total science	0.0102 ± 0.0013
Attractivity index	0.563
Test statistics	−5.996
Fraction of cited publications	0.5849 ± 0.0677
Expected citation rate	77.5
Mean impact factor	1.462
Mean citation rate	1.113 ± 0.182
Relative citation rate	0.761
Test statistics	−1.919

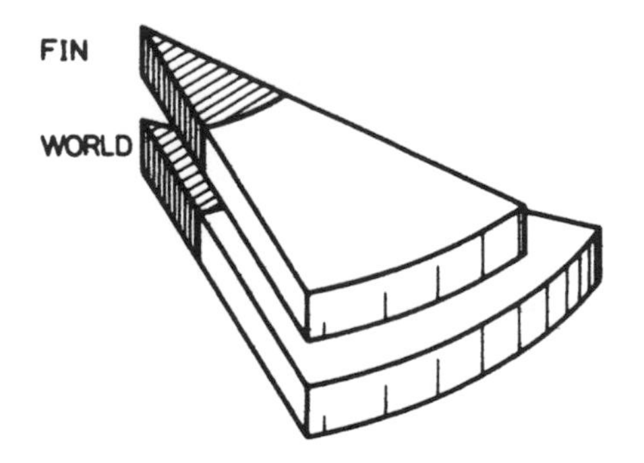

FIGURE 140. FIN-A. Proportion of chemistry and analytical chemistry papers in Finnish science publications.

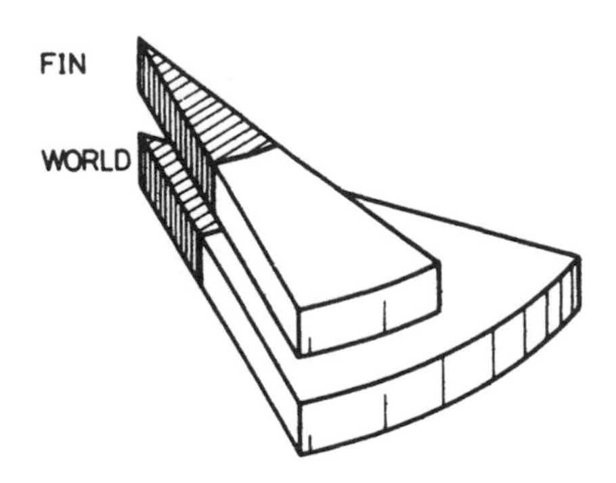

FIGURE 141. FIN-B. Citation to chemistry and analytical chemistry papers as a proportion of citations to Finnish science publications.

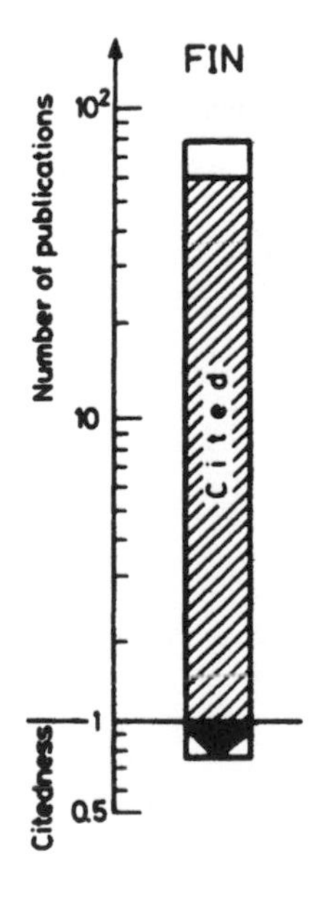

FIGURE 142. FIN-C. Number of Finnish analytical chemistry papers and the cited fraction.

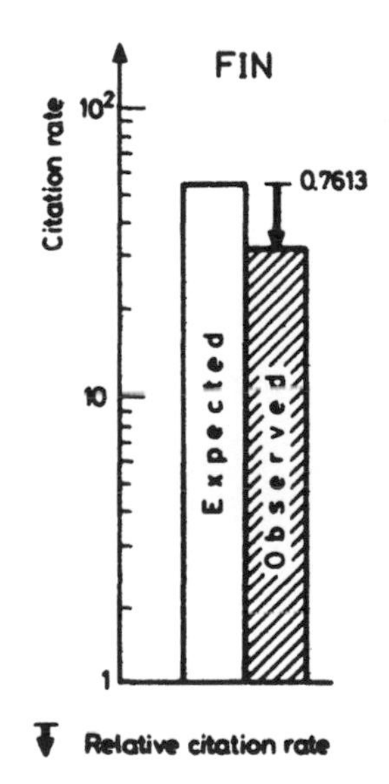

FIGURE 143. FIN-D. Expected and observed citation rates of Finnish analytical chemistry papers.

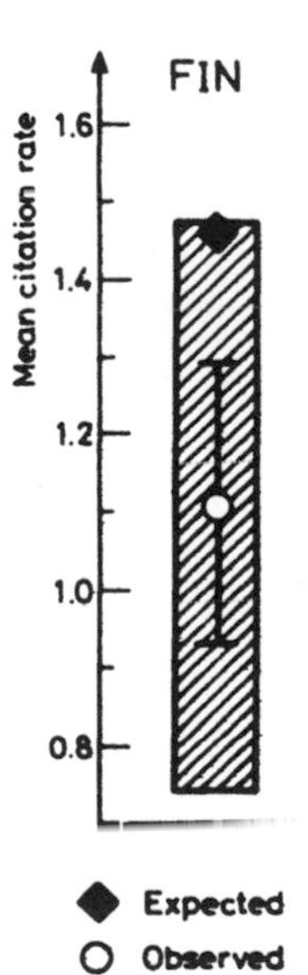

FIGURE 144. FIN-E. Expected and observed values of mean citation rate of Finnish analytical chemistry papers.

Table 81
SCIENTOMETRIC INDICATOR VALUES 1978—1980

France

Publication counts	
Analytical chemistry	631
Chemistry	5,708
Total science	35,995
Analytical chemistry publications as a proportion of publications	
In chemistry	0.1105 ± 0.0042
Activity index	0.914
Test statistics	−2.518
In total science	0.0175 ± 0.0007
Activity index	1.015
Test statistics	0.373
Citation rates	
Analytical chemistry	926
Chemistry	7,646
Total science	42,361
Citations to analytical chemistry publications as a proportion of citations	
To chemistry	0.1211 ± 0.0037
Attractivity index	0.888
Test statistics	−4.097
To total science	0.0219 ± 0.0007
Attractivity index	1.206
Test statistics	5.252
Fraction of cited publications	0.5737 ± 0.0197
Expected citation rate	928.8
Mean impact factor	1.472
Mean citation rate	1.468 ± 0.080
Relative citation rate	0.997
Test statistics	−0.056

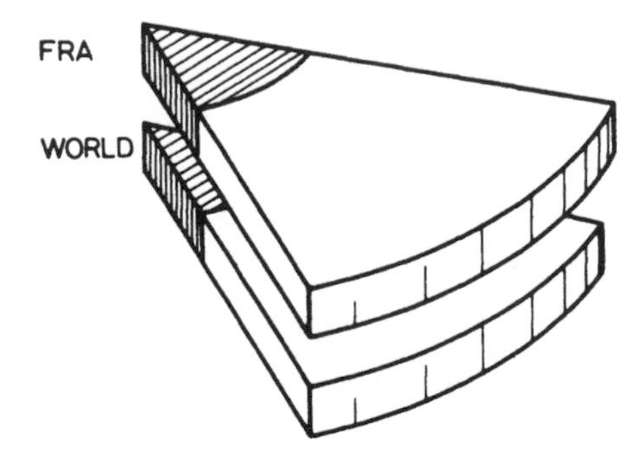

FIGURE 145. FRA-A. Proportion of chemistry and analytical chemistry papers in French science publications.

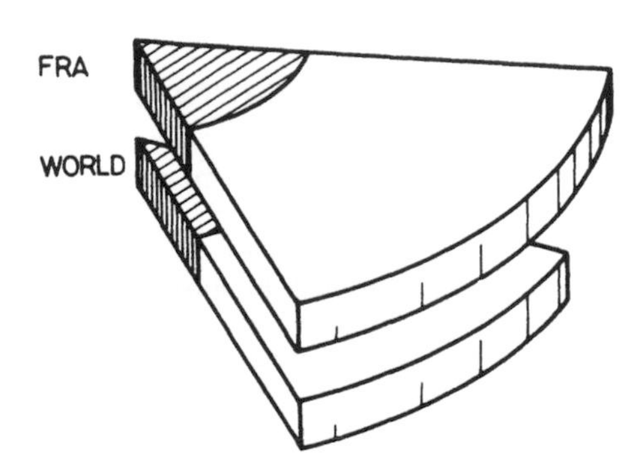

FIGURE 146. FRA-B. Citations to chemistry and analytical chemistry papers as a proportion of citations to French science publications.

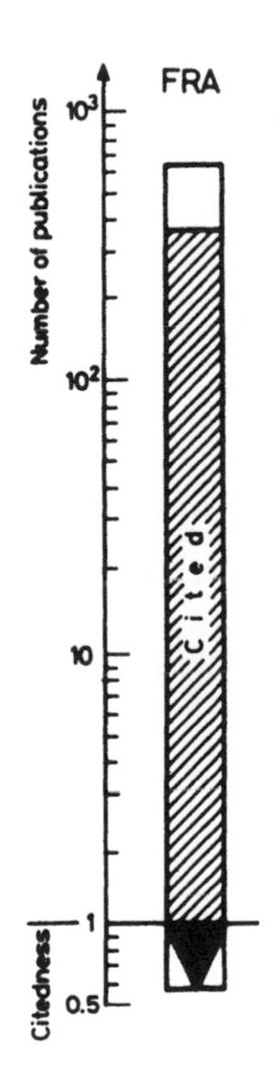

FIGURE 147. FRA-C. Number of French analytical chemistry papers and the cited fraction.

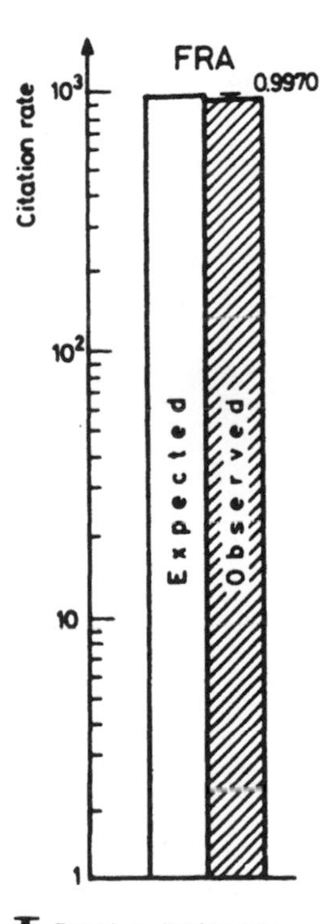

FIGURE 148. FRA-D. Expected and observed citation rates of French analytical chemistry papers.

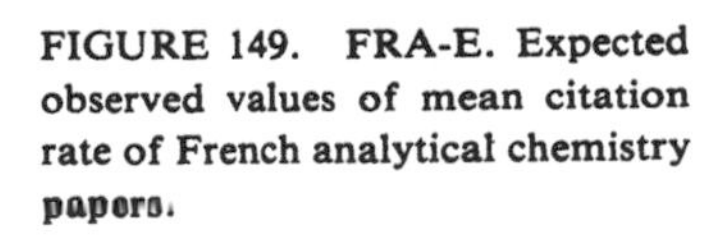
FIGURE 149. FRA-E. Expected observed values of mean citation rate of French analytical chemistry papers.

Table 82
SCIENTOMETRIC INDICATOR VALUES 1978—1980

East Germany (German Democratic Republic)

Publication counts	
Analytical chemistry	112
Chemistry	1828
Total science	6617
Analytical chemistry publications as a proportion of publications	
In chemistry	0.0613 ± 0.0056
Activity index	0.506
Test statistics	−10.648
In total science	0.0169 ± 0.0016
Activity index	0.980
Test statistics	−0.218
Citation rates	
Analytical chemistry	88
Chemistry	1194
Total science	3761
Citations to analytical chemistry publications as a proportion of citations	
To chemistry	0.0737 ± 0.0076
Attractivity index	0.540
Test statistics	−8.291
To total science	0.0234 ± 0.0025
Attractivity index	1.291
Test statistics	2.138
Fraction of cited publications	0.4821 ± 0.0472
Expected citation rate	113.7
Mean impact factor	1.015
Mean citation rate	0.786 ± 0.100
Relative citation rate	0.774
Test statistics	−2.305

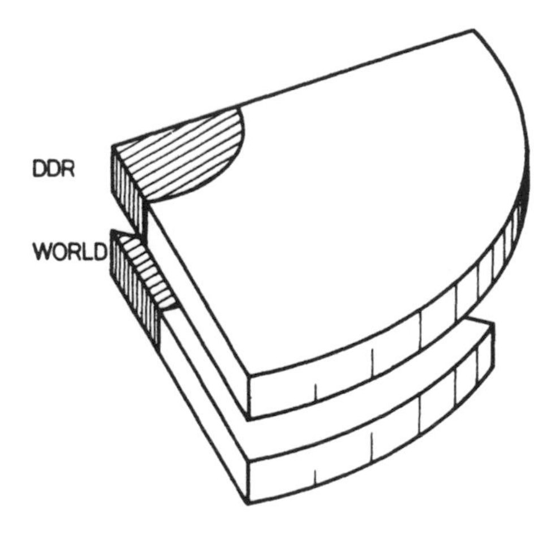

FIGURE 150. DDR-A. Proportion of chemistry and analytical chemistry papers in East German science publications.

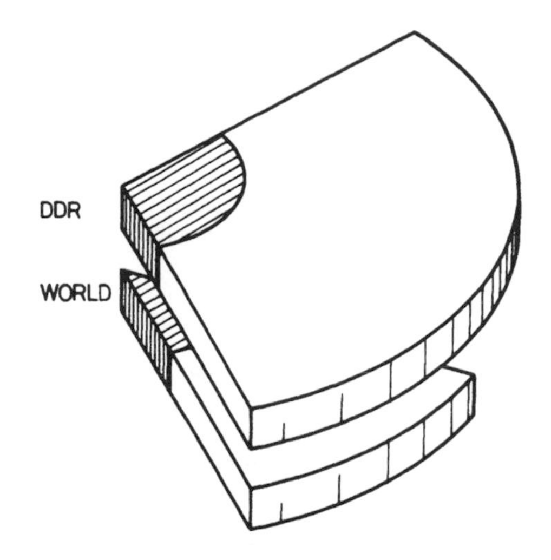

FIGURE 151. DDR-B. Citation to chemistry and analytical chemistry papers as a proportion of citations to East German science publications.

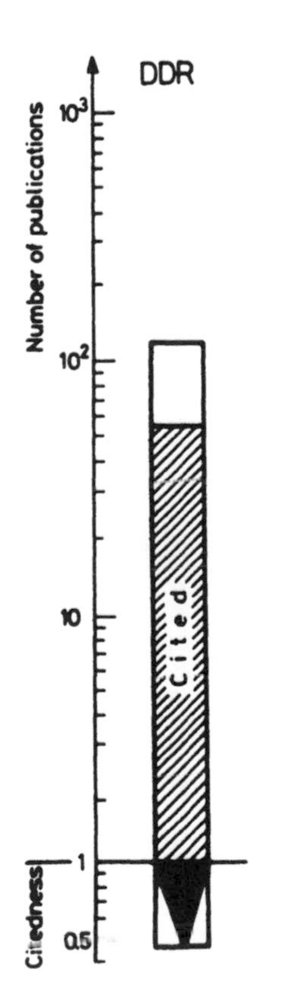

FIGURE 152. DDR-C. Number of East German analytical chemistry papers and the cited fraction.

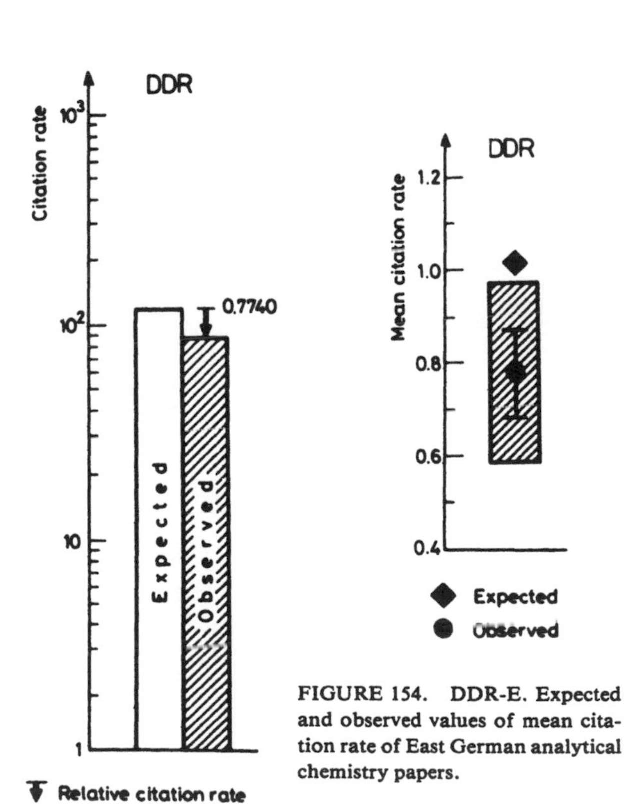

FIGURE 153. DDR-D. Expected and observed citation rates of East German analytical chemistry papers.

FIGURE 154. DDR-E. Expected and observed values of mean citation rate of East German analytical chemistry papers.

Table 83
SCIENTOMETRIC INDICATOR VALUES 1978—1980

Greece

Publication counts	
Analytical chemistry	32
Chemistry	135
Total science	1110
Analytical chemistry publications as a proportion of publications	
In chemistry	0.2370 ± 0.0366
Activity index	1.959
Test statistics	3.170
In total science	0.0288 ± 0.0050
Activity index	1.669
Test statistics	2.301
Citation rates	
Analytical chemistry	35
Chemistry	132
Total science	721
Citations to analytical chemistry publications as a proportion of citations	
To chemistry	0.2652 ± 0.0384
Attractivity index	1.944
Test statistics	3.351
To total science	0.0485 ± 0.0080
Attractivity index	2.678
Test statistics	3.800
Fraction of cited publications	0.6563 ± 0.0840
Expected citation rate	43.5
Mean impact factor	1.359
Mean citation rate	1.094 ± 0.190
Relative citation rate	0.805
Test statistics	−1.400

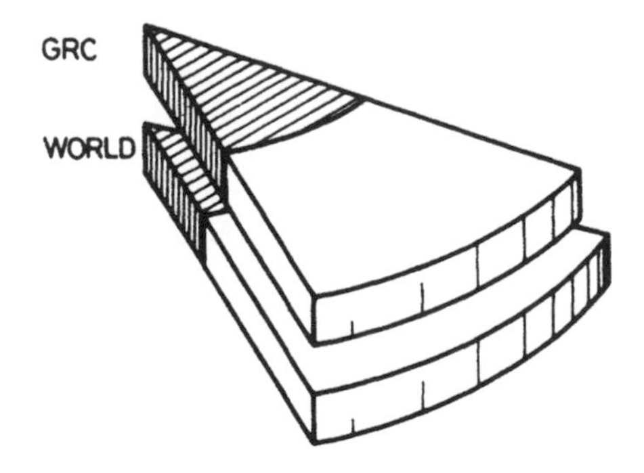

FIGURE 155. GRC-A. Proportion of chemistry and analytical chemistry papers in Greek science publications.

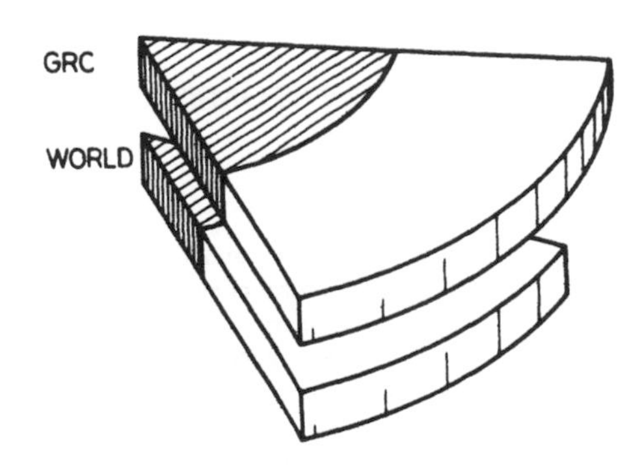

FIGURE 156. GRC-B. Citations to chemistry and analytical chemistry papers as a proportion of citations to Greek science publications.

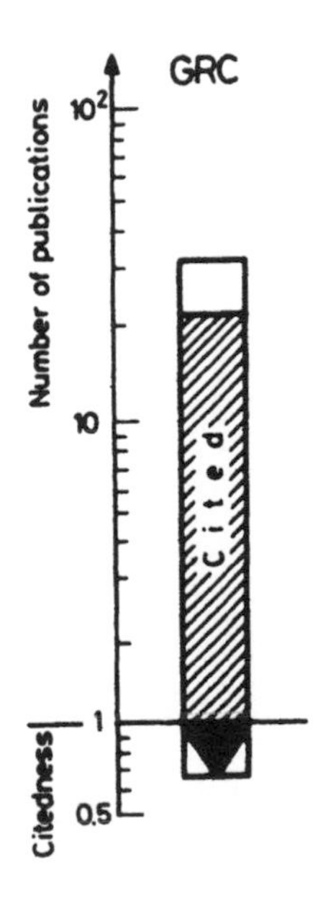

FIGURE 157. GRC-C. Number of Greek analytical chemistry papers and the cited fraction.

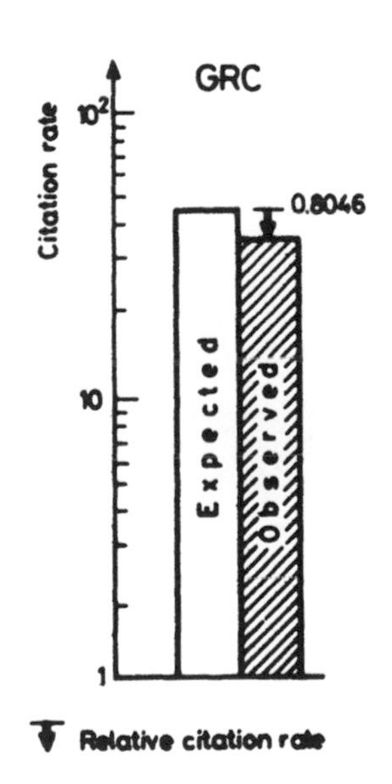

FIGURE 158. GRC-D. Expected and observed citation rates of Greek analytical chemistry papers.

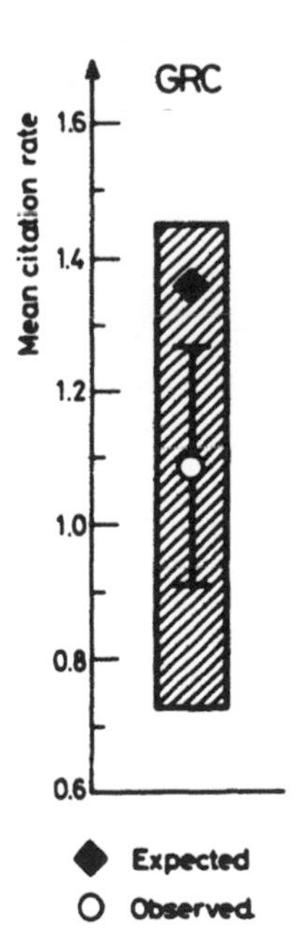

FIGURE 159. GRC-E. Expected and observed values of mean citation rate of Greek analytical chemistry papers.

Table 84
SCIENTOMETRIC INDICATOR VALUES 1978—1980

Hungary

Publication counts	
Analytical chemistry	172
Chemistry	1075
Total science	3868
Analytical chemistry publications as a proportion of publications	
In chemistry	0.1600 ± 0.0112
Activity index	1.322
Test statistics	3.488
In total science	0.0445 ± 0.0033
Activity index	2.574
Test statistics	8.205
Citation rates	
Analytical chemistry	162
Chemistry	790
Total science	2754
Citations to analytical chemistry publications as a proportion of citations	
To chemistry	0.2051 ± 0.0144
Attractivity index	1.503
Test statistics	4.780
To total science	0.0588 ± 0.0045
Attractivity index	3.245
Text statistics	9.076
Fraction of cited publications	0.4593 ± 0.0380
Expected citation rate	199.9
Mean impact factor	1.162
Mean citation rate	0.942 ± 0.110
Relative citation rate	0.810
Test statistics	−1.996

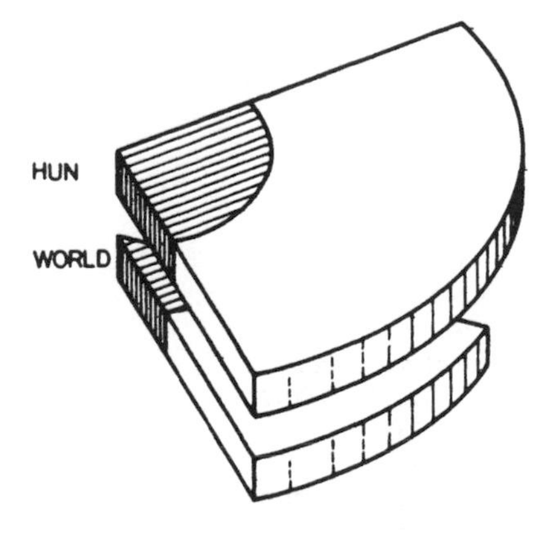

FIGURE 160. HUN-A. Proportion of chemistry and analytical chemistry papers in Hungarian science publications.

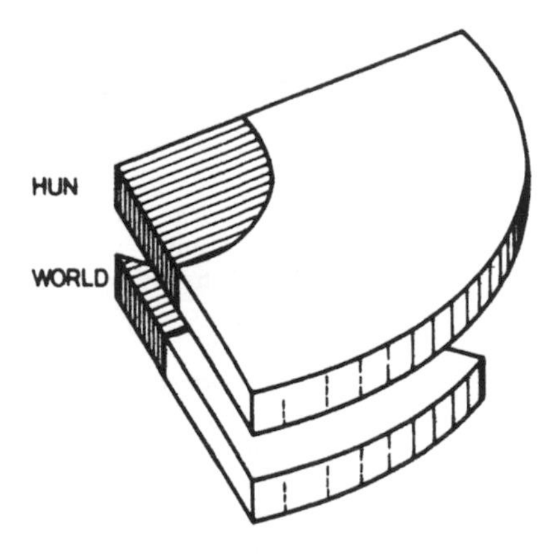

FIGURE 161. HUN-B. Citation to chemistry and analytical chemistry papers as a proportion of citations to Hungarian science publications.

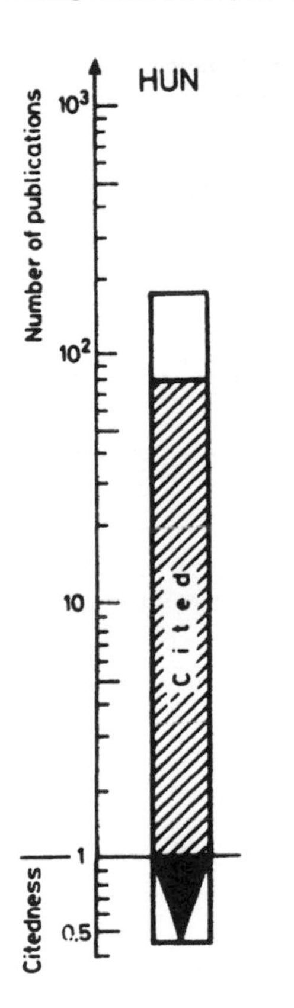

FIGURE 162. HUN-C. Number of Hungarian analytical chemistry papers and the cited fraction.

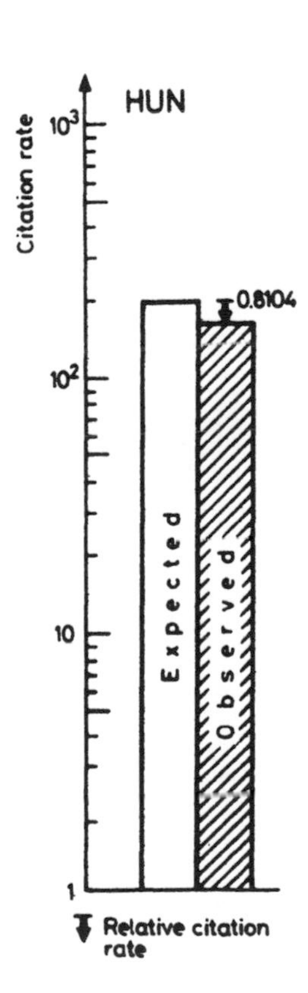

FIGURE 163. HUN-D. Expected and observed citation rates of Hungarian analytical chemistry papers.

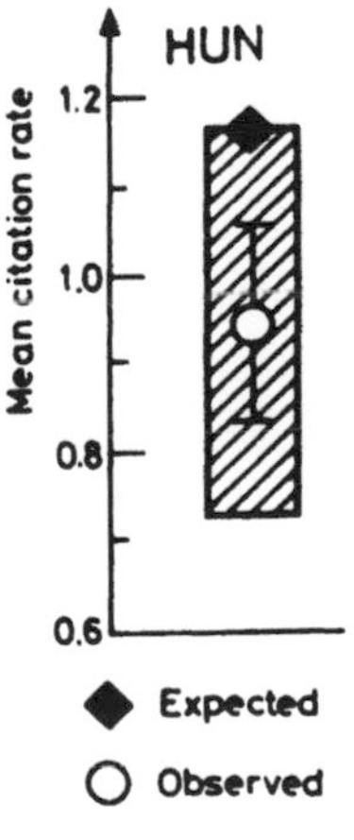

FIGURE 164. HUN-E. Expected and observed values of mean citation rate of Hungarian analytical chemistry papers.

Table 85
SCIENTOMETRIC INDICATOR VALUES 1978—1980

India

Publication counts	
Analytical chemistry	531
Chemistry	5,243
Total science	21,693
Analytical chemistry publications as a proportion of publications	
In chemistry	0.1013 ± 0.0042
Activity index	0.837
Test statistics	−4.733
In total science	0.0245 ± 0.0010
Activity index	1.417
Test statistics	6.868
Citation rates	
Analytical chemistry	308
Chemistry	2,536
Total science	8,802
Citations to analytical chemistry publications as a proportion of citations	
To chemistry	0.1215 ± 0.0065
Attractivity index	0.890
Test statistics	−2.304
To total science	0.0350 ± 0.0020
Attractivity index	1.930
Test statistics	8.610
Fraction of cited publications	0.3710 ± 0.0210
Expected citation rate	514.5
Mean impact factor	0.969
Mean citation rate	0.580 ± 0.041
Relative citation rate	0.599
Test statistics	−9.477

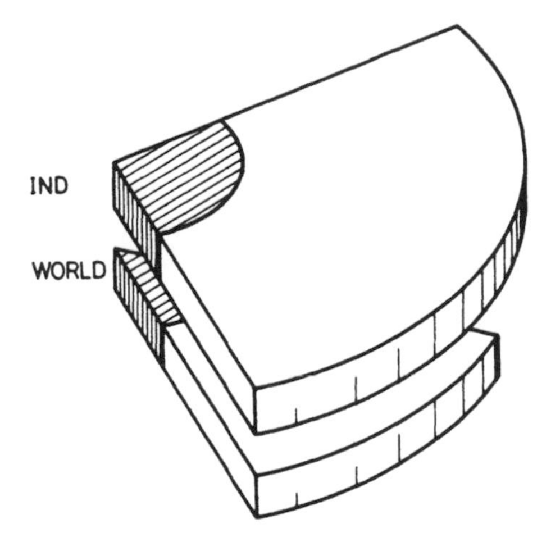

FIGURE 165. IND-A. Proportion of chemistry and analytical chemistry papers in Indian science publications.

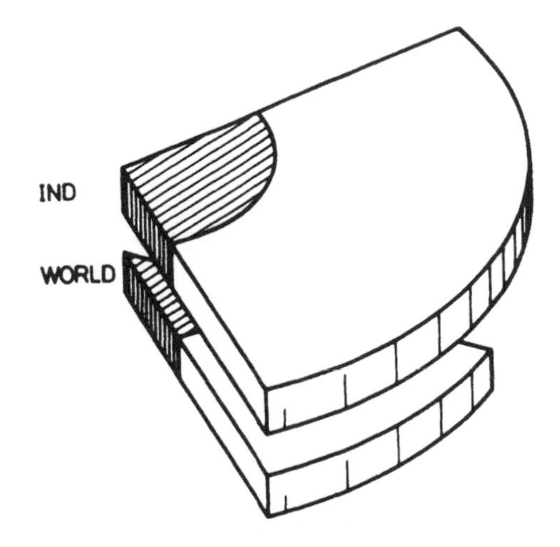

FIGURE 166. IND-B. Citations to chemistry and analytical chemistry papers as a proportion of citations to Indian science publications.

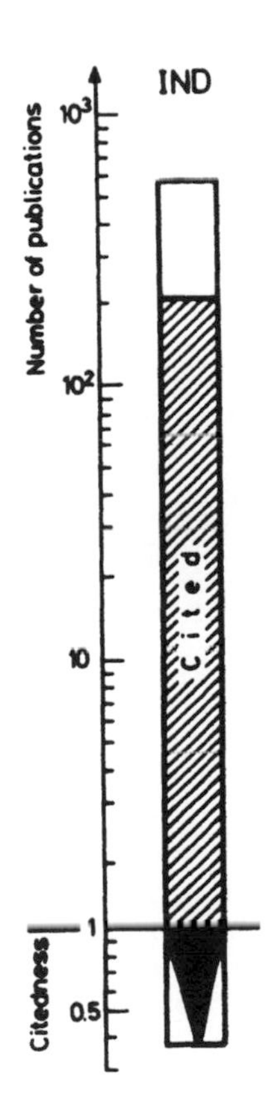

FIGURE 167. IND-C. Number of Indian analytical chemistry papers and the cited fraction.

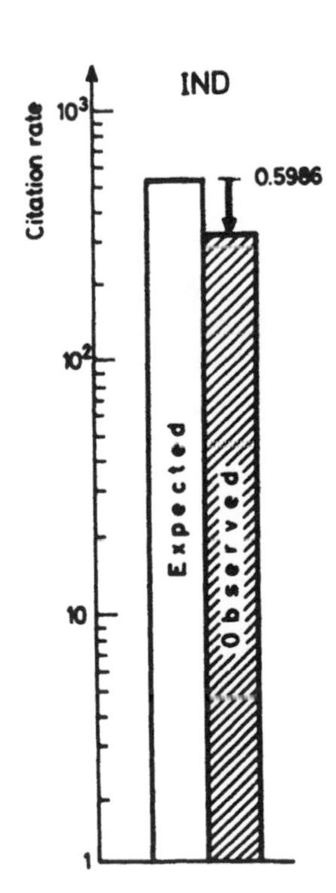

FIGURE 168. IND-D. Expected and observed citation rates of Indian analytical chemistry papers.

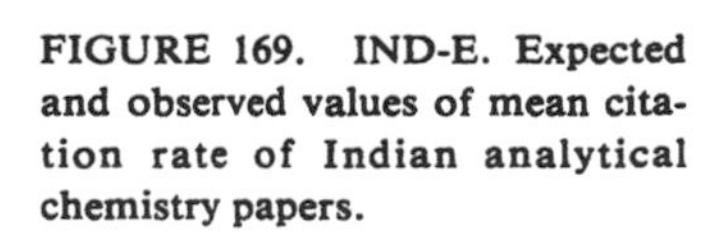

FIGURE 169. IND-E. Expected and observed values of mean citation rate of Indian analytical chemistry papers.

Table 86
SCIENTOMETRIC INDICATOR VALUES 1978—1980

Ireland

Publication counts	
Analytical chemistry	12
Chemistry	136
Total science	1287
Analytical chemistry publications as a proportion of publications	
In chemistry	0.0882 ± 0.0243
Activity index	0.729
Test statistics	−1.347
In total science	0.0093 ± 0.0027
Activity index	0.540
Test statistics	−2.967
Citation rates	
Analytical chemistry	26
Chemistry	217
Total science	885
Citations to analytical chemistry publications as a proportion of citations	
To chemistry	0.1198 ± 0.0220
Attractivity index	0.878
Test statistics	−0.752
To total science	0.0294 ± 0.0057
Attractivity index	1.621
Test statistics	1.982
Fraction of cited publications	0.6667 ± 0.1361
Expected citation rate	21.2
Mean impact factor	1.767
Mean citation rate	2.167 ± 0.787
Relative citation rate	1.226
Test statistics	0.508

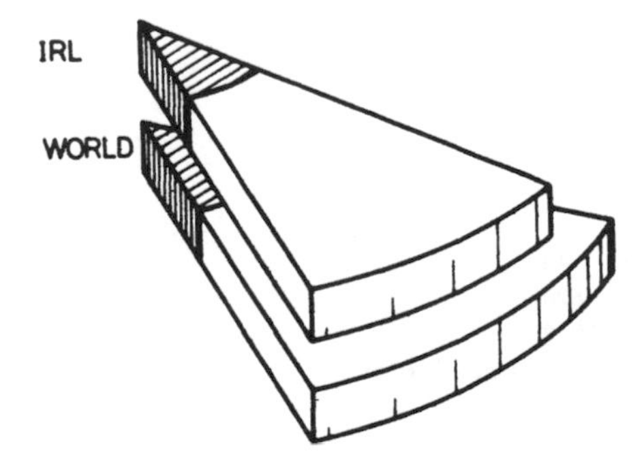

FIGURE 170. IRL-A. Proportion of chemistry and analytical chemistry papers in Irish science publications.

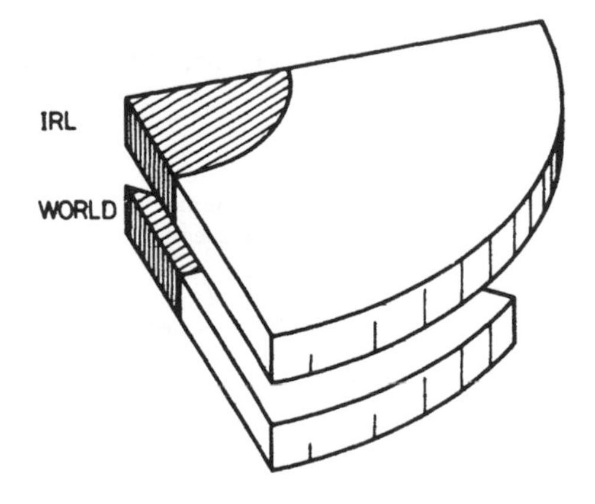

FIGURE 171. IRL-B. Citation to chemistry and analytical chemistry papers as a proportion of citations to Irish science publications.

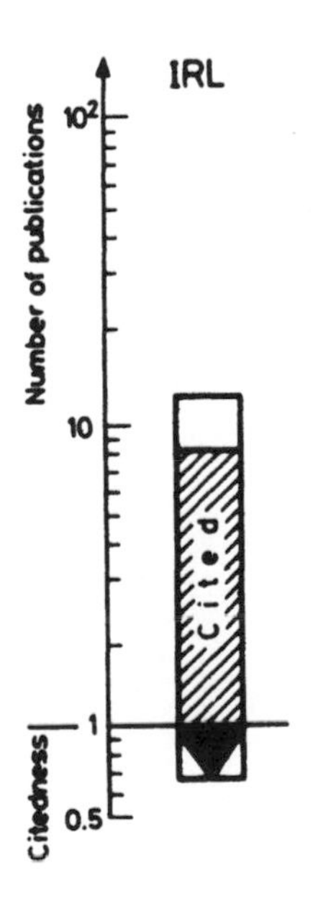

FIGURE 172. IRL-C. Number of Irish analytical chemistry papers and the cited fraction.

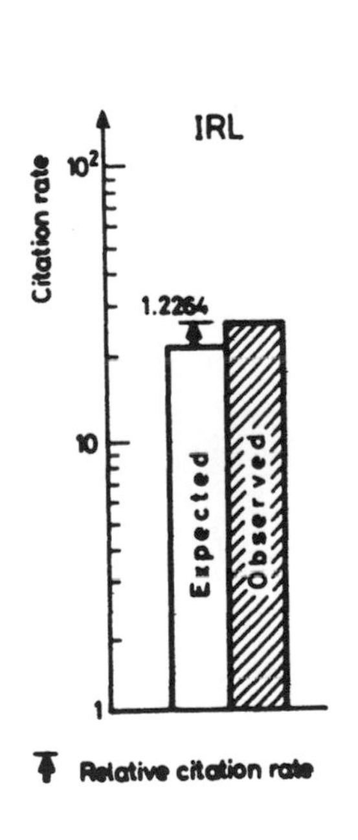

FIGURE 173. IRL-D. Expected and observed citation rates of Irish analytical chemistry papers.

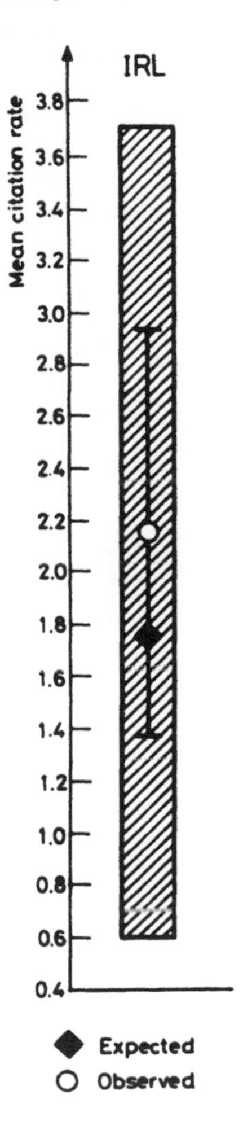

FIGURE 174. IRL-E. Expected and observed values of mean citation rate of Irish analytical chemistry papers.

Table 87
SCIENTOMETRIC INDICATOR VALUES 1978—1980

Israel

Publication counts	
Analytical chemistry	46
Chemistry	620
Total science	7018
Analytical chemistry publications as a proportion of publications	
In chemistry	0.0742 ± 0.0105
Activity index	0.613
Test statistics	−4.447
In total science	0.0066 ± 0.0010
Activity index	0.379
Test statistics	−11.127
Citation rates	
Analytical chemistry	37
Chemistry	931
Total science	9374
Citations to analytical chemistry publications as a proportion of citations	
To chemistry	0.0397 ± 0.0064
Attractivity index	0.291
Test statistics	−15.096
To total science	0.0039 ± 0.0006
Attractivity index	0.218
Test statistics	−21.898
Fraction of cited publications	0.4783 ± 0.0737
Expected citation rate	81.4
Mean impact factor	1.770
Mean citation rate	0.804 ± 0.162
Relative citation rate	0.455
Test statistics	−5.945

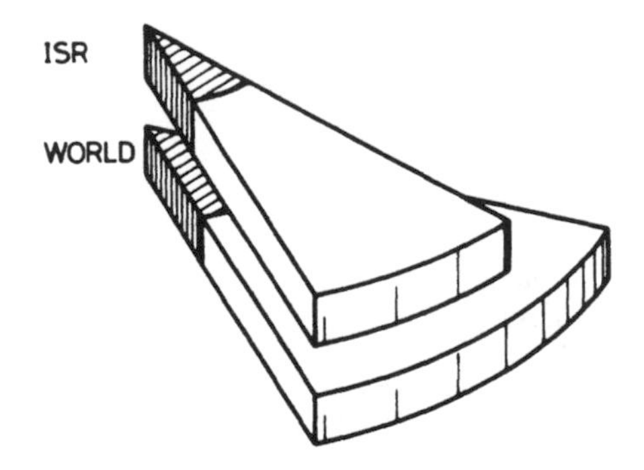

FIGURE 175. ISR-A. Proportion of chemistry and analytical chemistry papers in Israeli science publications.

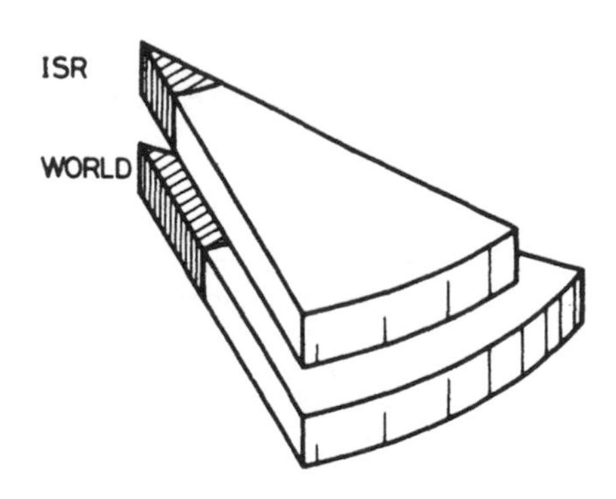

FIGURE 176. ISR-B. Citations to chemistry and analytical chemistry papers as a proportion of citations to Israeli science publications.

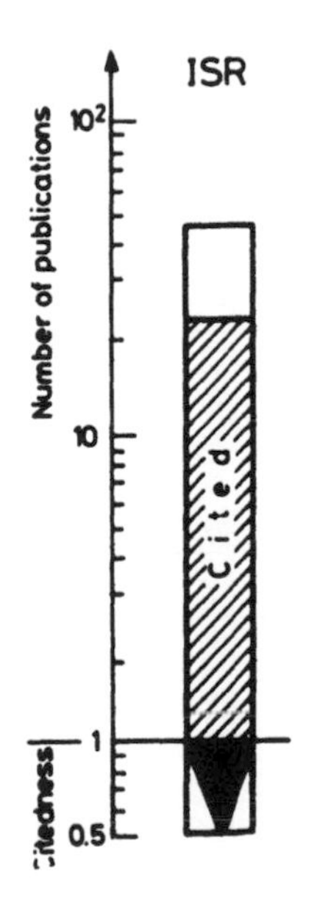

FIGURE 177. ISR-C. Number of Israeli analytical chemistry papers and the cited fraction.

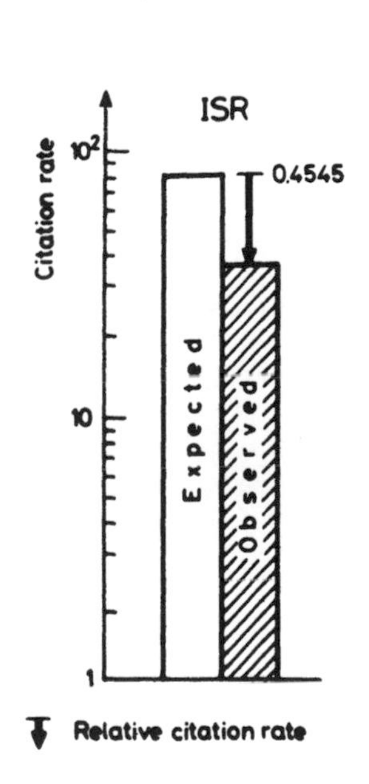

FIGURE 178. ISR-D. Expected and observed citation rates of Israeli analytical chemistry papers.

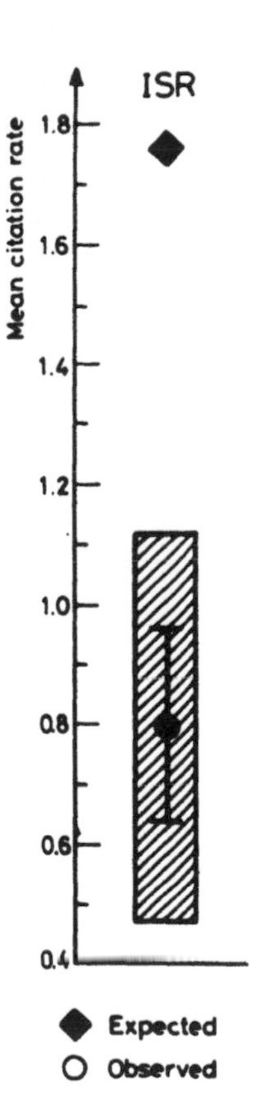

FIGURE 179. ISR-E. Expected and observed values of mean citation rate of Israeli analytical chemistry papers.

Table 88
SCIENTOMETRIC INDICATOR VALUES 1978—1980

Italy

Publication counts	
Analytical chemistry	267
Chemistry	3,039
Total science	13,235
Analytical chemistry publications as a proportion of publications	
In chemistry	0.0879 ± 0.0051
Activity index	0.726
Test statistics	−6.454
In total science	0.0202 ± 0.0012
Activity index	1.168
Test statistics	2.374
Citation rates	
Analytical chemistry	295
Chemistry	3,656
Total science	14,752
Citations to analytical chemistry publications as a proportion of citations	
To chemistry	0.0807 ± 0.0045
Attractivity index	0.592
Test statistics	−12.367
To total science	0.0200 ± 0.0012
Attractivity index	1.103
Test statistics	1.621
Fraction of cited publications	0.5805 ± 0.0302
Expected citation rate	448.0
Mean impact factor	1.678
Mean citation rate	1.105 ± 0.081
Relative citation rate	0.658
Test statistics	−7.066

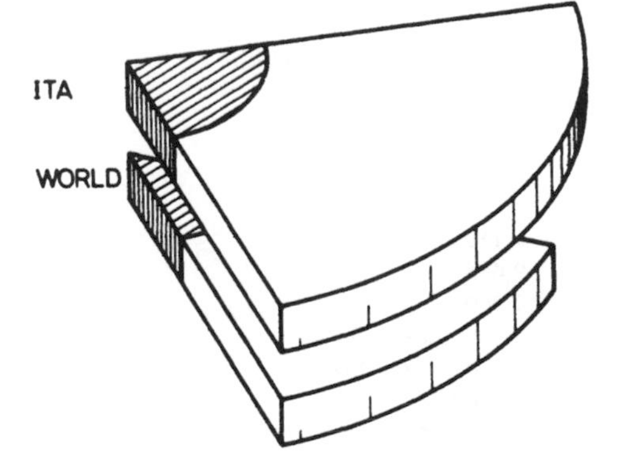

FIGURE 180. ITA-A. Proportion of chemistry and analytical chemistry papers in Italian science publications.

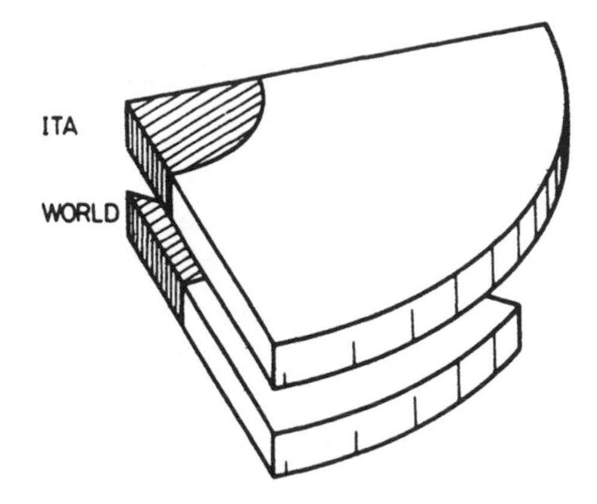

FIGURE 181. ITA-B. Citations to chemistry and analytical chemistry papers as a proportion of citations to Italian science publications.

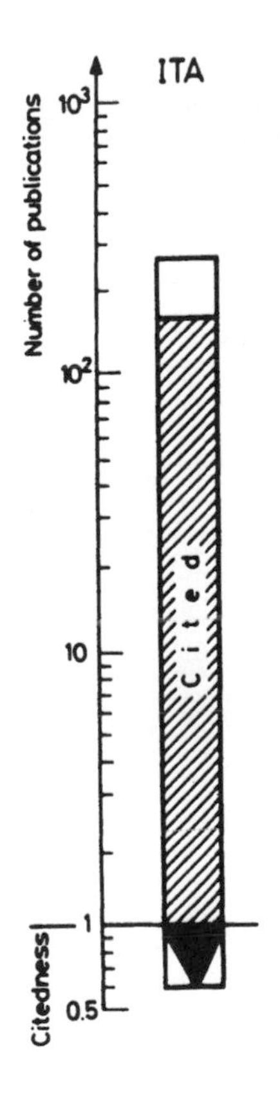

FIGURE 182. ITA-C. Number of Italian analytical chemistry papers and the cited fraction.

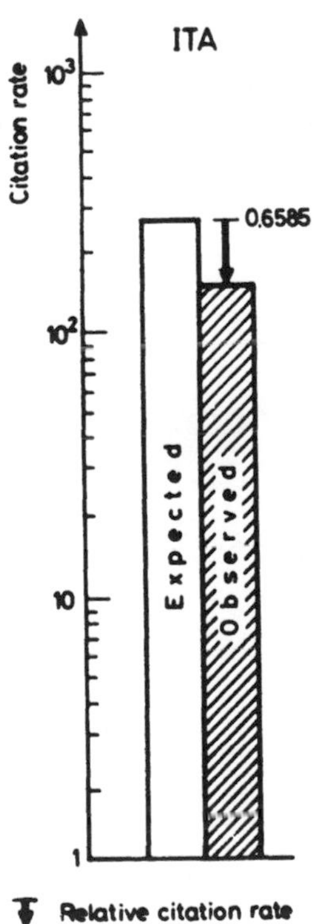

FIGURE 183. ITA-D. Expected and observed citation rates of Italian analytical chemistry papers.

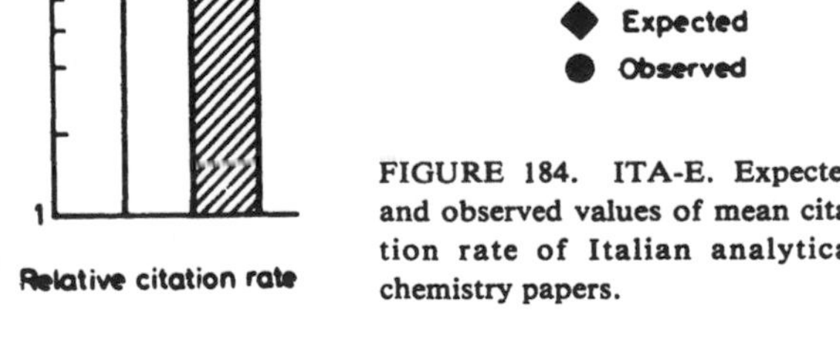

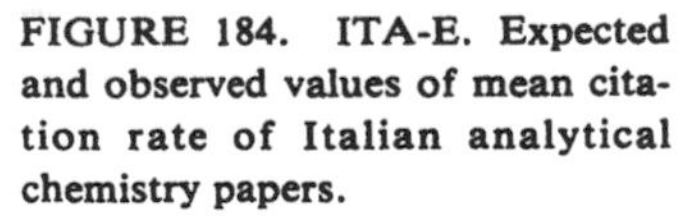

FIGURE 184. ITA-E. Expected and observed values of mean citation rate of Italian analytical chemistry papers.

Table 89
SCIENTOMETRIC INDICATOR VALUES 1978—1980

Japan

Publication counts	
Analytical chemistry	1,119
Chemistry	9,655
Total science	40,709
Analytical chemistry publications as a proportion of publications	
In chemistry	0.1159 ± 0.0033
Activity index	0.958
Test statistics	−1.565
In total science	0.0275 ± 0.0008
Activity index	1.591
Test statistics	12.606
Citation rates	
Analytical chemistry	1,360
Chemistry	11,955
Total science	47,764
Citations to analytical chemistry publications as a proportion of citations	
To chemistry	0.1138 ± 0.0029
Attractivity index	0.834
Test statistics	−7.795
To total science	0.0285 ± 0.0008
Attractivity index	1.571
Test statistics	13.593
Fraction of cited publications	0.5255 ± 0.0149
Expected citation rate	1,131.5
Mean impact factor	1.011
Mean citation rate	1.215 ± 0.052
Relative citation rate	1.202
Test statistics	3.933

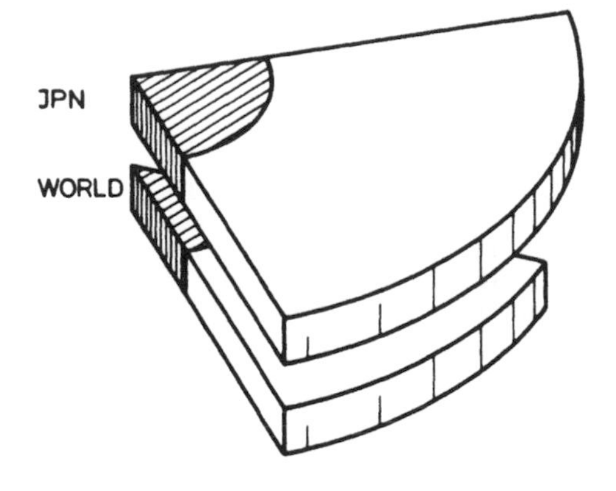

FIGURE 185. JPN-A. Proportion of chemistry and analytical chemistry papers in Japanese science publications.

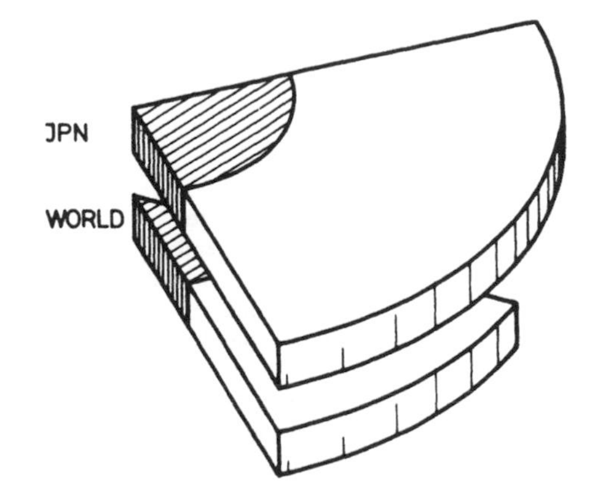

FIGURE 186. JPN-B. Citations to chemistry and analytical chemistry papers as a proportion of citations to Japanese science publications.

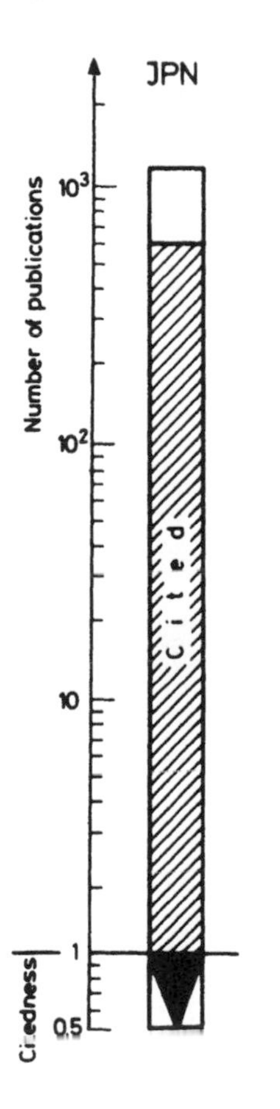

FIGURE 187. JPN-C. Number of Japanese analytical chemistry papers and the cited fraction.

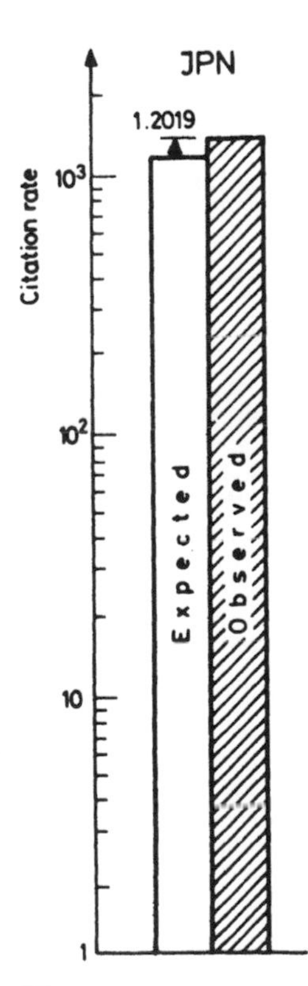

FIGURE 188. JPN-D. Expected and observed citation rates of Japanese analytical chemistry papers.

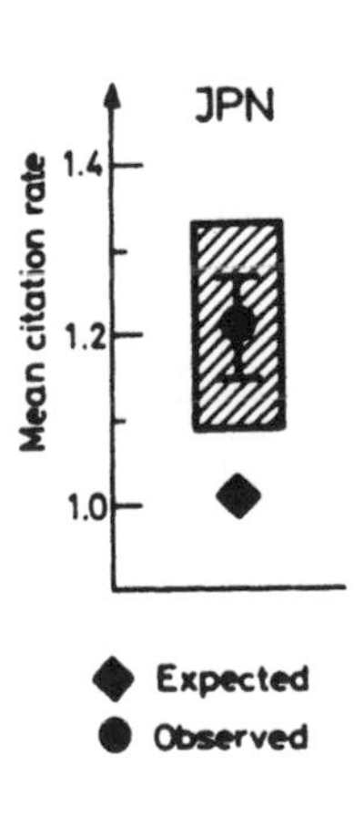

FIGURE 189. JPN-E. Expected and observed values of mean citation rate of Japanese analytical chemistry papers.

Table 90
SCIENTOMETRIC INDICATOR VALUES 1978—1980

Netherlands

Publication counts	
Analytical chemistry	303
Chemistry	1,594
Total science	10,141
Analytical chemistry publications as a proportion of publications:	
In chemistry	0.1901 ± 0.0098
Activity index	1.571
Test statistics	7.030
In total science	0.0299 ± 0.0017
Activity index	1.730
Test statistics	7.457
Citation rates	
Analytical chemistry	531
Chemistry	2,868
Total science	17,399
Citations to analytical chemistry publications as a proportion of citations	
To chemistry	0.1851 ± 0.0073
Attractivity index	1.357
Test statistics	6.722
To total science	0.0305 ± 0.0013
Attractivity index	1.683
Test statistics	9.501
Fraction of cited publications	0.6073 ± 0.0281
Expected citation rate	497.2
Mean impact factor	1.641
Mean citation rate	1.752 ± 0.135
Relative citation rate	1.068
Test statistics	0.826

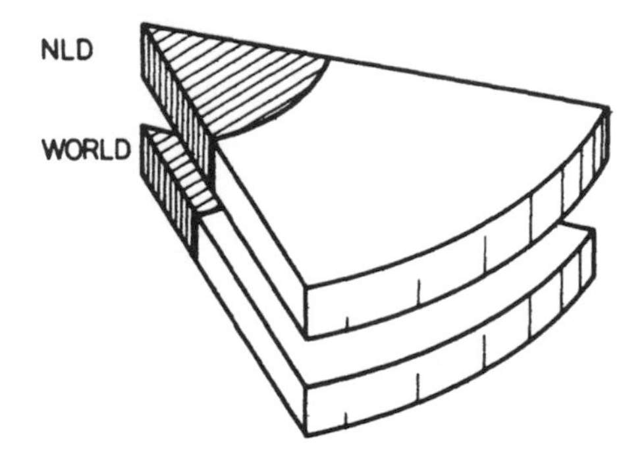

FIGURE 190. NLD-A. Proportion of chemistry and analytical chemistry papers in Dutch science publications.

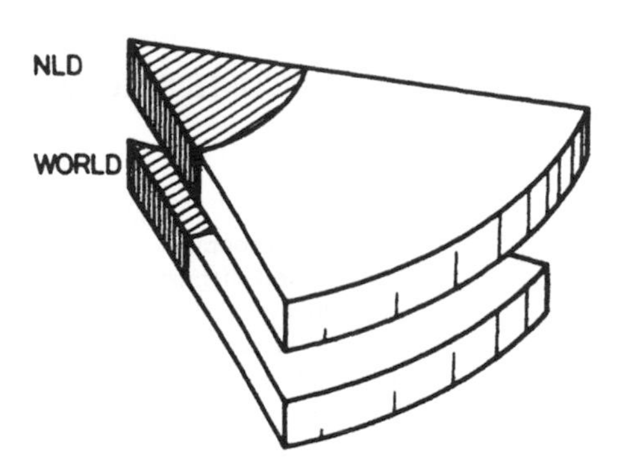

FIGURE 191. NLD-B. Citations to chemistry and analytical chemistry papers as a proportion of citations to Dutch science publications.

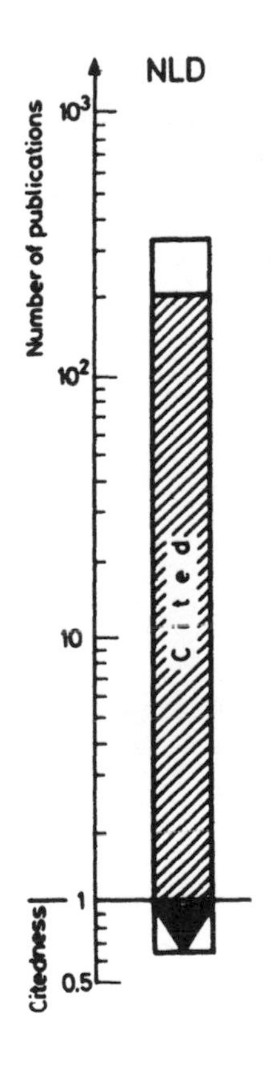

FIGURE 192. NLD-C. Number of Dutch analytical chemistry papers and the cited fraction.

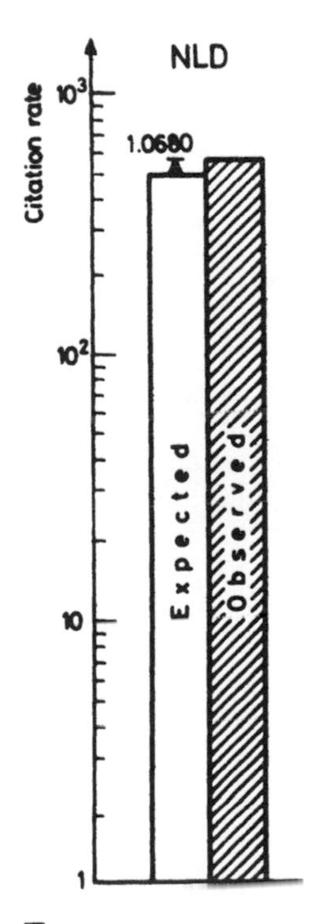

FIGURE 193. NLD-D. Expected and observed citation rates of Dutch analytical chemistry papers.

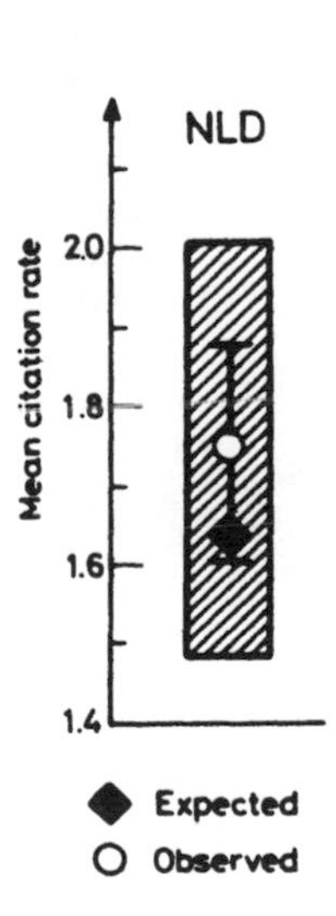

FIGURE 194. NLD-E. Expected and observed values of mean citation rate of Dutch analytical chemistry papers.

Table 91
SCIENTOMETRIC INDICATOR VALUES 1978—1980

New Zealand

Publication counts	
Analytical chemistry	32
Chemistry	246
Total science	3332
Analytical chemistry publications as a proportion of publications	
In chemistry	0.1301 ± 0.0214
Activity index	1.075
Test statistics	0.423
In total science	0.0096 ± 0.0017
Activity index	0.556
Test statistics	−4.539
Citation rates	
Analytical chemistry	93
Chemistry	386
Total science	3021
Citations to analytical chemistry publications as a proportion of citations	
To chemistry	0.2409 ± 0.0218
Attractivity index	1.766
Test statistics	4.803
To total science	0.0308 ± 0.0031
Attractivity index	1.698
Test statistics	4.027
Fraction of cited publications	0.6875 ± 0.0819
Expected citation rate	60.9
Mean impact factor	1.903
Mean citation rate	2.906 ± 0.677
Relative citation rate	1.527
Test statistics	1.482

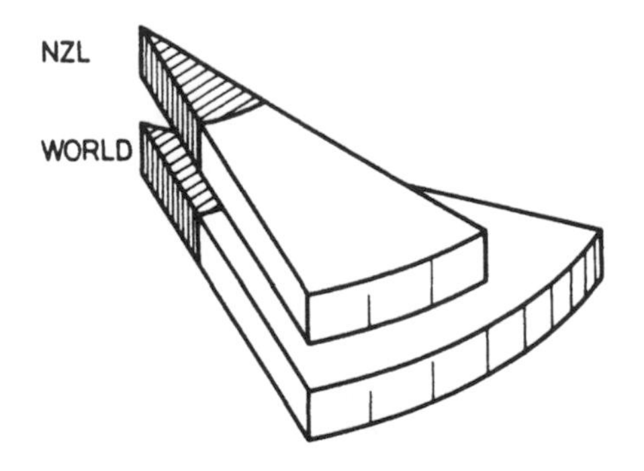

FIGURE 195. NZL-A. Proportion of chemistry and analytical chemistry papers in New Zealand science publications.

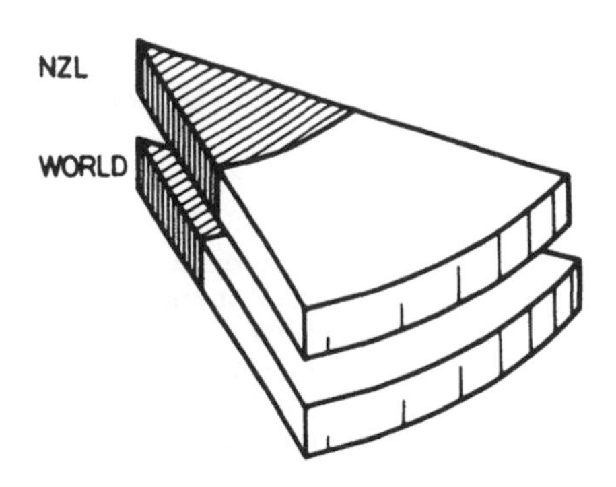

FIGURE 196. NZL-B. Citations to chemistry and analytical chemistry papers as a proportion of citations to New Zealand science publications.

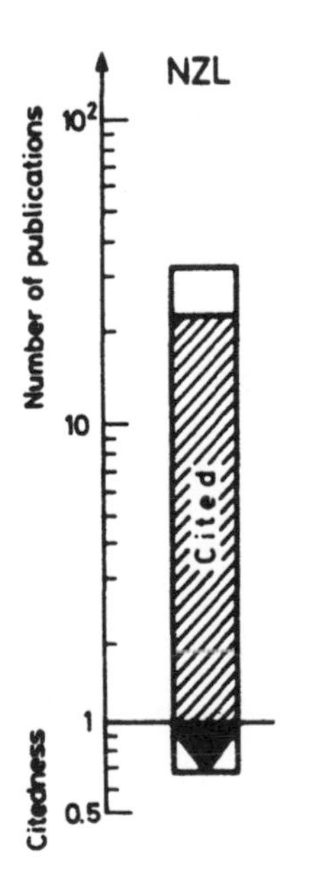

FIGURE 197. NZL-C. Number of New Zealand analytical chemistry papers and the cited fraction.

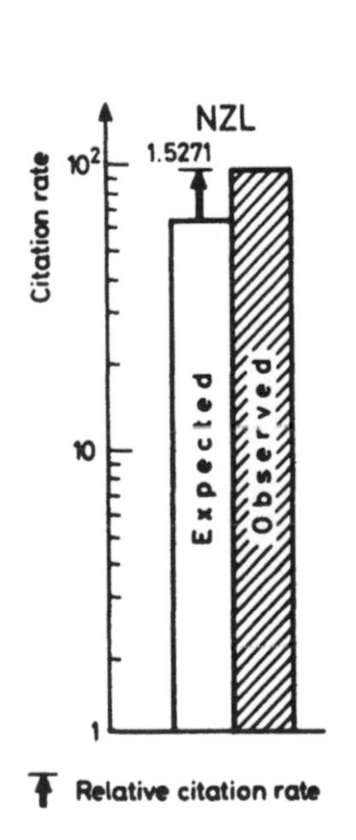

FIGURE 198. NZL-D. Expected and observed citation rates of New Zealand analytical chemistry papers.

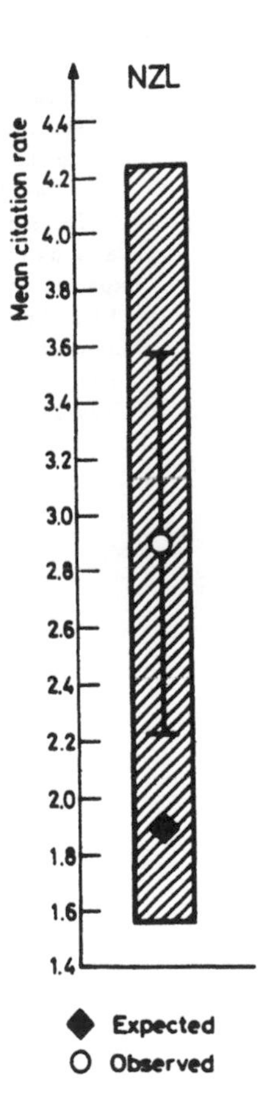

FIGURE 199. NZL-E. Expected and observed values of mean citation rate of New Zealand analytical chemistry papers.

Table 92
SCIENTOMETRIC INDICATOR VALUES 1978—1980

Norway

Publication counts	
Analytical chemistry	74
Chemistry	370
Total science	3612
Analytical chemistry publications as a proportion of publications	
In chemistry	0.2000 ± 0.0208
Activity index	1.653
Test statistics	3.799
In total science	0.0205 ± 0.0024
Activity index	1.186
Test statistics	1.364
Citation rates	
Analytical chemistry	117
Chemistry	489
Total science	4832
Citations to analytical chemistry publications as a proportion of citations	
To chemistry	0.2393 ± 0.0193
Attractivity index	1.754
Test statistics	5.332
To total science	0.0242 ± 0.0022
Attractivity index	1.336
Test statistics	2.752
Fraction of cited publications	0.6622 ± 0.0550
Expected citation rate	122.9
Mean impact factor	1.661
Mean citation rate	1.581 ± 0.209
Relative citation rate	0.952
Test statistics	−0.382

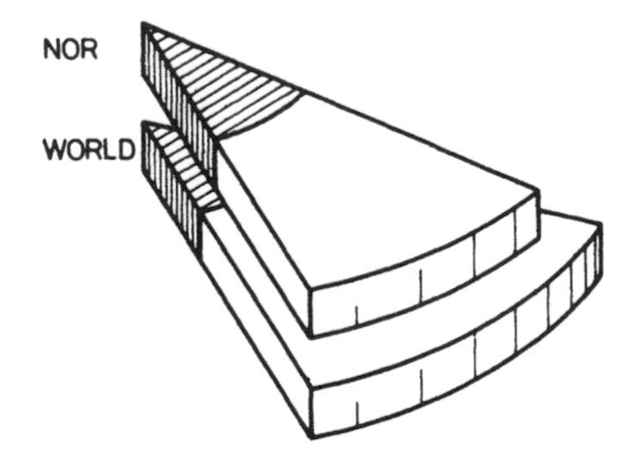

FIGURE 200. NOR-A. Proportion of chemistry and analytical chemistry papers in Norwegian science publications.

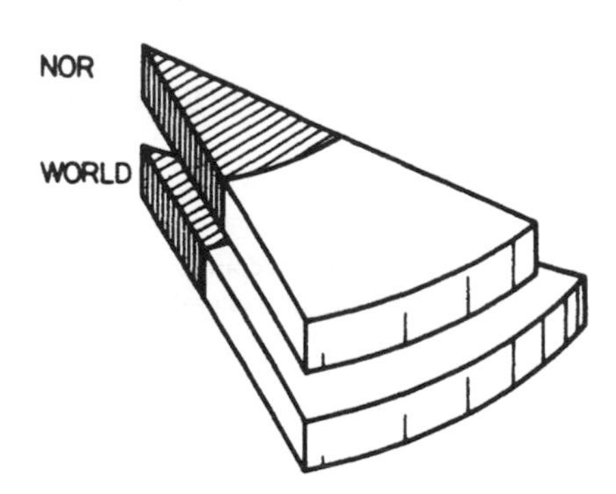

FIGURE 201. NOR-B. Citations to chemistry and analytical chemistry papers as a proportion of citations to Norwegian science publications.

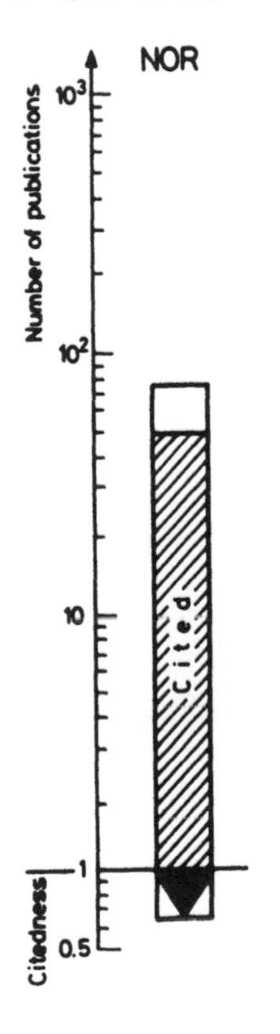

FIGURE 202. NOR-C. Number of Norwegian analytical chemistry papers and the cited fraction.

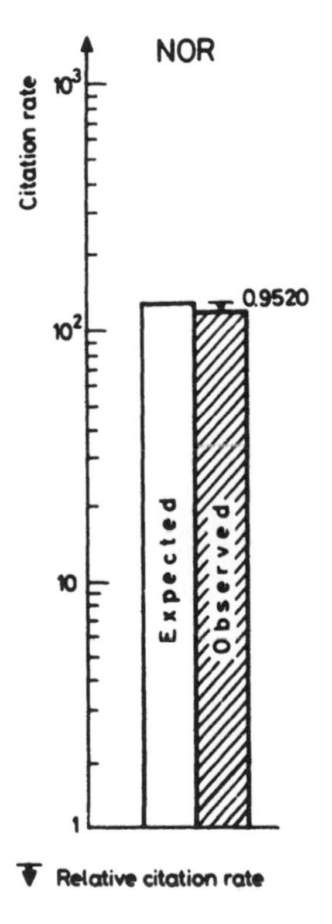

FIGURE 203. NOR-D. Expected and observed citation rates of Norwegian analytical chemistry papers.

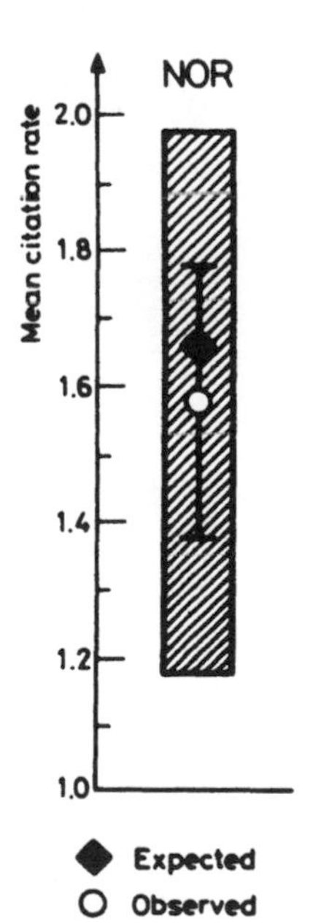

FIGURE 204. NOR-E. Expected and observed values of mean citation rate of Norwegian analytical chemistry papers.

Table 93
SCIENTOMETRIC INDICATOR VALUES 1978—1980

Pakistan

Publication counts	
Analytical chemistry	29
Chemistry	47
Total science	243
Analytical chemistry publications as a proportion of publications	
In chemistry	0.6170 ± 0.0709
Activity index	5.099
Test statistics	6.995
In total science	0.1193 ± 0.0208
Activity index	6.909
Test statistics	4.908
Citation rates	
Analytical chemistry	20
Chemistry	29
Total science	128
Citations to analytical chemistry publications as a proportion of citations	
To chemistry	0.6897 ± 0.0859
Attractivity index	5.056
Test statistics	6.440
To total science	0.1563 ± 0.0321
Attractivity index	8.619
Test statistics	4.304
Fraction of cited publications	0.5172 ± 0.0928
Expected citation rate	29.6
Mean impact factor	1.021
Mean citation rate	0.690 ± 0.154
Relative citation rate	0.676
Test statistics	−2.147

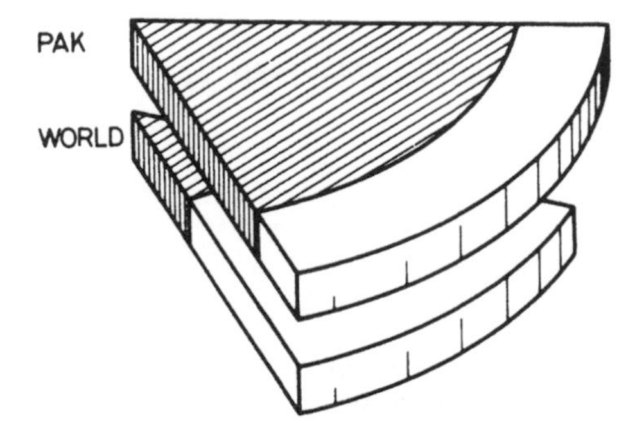

FIGURE 205. PAK-A. Proportion of chemistry and analytical chemistry papers in Pakistani science publications.

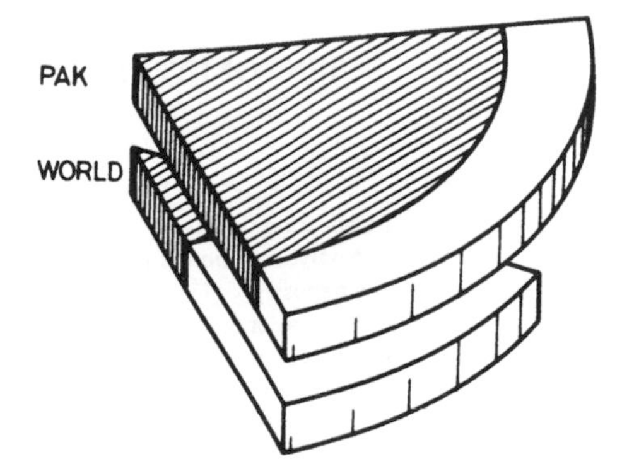

FIGURE 206. PAK-B. Citations to chemistry and analytical chemistry papers as a proportion of citations to Pakistani science publications.

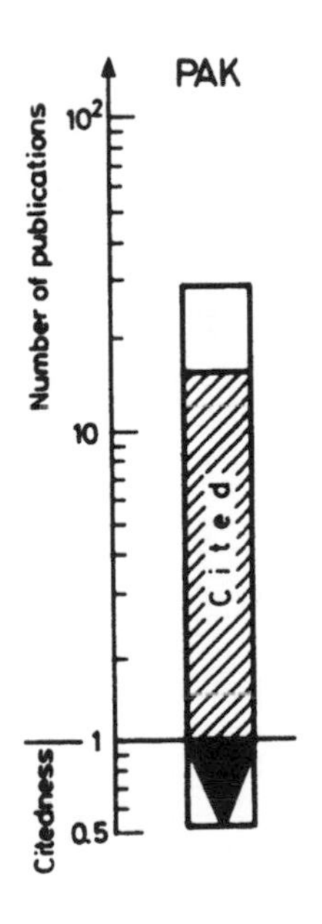

FIGURE 207. PAK-C. Number of Pakistani analytical chemistry papers and the cited fraction.

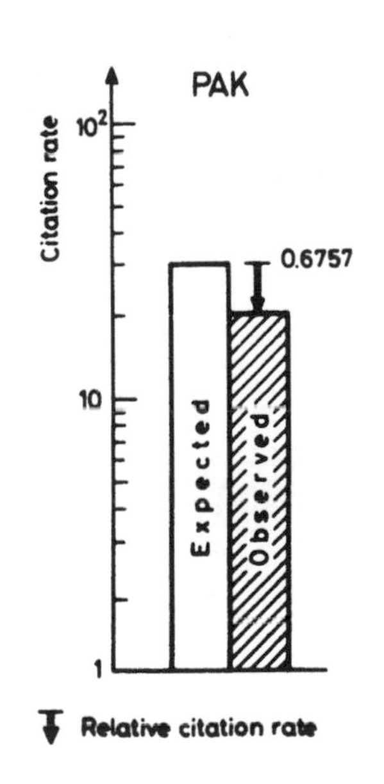

FIGURE 208. PAK-D. Expected and observed citation rates of Pakistani analytical chemistry papers.

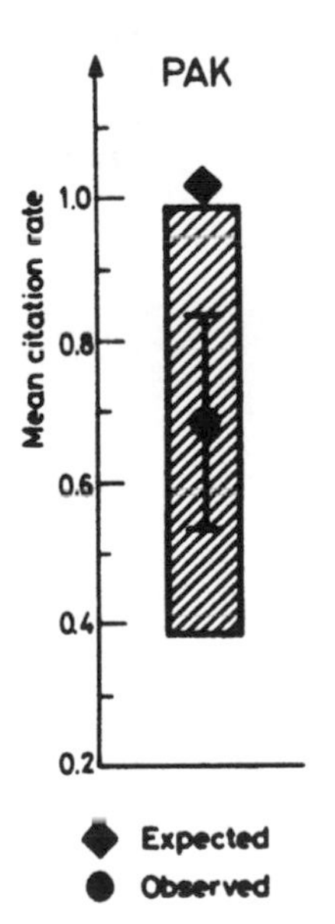

FIGURE 209. PAK-E. Expected and observed values of mean citation rate of Pakistani analytical chemistry papers.

Table 94
SCIENTOMETRIC INDICATOR VALUES 1978—1980

Poland

Publication counts	
Analytical chemistry	503
Chemistry	2561
Total science	8206
Analytical chemistry publications as a proportion of publications	
In chemistry	0.1964 ± 0.0079
Activity index	1.623
Test statistics	9.606
In total science	0.0613 ± 0.0026
Activity index	3.549
Test statistics	16.626
Citation rates	
Analytical chemistry	317
Chemistry	1465
Total science	4747
Citations to analytical chemistry publications as a proportion of citations	
To chemistry	0.2164 ± 0.0108
Attractivity index	1.586
Test statistics	7.435
To total science	0.0668 ± 0.0036
Attractivity index	3.684
Test statistics	13.427
Fraction of cited publications	0.3877 ± 0.0217
Expected citation rate	401.9
Mean impact factor	0.799
Mean citation rate	0.630 ± 0.045
Relative citation rate	0.789
Test statistics	−3.747

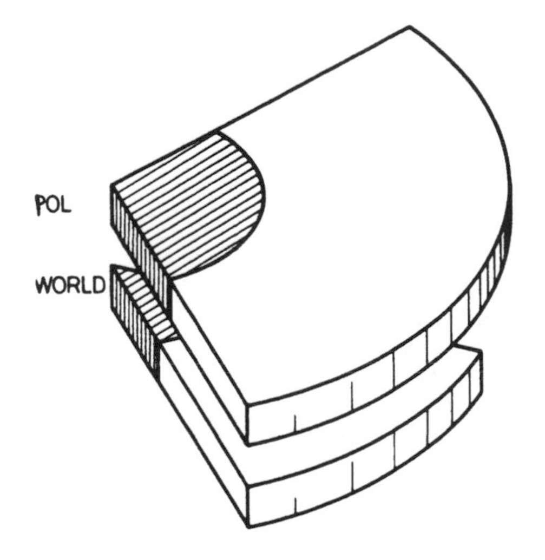

FIGURE 210. POL-A. Proportion of chemistry and analytical chemistry papers in Polish science publications.

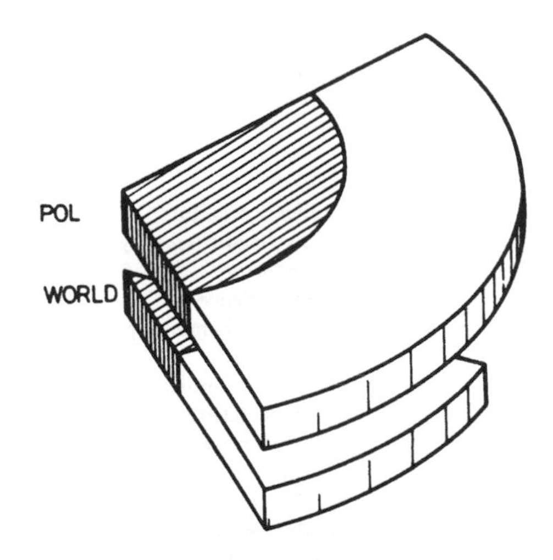

FIGURE 211. POL-B. Citations to chemistry and analytical chemistry papers as a proportion of citations to Polish science publications.

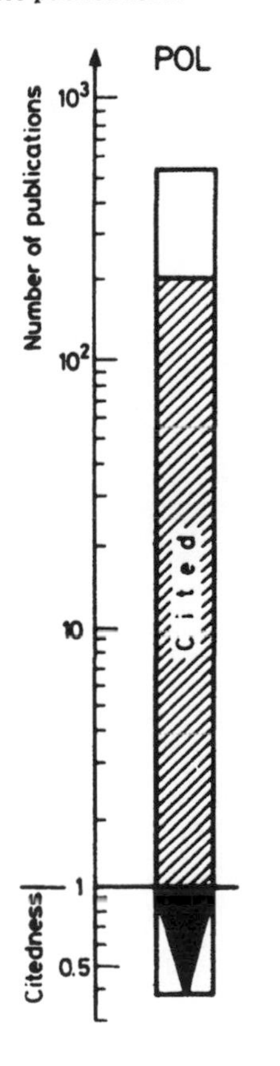

FIGURE 212. POL-C. Number of Polish analytical chemistry papers and the cited fraction.

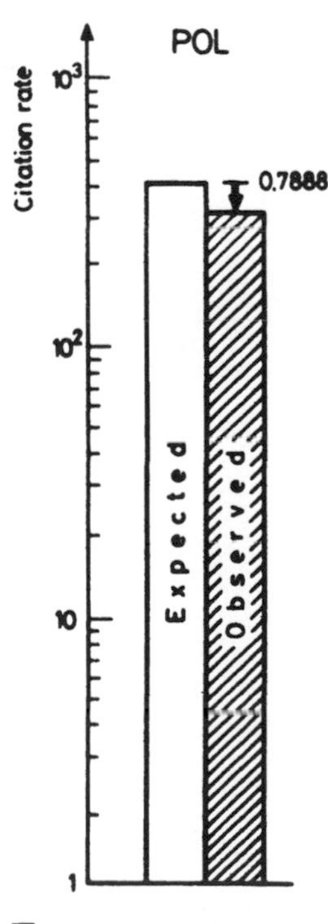

FIGURE 213. POL-D. Expected and observed citation rates of Polish analytical chemistry papers.

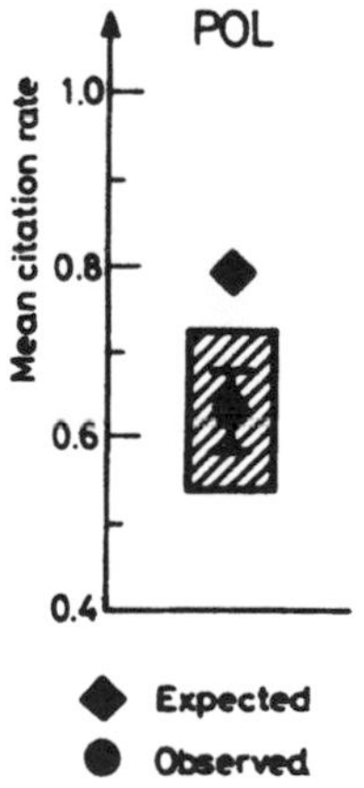

FIGURE 214. POL-E. Expected and observed values of mean citation rate of Polish analytical chemistry papers.

Table 95
SCIENTOMETRIC INDICATOR VALUES 1978—1980

Romania

Publication counts	
Analytical chemistry	42
Chemistry	428
Total science	1452
Analytical chemistry publications as a proportion of publications	
In chemistry	0.0981 ± 0.0144
Activity index	0.811
Test statistics	−1.590
In total science	0.0289 ± 0.0044
Activity index	1.675
Test statistics	2.649
Citation rates	
Analytical chemistry	14
Chemistry	142
Total science	512
Citations to analytical chemistry publications as a proportion of citations	
To chemistry	0.0986 ± 0.0250
Attractivity index	0.723
Test statistics	−1.511
To total science	0.0273 ± 0.0072
Attractivity index	1.508
Test statistics	1.279
Fraction of cited publications	0.2619 ± 0.0678
Expected citation rate	38.5
Mean impact factor	0.917
Mean citation rate	0.333 ± 0.098
Relative citation rate	0.364
Test statistics	−5.971

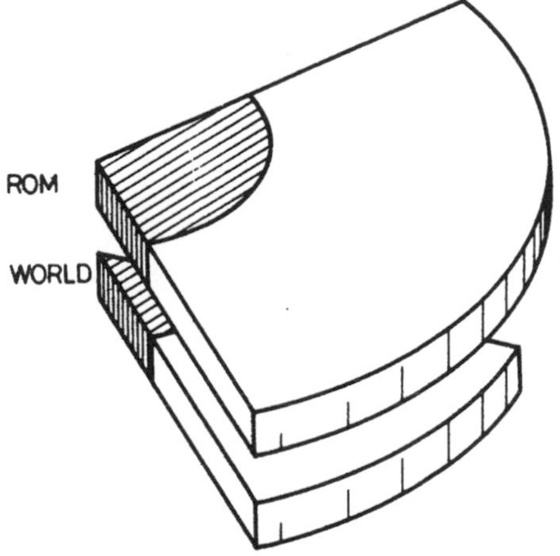

FIGURE 215. ROM-A. Proportion of chemistry and analytical chemistry papers in Romanian science publications.

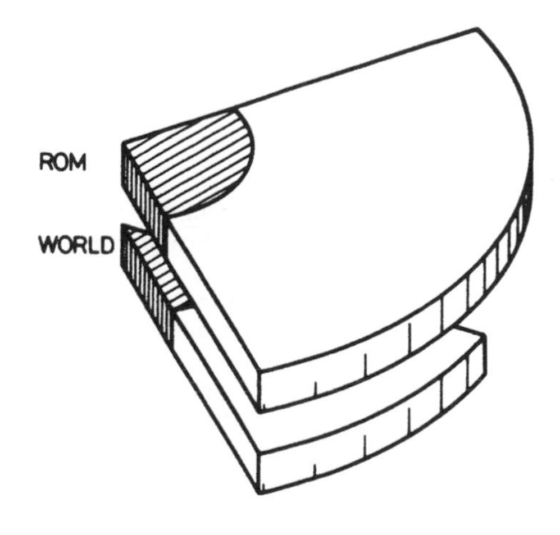

FIGURE 216. ROM-B. Citations to chemistry and analytical chemistry papers as a proportion of citations to Romanian science publications.

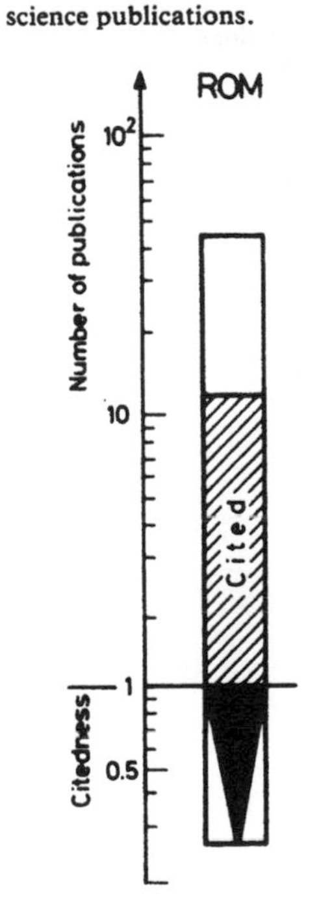

FIGURE 217. ROM-C. Number of Romanian analytical chemistry papers and the cited fraction.

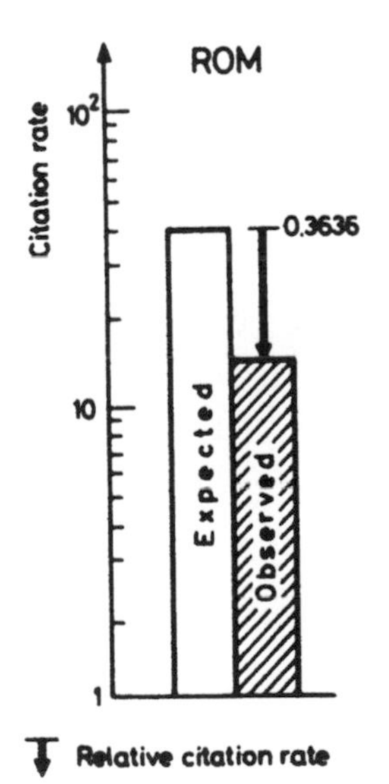

FIGURE 218. ROM-D. Expected and observed citation rates of Romanian analytical chemistry papers.

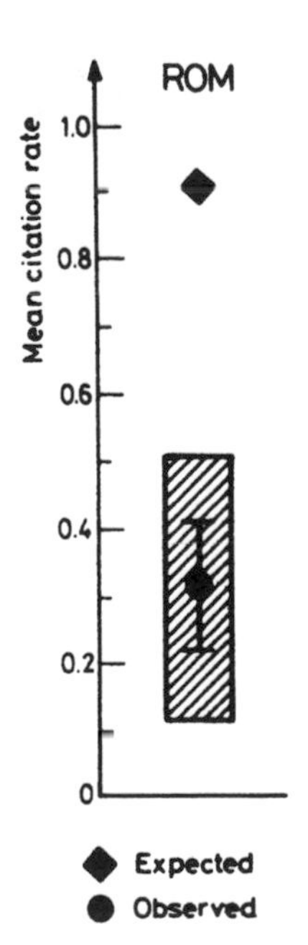

FIGURE 219. ROM-E. Expected and observed values of mean citation rate of Romanian analytical chemistry papers.

Table 96
SCIENTOMETRIC INDICATOR VALUES 1978—1980

Scotland

Publication counts	
Analytical chemistry	91
Chemistry	771
Total science	7,334
Analytical chemistry publications as a proportion of publications	
In chemistry	0.1180 ± 0.0116
Activity index	0.975
Test statistics	−0.256
In total science	0.0124 ± 0.0013
Activity index	0.718
Test statistics	−3.763
Citation rates	
Analytical chemistry	158
Chemistry	1,227
Total science	10,515
Citations to analytical chemistry publications as a proportion of citations	
To chemistry	0.1288 ± 0.0096
Attractivity index	0.944
Test statistics	−0.798
To total science	0.0150 ± 0.0012
Attractivity index	0.829
Test statistics	−2.615
Fraction of cited publications	0.7253 ± 0.0468
Expected citation rate	140.6
Mean impact factor	1.545
Mean citation rate	1.736 ± 0.183
Relative citation rate	1.124
Test statistics	1.044

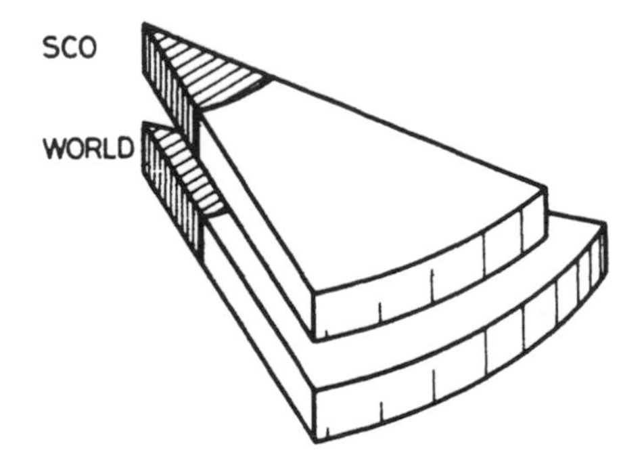

FIGURE 220. SCO-A. Proportion of chemistry and analytical chemistry papers in Scottish science publications.

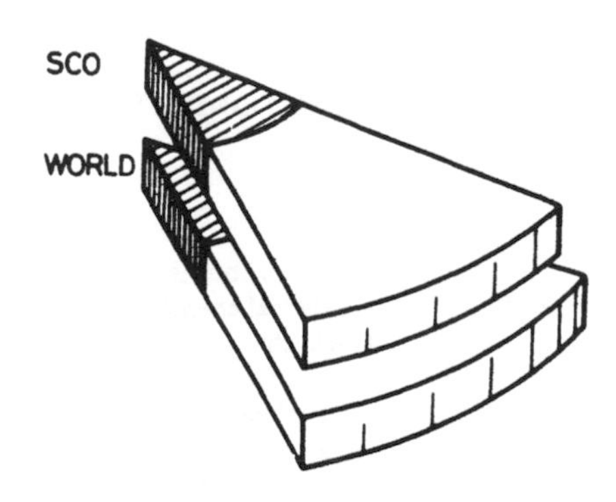

FIGURE 221. SCO-B. Citations to chemistry and analytical chemistry papers as a proportion of citations to Scottish science publications.

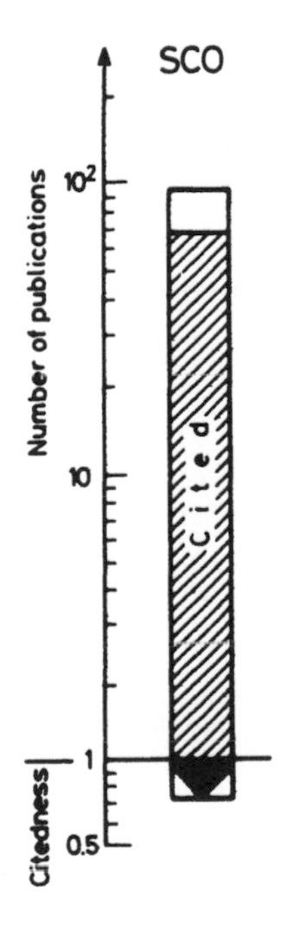

FIGURE 222. SCO-C. Number of Scottish analytical chemistry papers and the cited fraction.

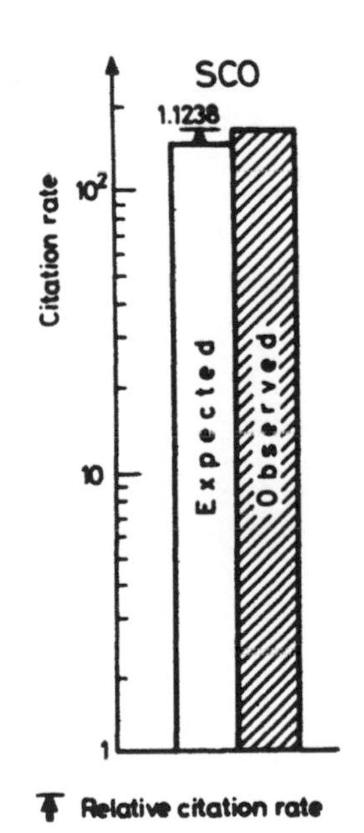

FIGURE 223. SCO-D. Expected and observed citation rates of Scottish analytical chemistry papers.

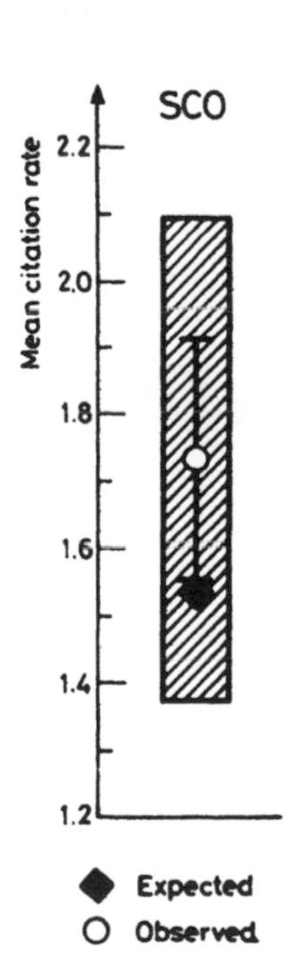

FIGURE 224. SCO-E. Expected and observed values of mean citation rate of Scottish analytical chemistry papers.

Table 97
SCIENTOMETRIC INDICATOR VALUES 1978—1980

South Africa

Publication counts	
Analytical chemistry	59
Chemistry	405
Total science	3956
Analytical chemistry publications as a proportion of publications	
In chemistry	0.1457 ± 0.0175
Activity index	1.204
Test statistics	1.408
In total science	0.0149 ± 0.0019
Activity index	0.863
Test statistics	−1.224
Citation rates	
Analytical chemistry	75
Chemistry	450
Total science	3078
Citations to analytical chemistry publications as a proportion of citations	
To chemistry	0.1667 ± 0.0176
Attractivity index	1.222
Test statistics	1.723
To total science	0.0244 ± 0.0028
Attractivity index	1.344
Test statistics	2.245
Fraction of cited publications	0.7119 ± 0.0590
Expected citation rate	88.7
Mean impact factor	1.503
Mean citation rate	1.271 ± 0.150
Relative citation rate	0.846
Test statistics	−1.546

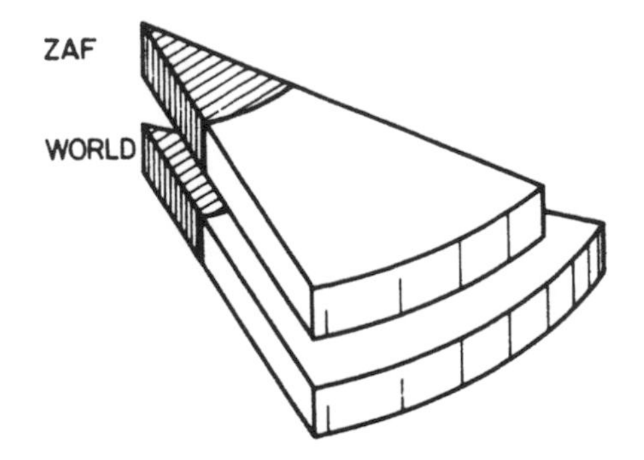

FIGURE 225. ZAF-A. Proportion of chemistry and analytical chemistry papers in South African science publications.

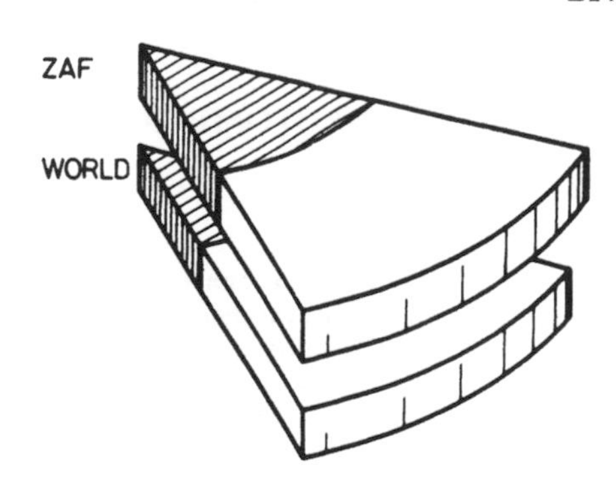

FIGURE 226. ZAF-B. Citations to chemistry and analytical chemistry papers as a proportion of citations to South African science publications.

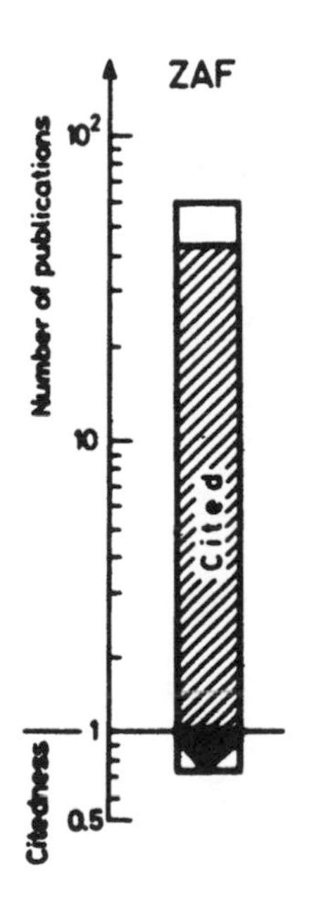

FIGURE 227. ZAF-C. Number of South African analytical chemistry papers and the cited fraction.

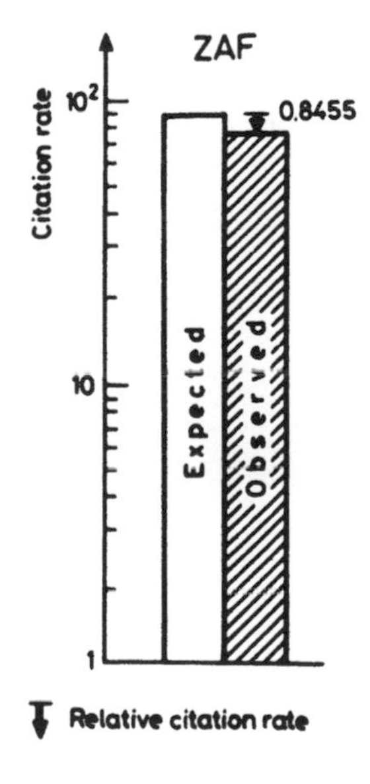

FIGURE 228. ZAF-D. Expected and observed citation rates of South African analytical chemistry papers.

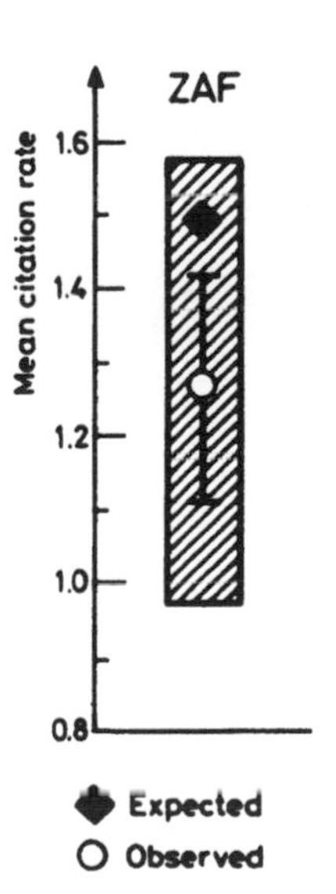

FIGURE 229. ZAF-E. Expected and observed values of mean citation rate of South African analytical chemistry papers.

Table 98
SCIENTOMETRIC INDICATOR VALUES 1978—1980

Spain

Publication counts	
Analytical chemistry	94
Chemistry	1081
Total science	3495
Analytical chemistry publications as a proportion of publications	
In chemistry	0.0870 ± 0.0086
Activity index	0.719
Test statistics	−3.972
In total science	0.0269 ± 0.0027
Activity index	1.557
Test statistics	3.517
Citation rates	
Analytical chemistry	100
Chemistry	742
Total science	2565
Citations to analytical chemistry publications as a proportion of citations	
To chemistry	0.1348 ± 0.0125
Attractivity index	0.988
Test statistics	−0.130
To total science	0.0390 ± 0.0038
Attractivity index	2.151
Test statistics	5.457
Fraction of cited publications	0.5426 ± 0.0514
Expected citation rate	123.5
Mean impact factor	1.314
Mean citation rate	1.064 ± 0.143
Relative citation rate	0.810
Test statistics	−1.752

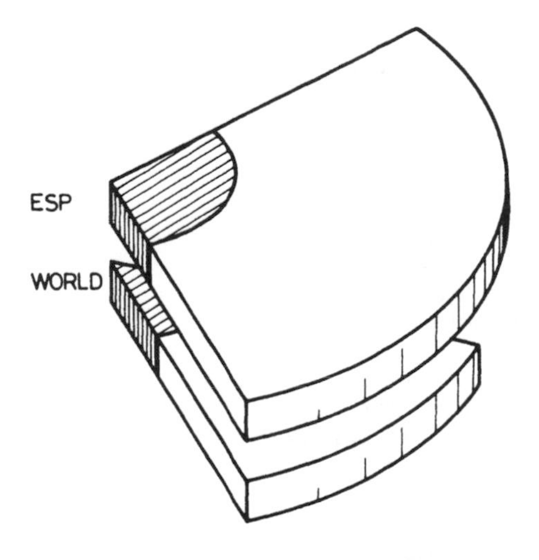

FIGURE 230. ESP-A. Proportion of chemistry and analytical chemistry papers in Spanish science publications.

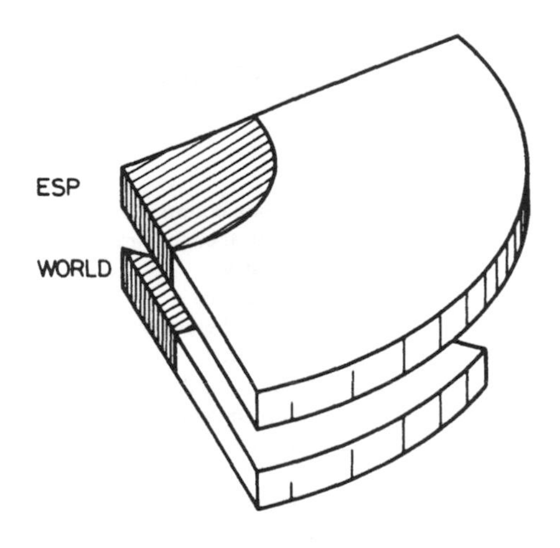

FIGURE 231. ESP-B. Citations to chemistry and analytical chemistry papers as a proportion of citations to Spanish science publications.

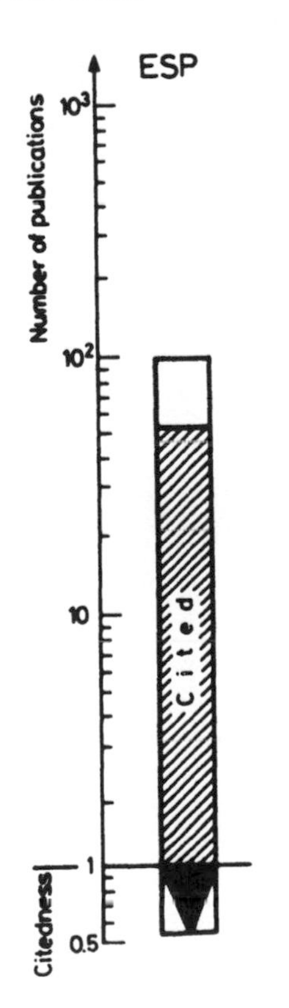

FIGURE 232. ESP-C. Number of Spanish analytical chemistry papers and the cited fraction.

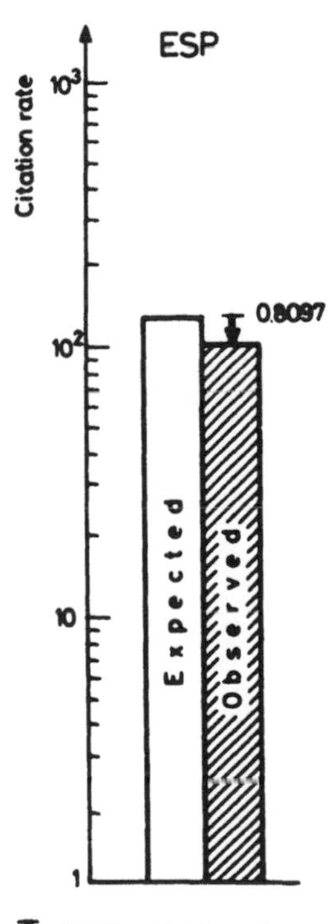

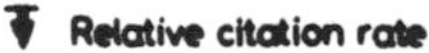

FIGURE 233. ESP-D. Expected and observed citation rates of Spanish analytical chemistry papers.

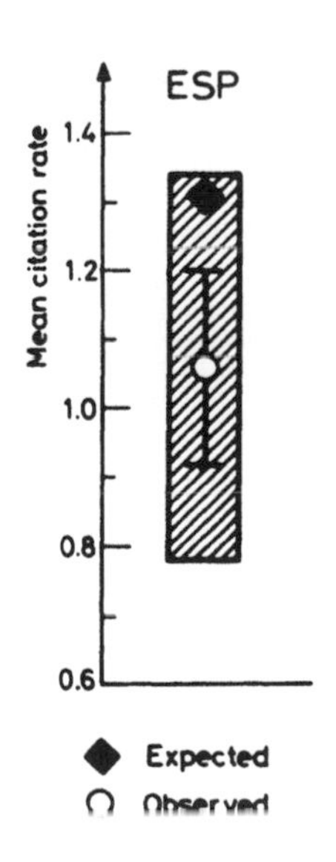

FIGURE 234. ESP-E. Expected and observed values of mean citation rate of Spanish analytical chemistry papers.

Table 99
SCIENTOMETRIC INDICATOR VALUES 1978—1980

Sweden

Publication counts	
Analytical chemistry	207
Chemistry	944
Total science	10,655
Analytical chemistry publications as a proportion of publications	
In chemistry	0.2193 ± 0.0135
Activity index	1.812
Test statistics	7.298
In total science	0.0194 ± 0.0013
Activity index	1.125
Test statistics	1.612
Citation rates	
Analytical chemistry	542
Chemistry	1,605
Total science	19,179
Citations to analytical chemistry publications as a proportion of citations	
To chemistry	0.3377 ± 0.0118
Attractivity index	2.476
Test statistics	17.052
To total science	0.0283 ± 0.0012
Attractivity index	1.559
Test statistics	8.467
Fraction of cited publications	0.7440 ± 0.0303
Expected citation rate	387.1
Mean impact factor	1.870
Mean citation rate	2.618 ± 0.204
Relative citation rate	1.400
Test statistics	3.670

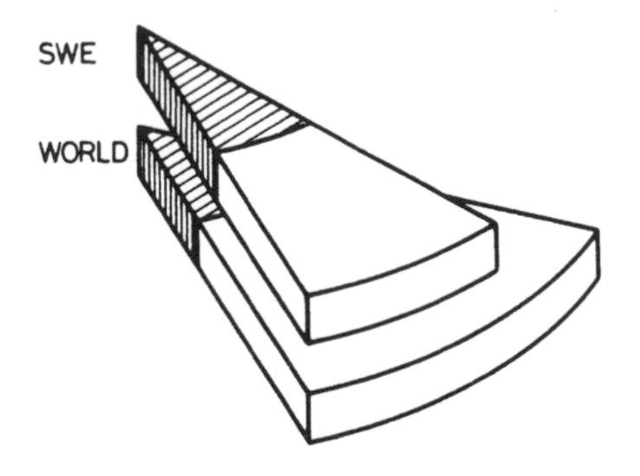

FIGURE 235. SWE-A. Proportion of chemistry and analytical chemistry papers in Swedish science publications.

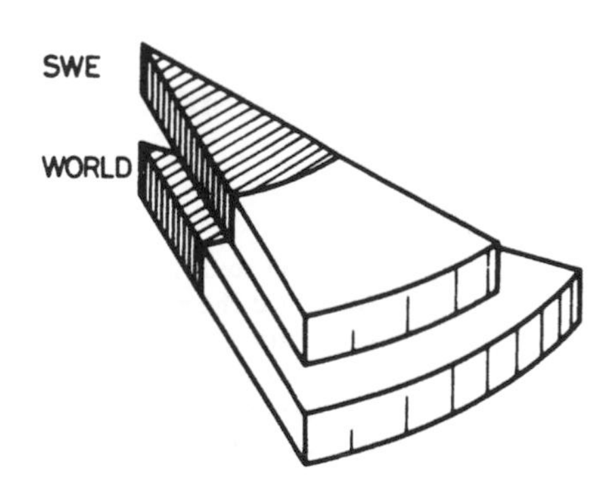

FIGURE 236. SWE-B. Citations to chemistry and analytical chemistry papers as a proportion of citations to Swedish science publications.

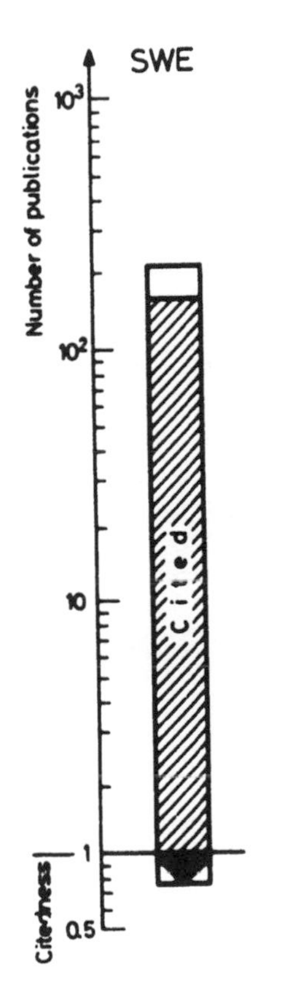

FIGURE 237. SWE-C. Number of Swedish analytical chemistry papers and the cited fraction.

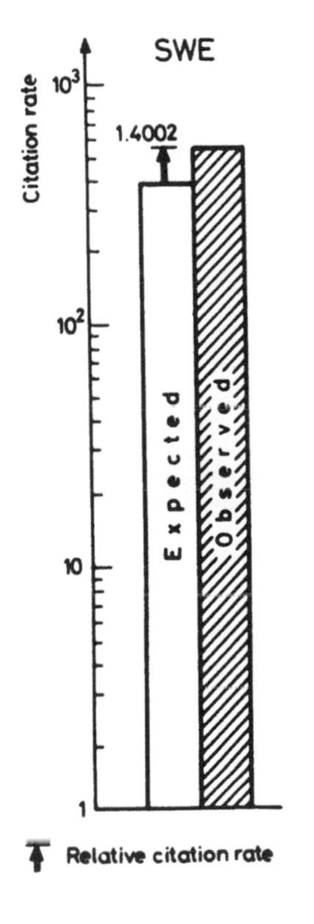

FIGURE 238. SWE-D. Expected and observed citation rates of Swedish analytical chemistry papers.

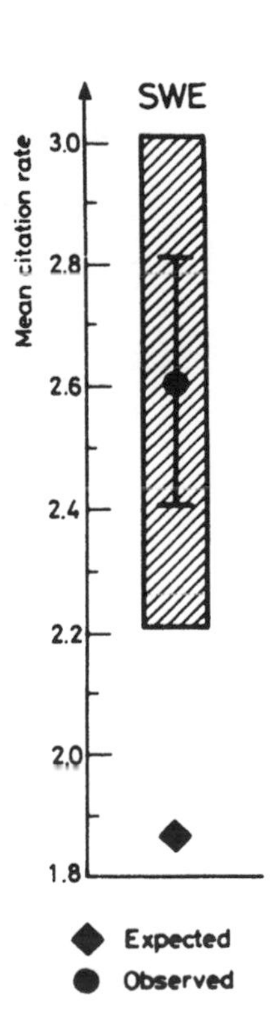

FIGURE 239. SWE-E. Expected and observed values of mean citation rate of Swedish analytical chemistry papers.

Table 100
SCIENTOMETRIC INDICATOR VALUES 1978—1980

Switzerland

Publication counts	
Analytical chemistry	131
Chemistry	1,227
Total science	9,150
Analytical chemistry publications as a proportion of publications	
In chemistry	0.1068 ± 0.0088
Activity index	0.882
Test statistics	−1.615
In total science	0.0143 ± 0.0012
Activity index	0.829
Test statistics	−2.380
Citation rates	
Analytical chemistry	258
Chemistry	2,547
Total science	17,962
Citations to analytical chemistry publications as a proportion of citations	
To chemistry	0.1013 ± 0.0060
Attractivity index	0.743
Test statistics	−5.871
To total science	0.0144 ± 0.0009
Attractivity index	0.792
Test statistics	−4.241
Fraction of cited publications	0.6336 ± 0.0421
Expected citation rate	217.9
Mean impact factor	1.663
Mean citation rate	1.969 ± 0.226
Relative citation rate	1.184
Test statistics	1.354

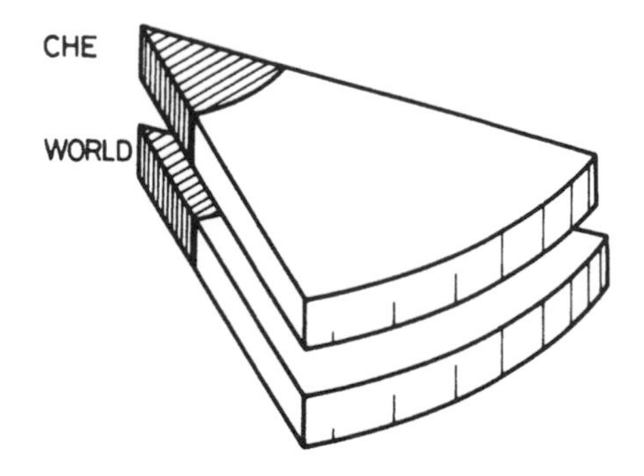

FIGURE 240. CHE-A. Proportion of chemistry and analytical chemistry papers in Swiss science publications.

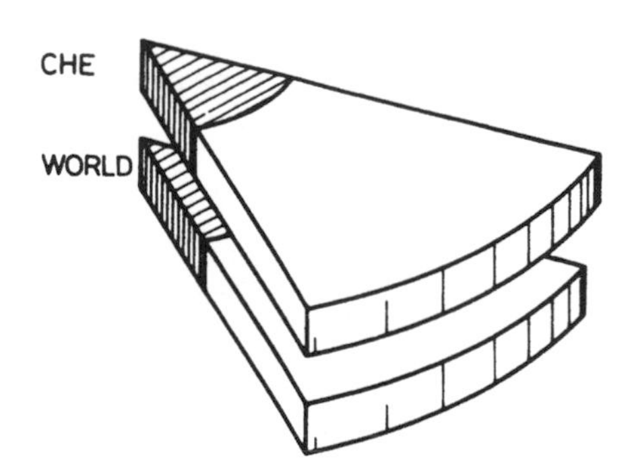

FIGURE 241. CHE-B. Citations to chemistry and analytical chemistry papers as a proportion of citations to Swiss science publications.

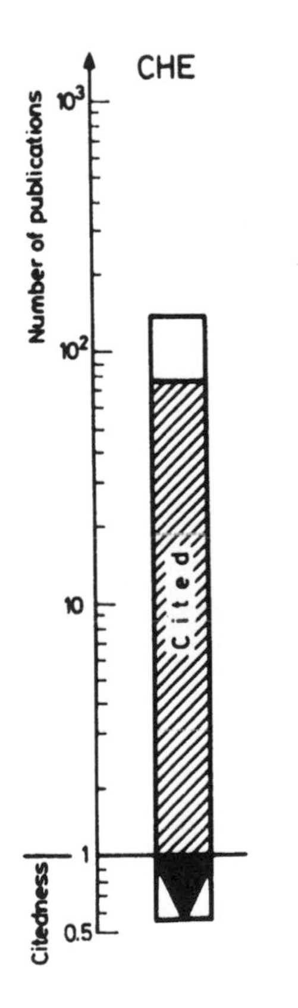

FIGURE 242. CHE-C. Number of Swiss analytical chemistry papers and the cited fraction.

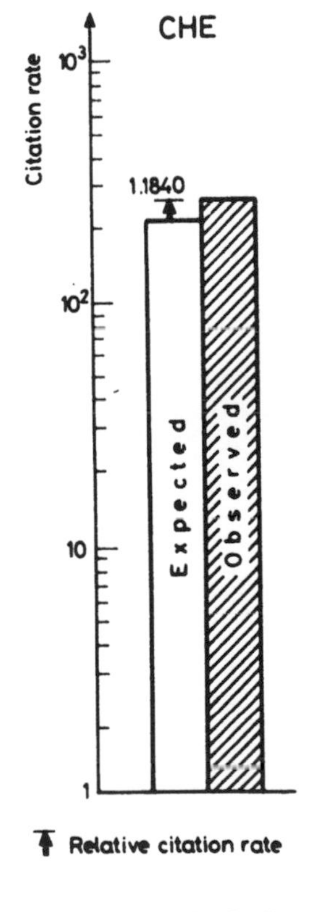

FIGURE 243. CHE-D. Expected and observed citation rates of Swiss analytical chemistry papers.

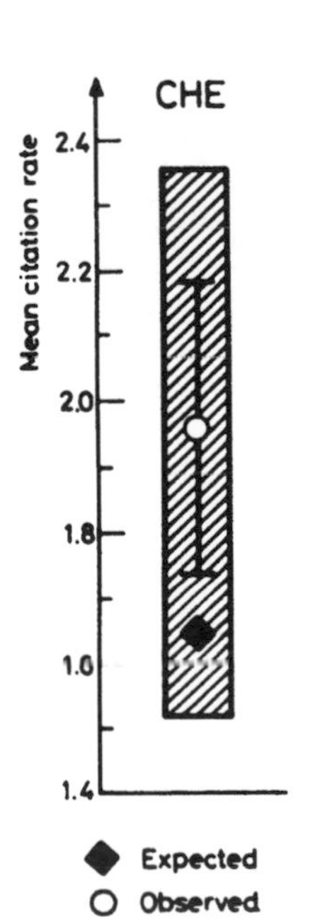

FIGURE 244. CHE-E. Expected and observed values of mean citation rate of Swiss analytical chemistry papers.

Table 101
SCIENTOMETRIC INDICATOR VALUES 1978—1980

U.S.S.R.

Publication counts	
Analytical chemistry	984
Chemistry	16,890
Total science	56,835
Analytical chemistry publications as a proportion of publications	
In chemistry	0.0583 ± 0.0018
Activity index	0.481
Test statistics	−34.810
In total science	0.0173 ± 0.0005
Activity index	1.002
Test statistics	0.075
Citation rates	
Analytical chemistry	252
Chemistry	4,893
Total science	17,777
Citations to analytical chemistry publications as a proportion of citations	
To chemistry	0.0515 ± 0.0032
Attractivity index	0.378
Test statistics	−26.868
To total science	0.0142 ± 0.0009
Attractivity index	0.782
Test statistics	−4.459
Fraction of cited publications	0.1291 ± 0.0107
Expected citation rate	274.5
Mean impact factor	0.279
Mean citation rate	0.256 ± 0.028
Relative citation rate	0.918
Test statistics	−0.806

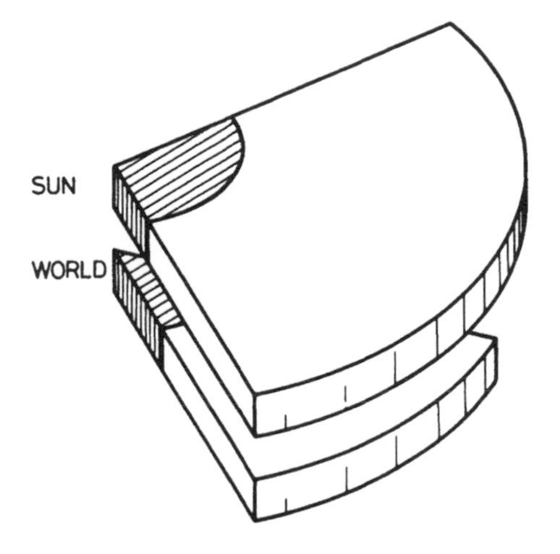

FIGURE 245. SUN-A. Proportion of chemistry and analytical chemistry papers in Soviet science publications.

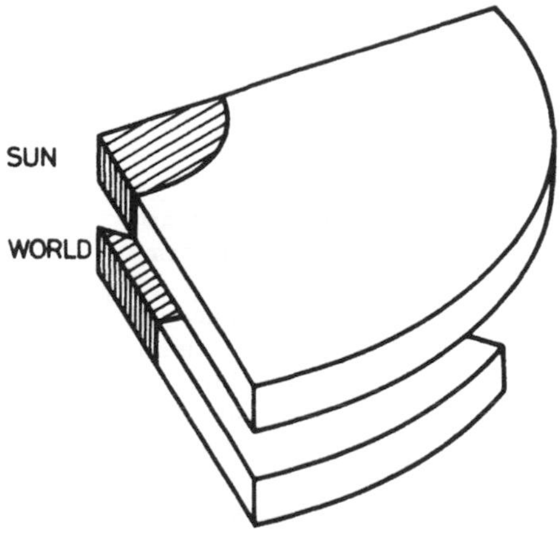

FIGURE 246. SUN-B. Citations to chemistry and analytical chemistry papers as a proportion of citations to Soviet science publications.

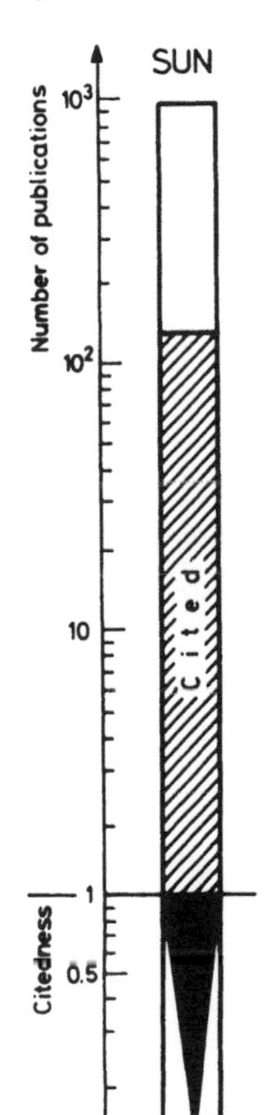

FIGURE 247. SUN-C. Number of Soviet analytical chemistry papers and the cited fraction.

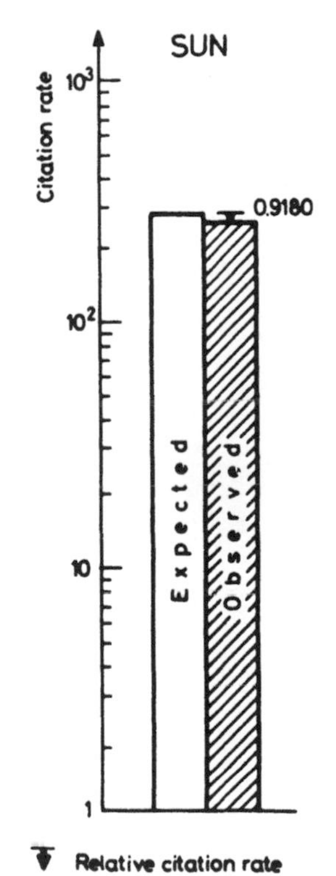

FIGURE 248. SUN-D. Expected and observed citation rates of Soviet analytical chemistry papers.

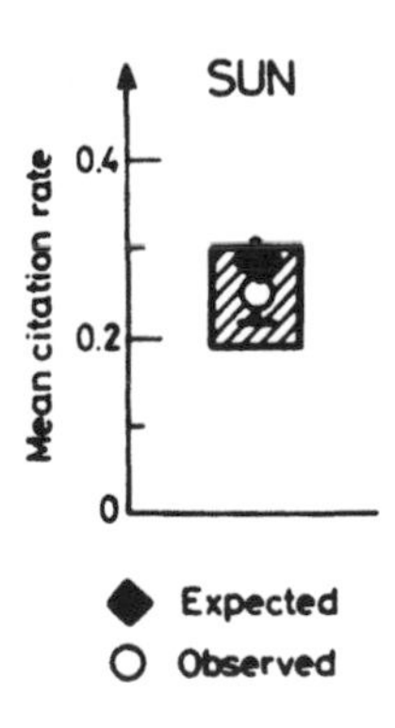

FIGURE 249. SUN-E. Expected and observed values of mean citation rate of Soviet analytical chemistry papers.

Table 102
SCIENTOMETRIC INDICATOR VALUES 1978—1980

U.S.

Publication counts	
Analytical chemistry	3,205
Chemistry	20,517
Total science	255,071
Analytical chemistry publications as a proportion of publications	
In chemistry	0.1562 ± 0.0025
Activity index	1.291
Test statistics	13.893
In total science	0.0126 ± 0.0002
Activity index	0.727
Test statistics	−21.343
Citation rates	
Analytical chemistry	7,439
Chemistry	46,591
Total science	494,648
Citations to analytical chemistry publications as a proportion of citations	
To chemistry	0.1597 ± 0.0017
Attractivity index	1.171
Test statistics	13.712
To total science	0.0150 ± 0.0002
Attractivity index	0.830
Test statistics	−17.855
Fraction of cited publications	0.7173 ± 0.0080
Expected citation rate	6,785.0
Mean impact factor	2.117
Mean citation rate	2.321 ± 0.047
Relative citation rate	1.096
Test statistics	4.344

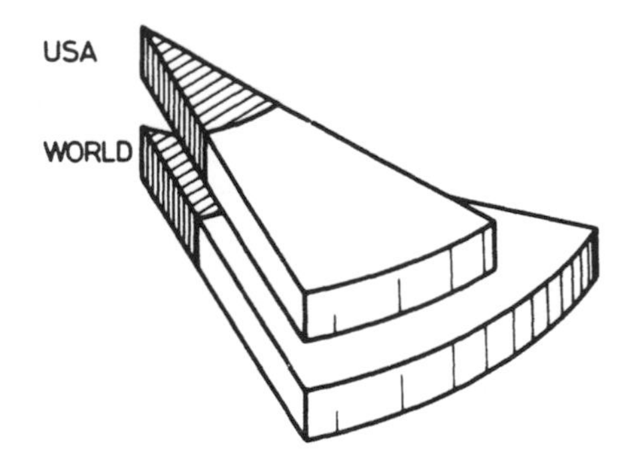

FIGURE 250. USA-A. Proportion of chemistry and analytical chemistry papers in U.S. science publications.

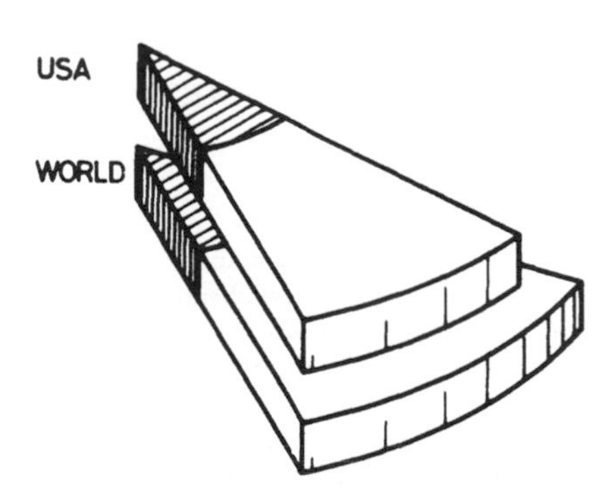

FIGURE 251. USA-B. Citations to chemistry and analytical chemistry papers as a proportion of citations to U.S. science publications.

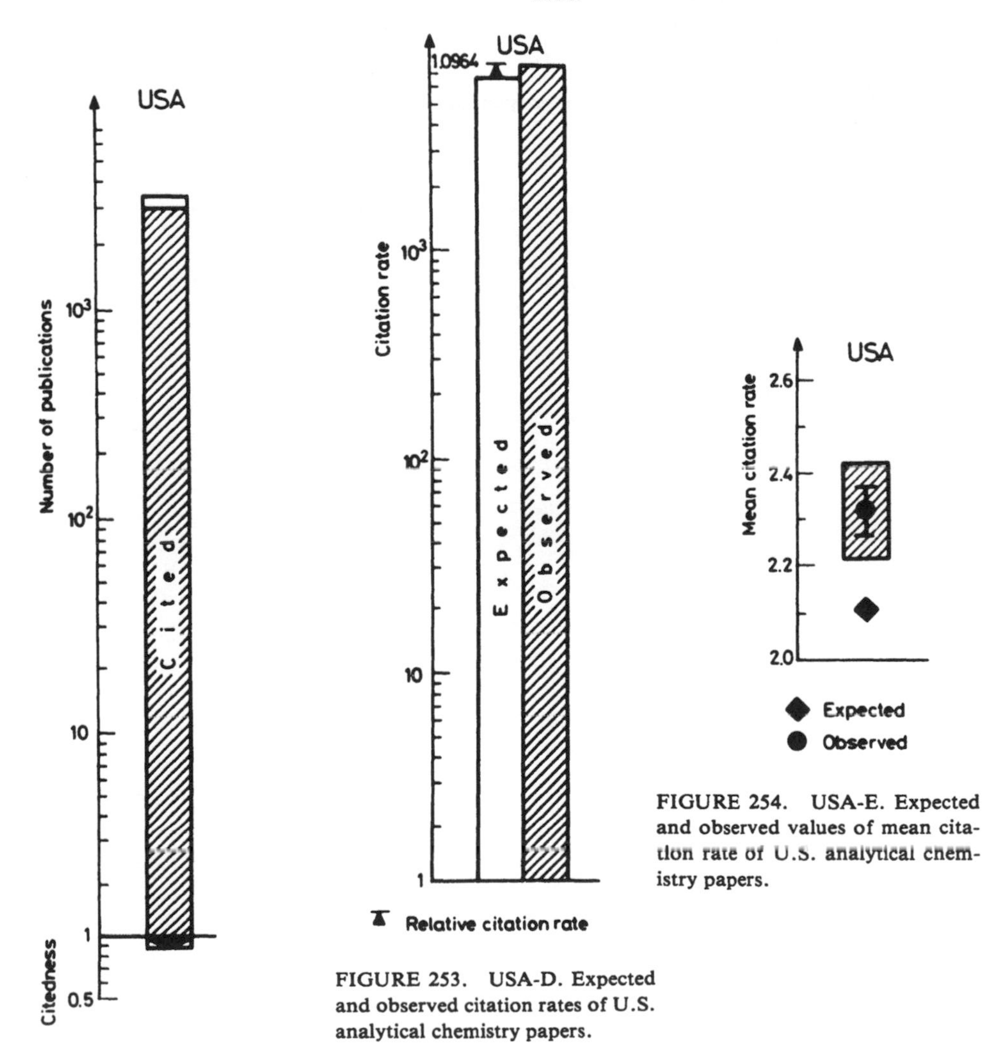

FIGURE 252. USA-C. Number of U.S. analytical chemistry papers and the cited fraction.

FIGURE 253. USA-D. Expected and observed citation rates of U.S. analytical chemistry papers.

FIGURE 254. USA-E. Expected and observed values of mean citation rate of U.S. analytical chemistry papers.

Table 103
SCIENTOMETRIC INDICATOR VALUES 1978—1980

Wales

Publication counts	
Analytical chemistry	49
Chemistry	289
Total science	2439
Analytical chemistry publications as a proportion of publications	
In chemistry	0.1696 ± 0.0221
Activity index	1.401
Test statistics	2.200
In total science	0.0201 ± 0.0028
Activity index	1.163
Test statistics	0.992
Citation rates	
Analytical chemistry	103
Chemistry	485
Total science	2827
Citations to analytical chemistry publications as a proportion of citations	
To chemistry	0.2124 ± 0.0186
Attractivity	1.557
Test statistics	4.091
To total science	0.0364 ± 0.0035
Attractivity index	2.010
Test statistics	5.195
Fraction of cited publications	0.7551 ± 0.0614
Expected citation rate	83.4
Mean impact factor	1.702
Mean citation rate	2.102 ± 0.302
Relative citation rate	1.235
Test statistics	1.324

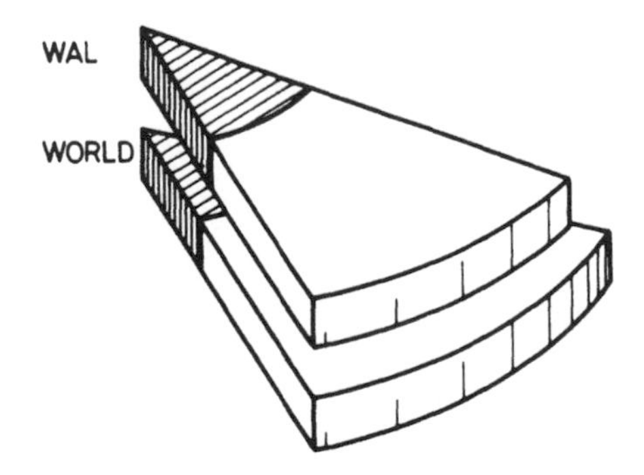

FIGURE 255. WAL-A. Proportion of chemistry and analytical chemistry papers in Welsh science publications.

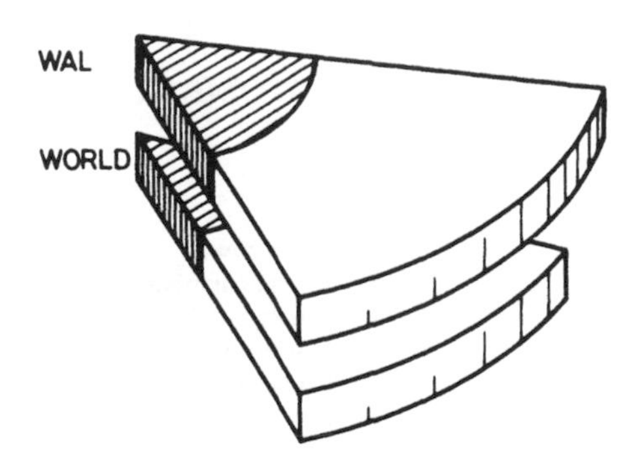

FIGURE 256. WAL-B. Citations to chemistry and analytical chemistry papers as a proportion of citations to Welsh science publications.

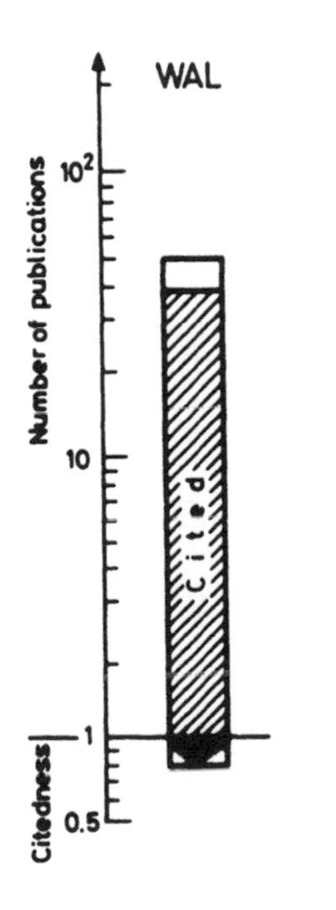

FIGURE 257. WAL-C. Number of Welsh analytical chemistry papers and the cited fraction.

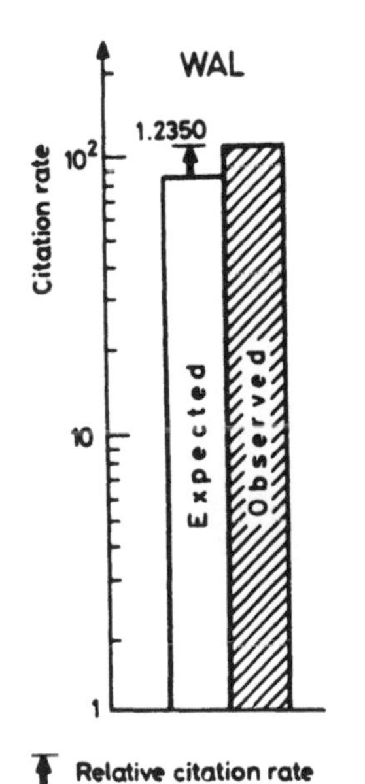

FIGURE 258. WAL-D. Expected and observed citation rates of Welsh analytical chemistry papers.

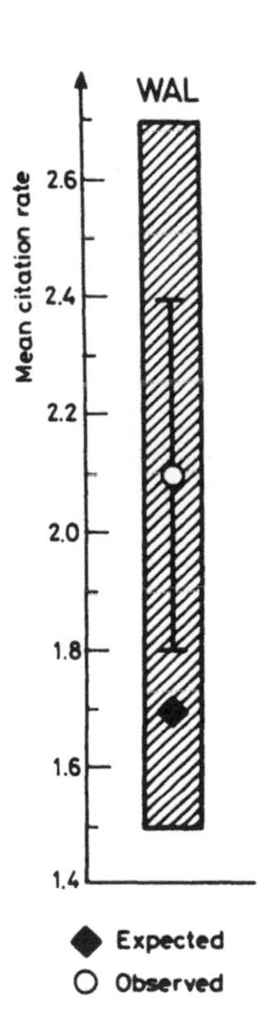

FIGURE 259. WAL-E. Expected and observed values of mean citation rate of Welsh analytical chemistry papers.

Table 104
SCIENTOMETRIC INDICATOR VALUES 1978—1980

Yugoslavia

Publication counts	
Analytical chemistry	87
Chemistry	447
Total science	1694
Analytical chemistry publications as a proportion of publications	
In chemistry	0.1946 ± 0.0187
Activity index	1.609
Test statistics	3.932
In total science	0.0514 ± 0.0054
Activity index	2.973
Test statistics	6.356
Citation rates	
Analytical chemistry	91
Chemistry	361
Total science	1089
Citations to analytical chemistry publications as a proportion of citations	
To chemistry	0.2521 ± 0.0229
Attractivity index	1.848
Test statistics	5.062
To total science	0.0836 ± 0.0084
Attractivity index	4.609
Test statistics	7.803
Fraction of cited publications	0.5057 ± 0.0536
Expected citation rate	100.8
Mean impact factor	1.159
Mean citation rate	1.046 ± 0.159
Relative citation rate	0.903
Test statistics	−0.708

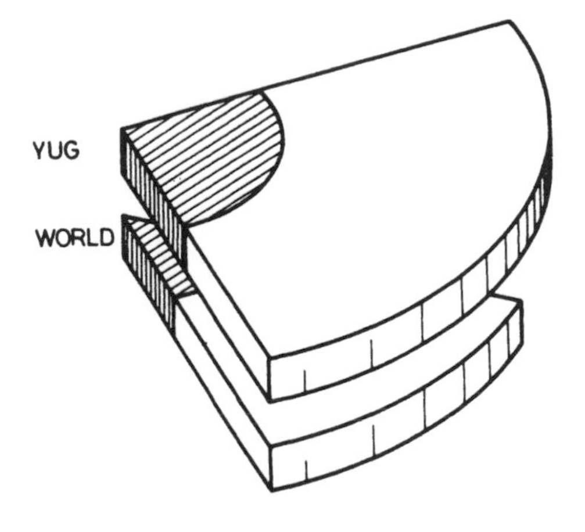

FIGURE 260. YUG-A. Proportion of chemistry and analytical chemistry papers in Yugoslavian science publications.

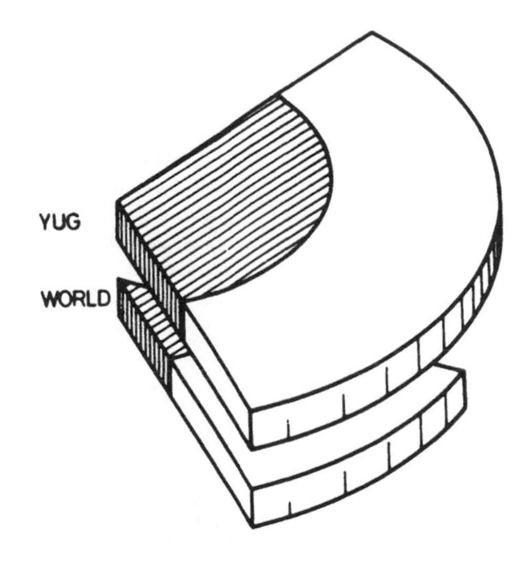

FIGURE 261. YUG-B. Citations to chemistry and analytical chemistry papers as a proportion of citations to Yugoslavian science publications.

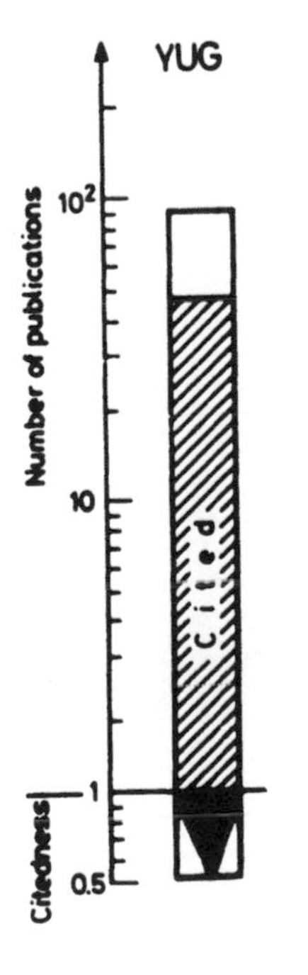

FIGURE 262. YUG-C. Number of Yugoslavian analytical chemistry papers and the cited fraction.

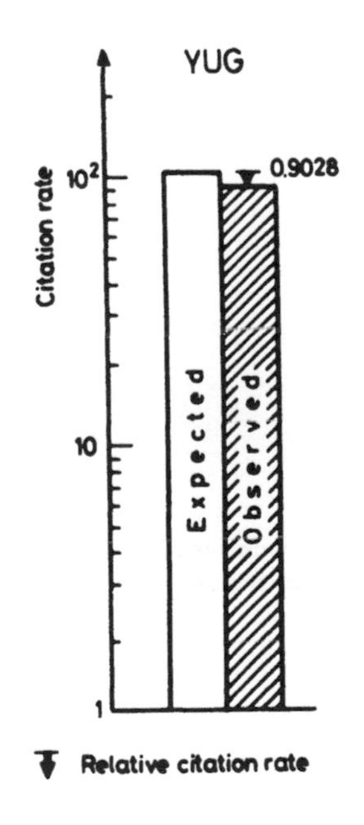

FIGURE 263. YUG-D. Expected and observed citation rates of Yugoslavian analytical chemistry papers.

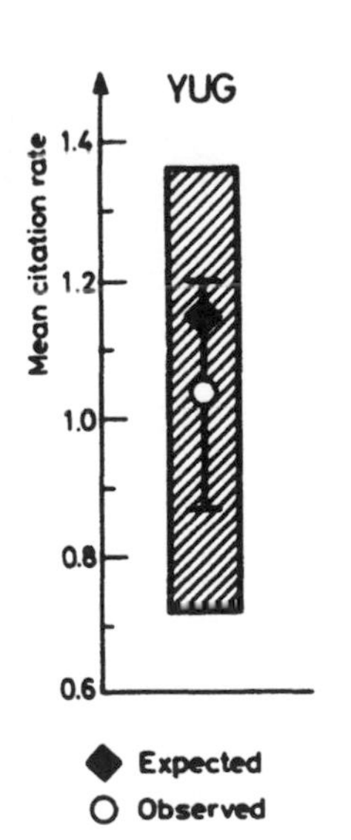

FIGURE 264. YUG-E. Expected and observed values of mean citation rate of Yugoslavian analytical chemistry papers.

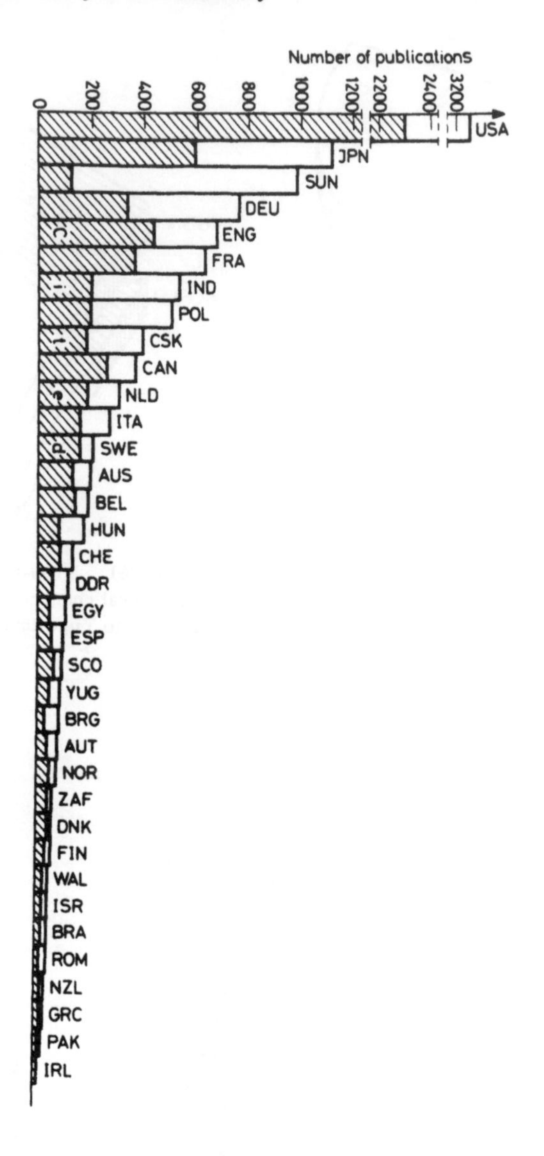

FIGURE 265. Number of analytical chemistry papers and the cited fraction.

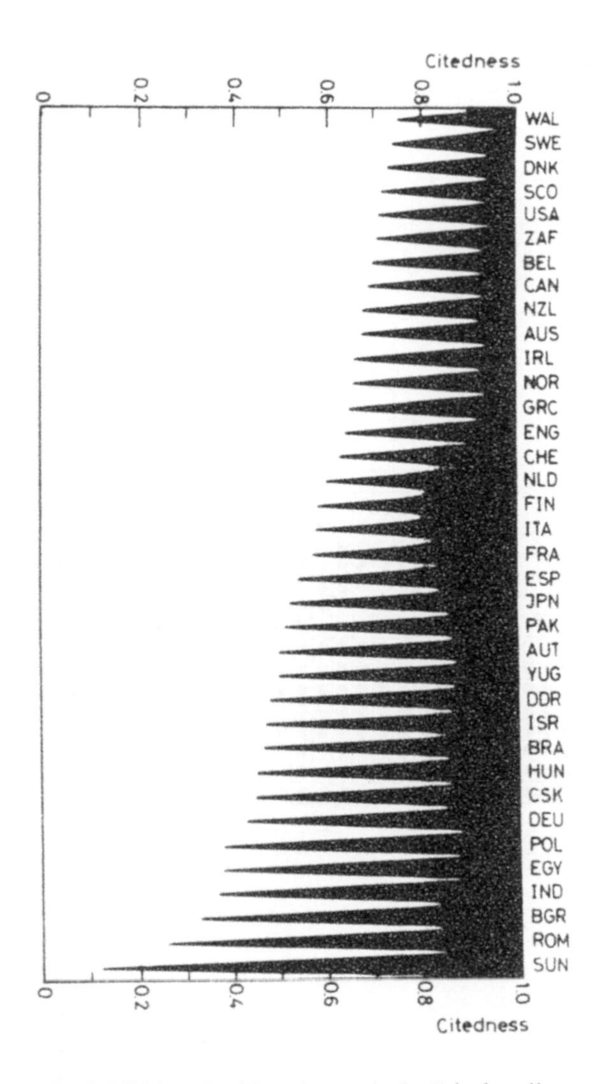

FIGURE 266. Ranking of countries by "citedness".

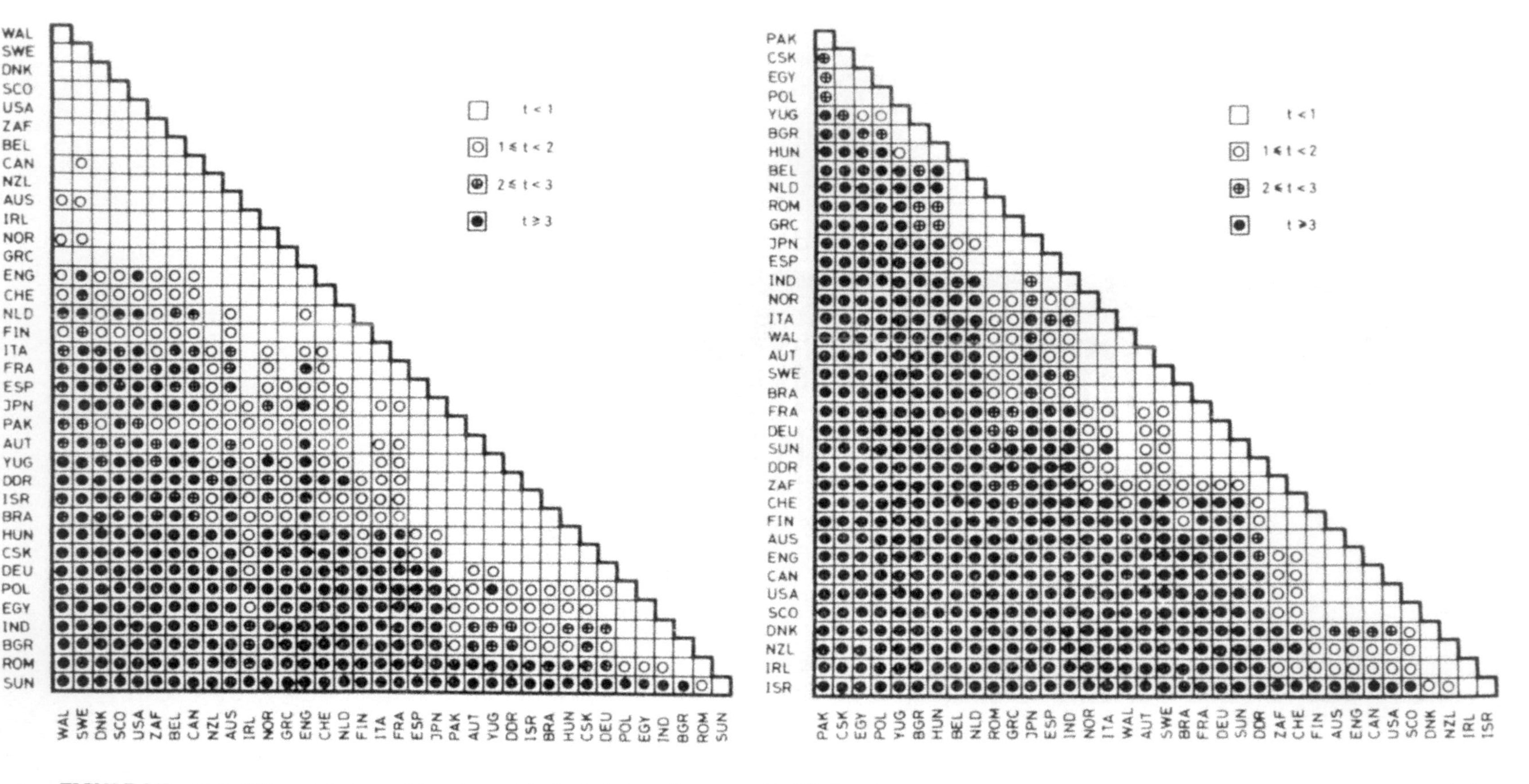

FIGURE 267. Reliability matrix for ranking of countries by "citedness".

FIGURE 268. Reliability matrix for ranking of countries by proportion of analytical chemistry papers.

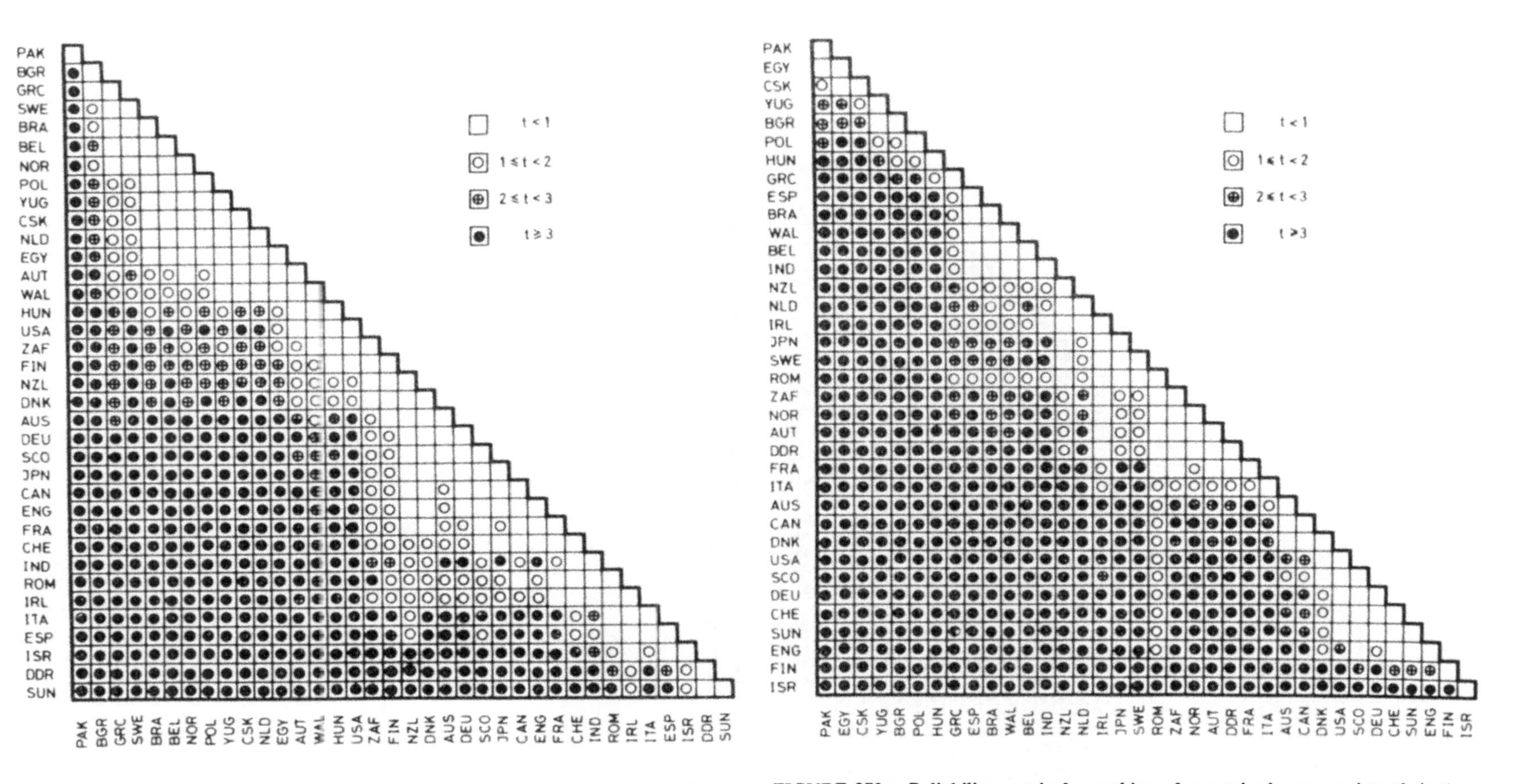

FIGURE 269. Reliability matrix for ranking of countries by proportion of analytical chemistry papers within chemistry.

FIGURE 270. Reliability matrix for ranking of countries by proportion of citations to analytical chemistry papers in citations to total science papers.

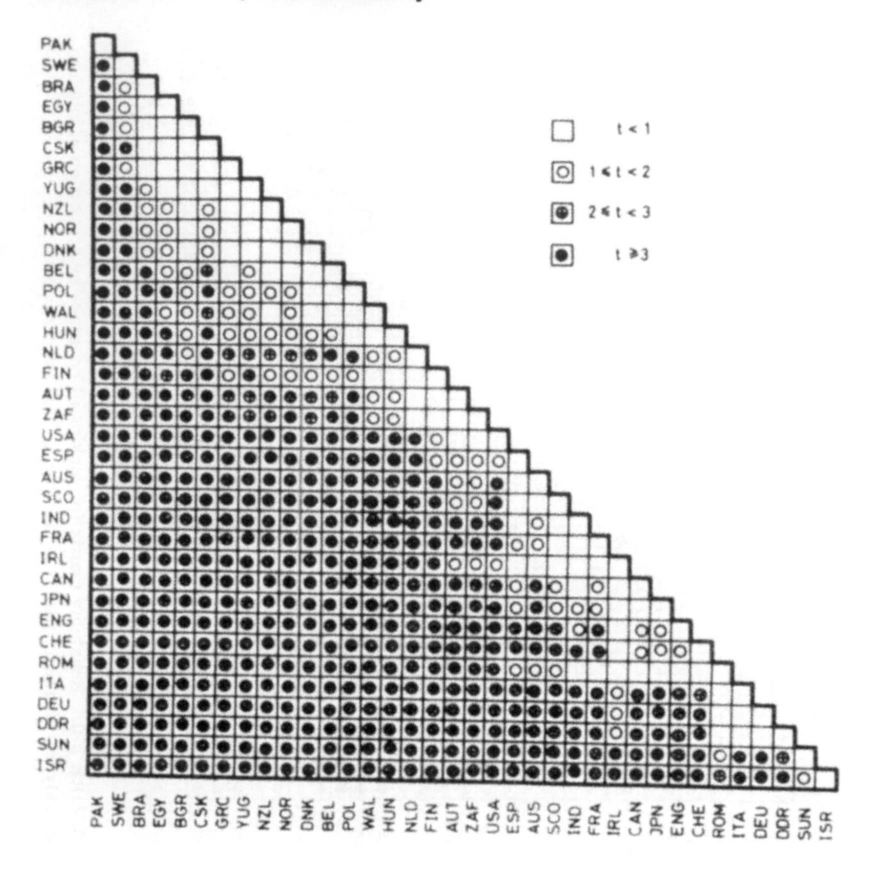

FIGURE 271. Reliability matrix for ranking of countries by proportion of citations to analytical chemistry papers in citations to chemistry papers.

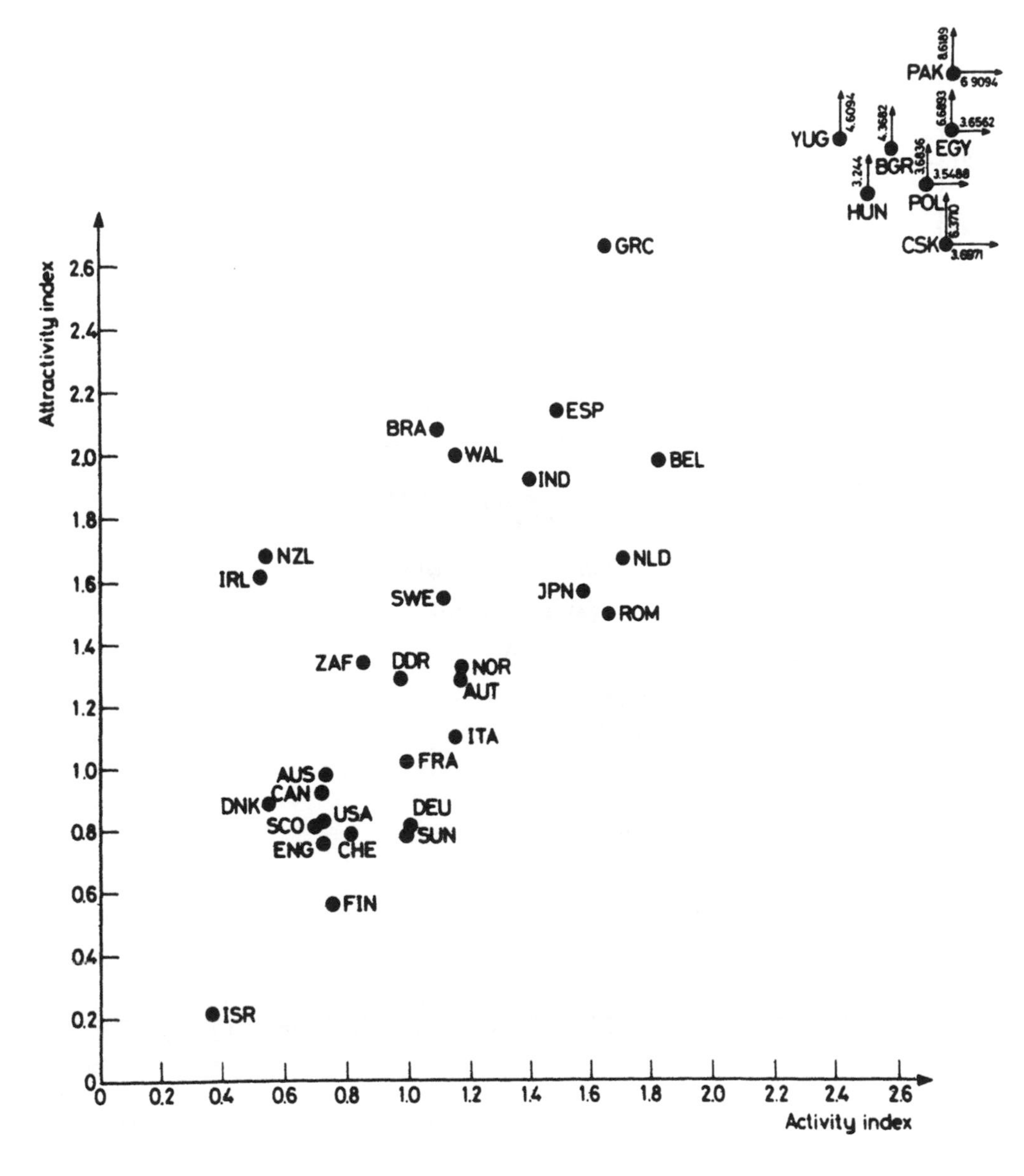

FIGURE 272. Activity and attractivity indexes of analytical chemistry in total science.

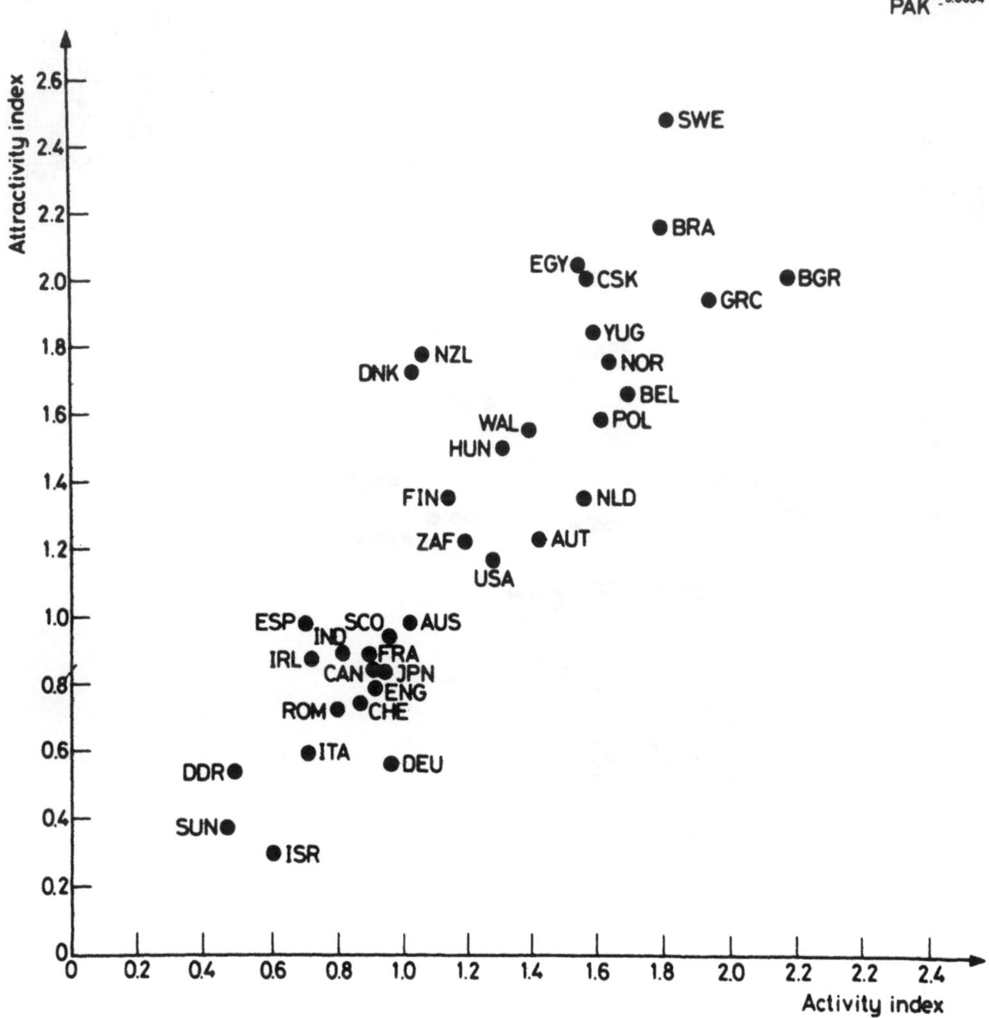

FIGURE 273. Activity and attractivity indexes of analytical chemistry within chemistry.

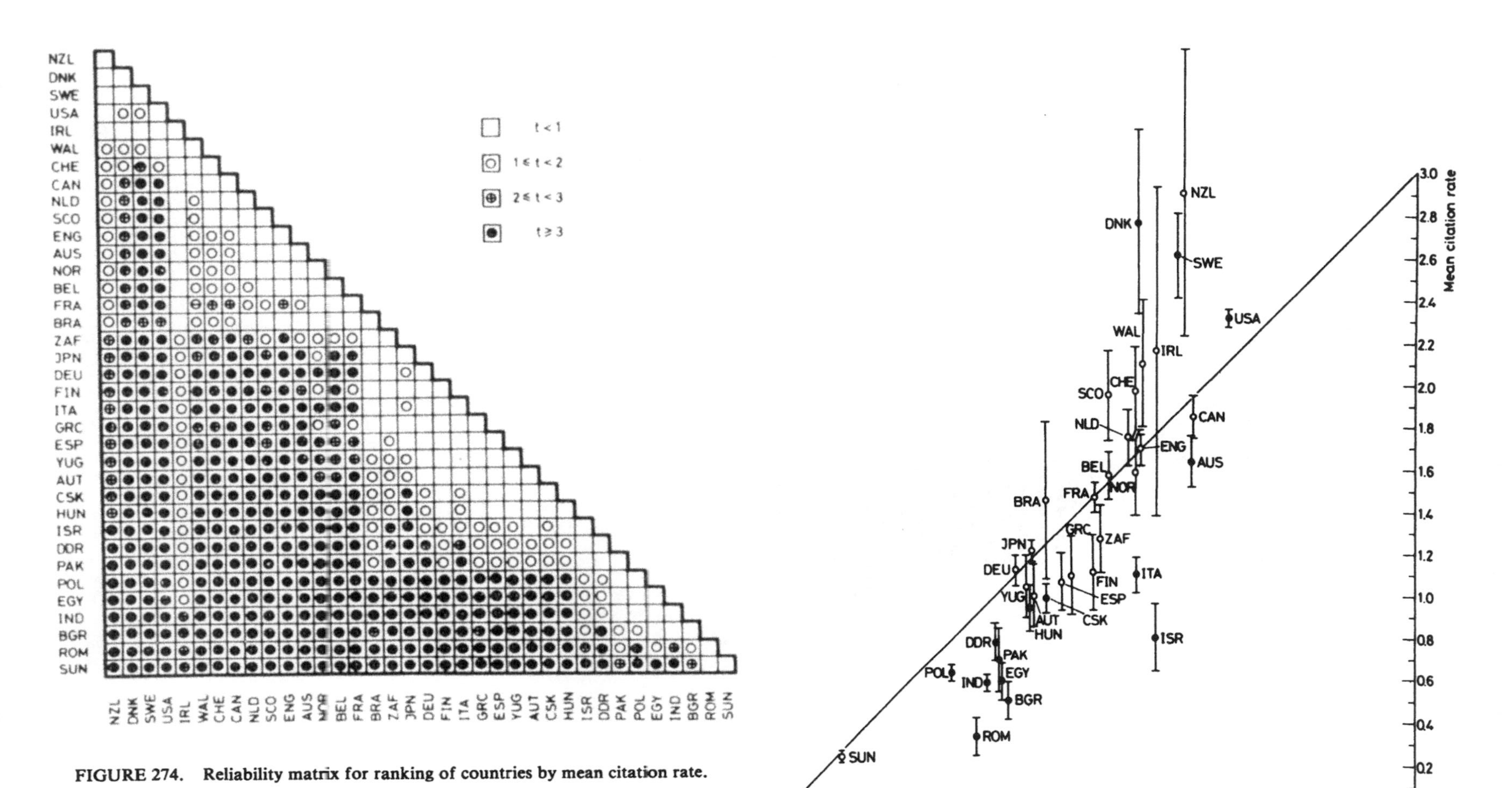

FIGURE 274. Reliability matrix for ranking of countries by mean citation rate.

FIGURE 275. Mean citation rates and mean impact factors.

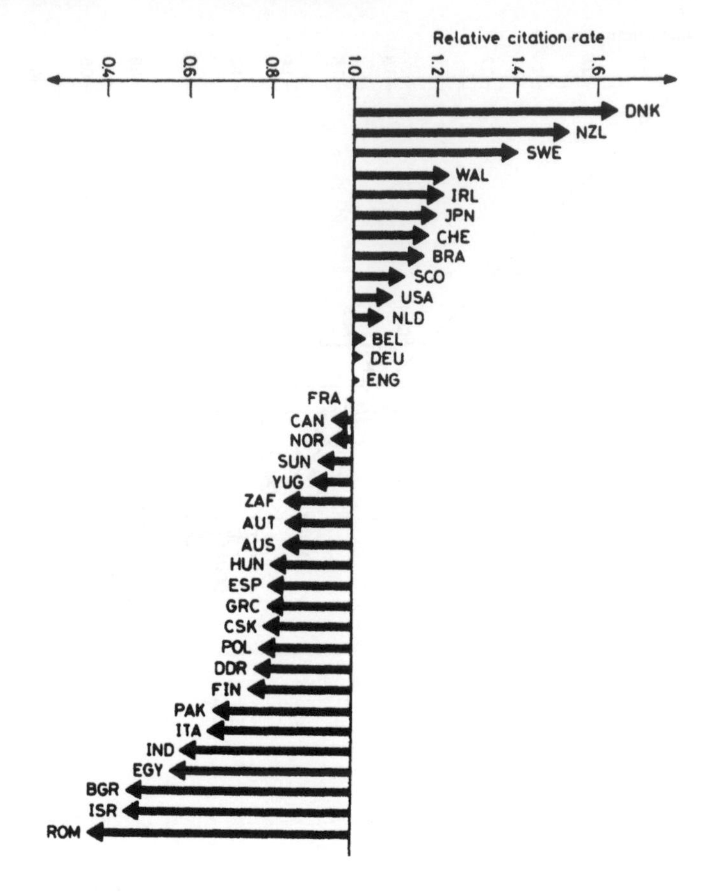

FIGURE 276. Ranking of countries by relative citation rate.

APPENDIX

LIST OF CHEMISTRY JOURNAL TITLES

Accounts of Chemical Research
Acta Chemica Scandinavica, Series A — Physical and Inorganic Chemistry
Acta Chemica Scandinavica, Series B — Organic Chemistry and Biochemistry
Acta Chimica Academiae Scientiarum Hungaricae
Acta Crystallographica, Section A
Acta Crystallographica, Section B
Advances in Colloid and Interface Science
Advances in Urethane Science and Technology
Anales de la Association Quimica Argentina
Anales de Quimica
Analusis
Analyst
Analytica Chimica Acta
Analytica Chimica Acta — Computer Techniques and Optimization
Analytical Chemistry
Analytical Letters
Angewandte Chemie — International Edition in English
Angewandte Makromolekulare Chemie
Annales de Chimie — Sciences des Matériaux
Annali di Chimica
Annual Reports on the Progress of Chemistry, Section A — Physical and Inorganic
Annual Reports on the Progress of Chemistry, Section B — Organic
Annual Review of Physical Chemistry
Applications of Surface Science
Applied Spectroscopy
Applied Spectroscopy Reviews
Australian Journal of Chemistry
Berichte der Bunsengesellschaft für Physikalische Chemie
Bulletin de l'Académie Polonaise des Sciences — Séries des Sciences Chimiques
Bulletin de la Société Chimique Belges
Bulletin de la Société Chimique de France, Part I
Bulletin de la Société Chimique de France, Part II
Bulletin of the Academy of Sciences of the USSR, Division of Chemical Sciences
Bulletin of the Chemical Society of Japan
Bunseki Kagaku
Canadian Journal of Chemistry
Canadian Journal of Spectroscopy
Carbohydrate Research
Carbon
Catalysis Reviews
Chemical Technology
Chemia Analyticzna
Chemica Scripta
Chemical, Biomedical and Environmental Instrumentation
Chemical Reviews
Chemical Senses and Flavour
Chemical Society Reviews

Chemicke Listy
Chemicke Zvesti
Chemiker-Zeitung
Chemische Berichte
Chemische Technik
Chemistry and Industry
Chemistry in Britain
Chemistry Letters
Chemosphere
Chimia
Chimica e l'Industria
Chromatographia
Collection of Czechoslovak Chemical Communications
Colloid and Polymer Sciences
Colloid Journal of the USSR
Color Research and Application
Comptes Rendus Hebdomadaires des Séances de l'Académie des Sciences
Coordination Chemistry Reviews
Critical Reviews in Analytical Chemistry
Croatica Chemica Acta
Denki Kagaku
Electrochimica Acta
European Polymer Journal
Faraday Discussions of the Chemical Society
Farmaco — Edizione Pratica
Farmaco — Edizione Scientifica
Finnish Chemical Letters
Fluoride
Gazetta Chimica Italiana
Grasas y Aceites
Helvetica Chimica Acta
Heterocycles
High Energy Chemistry
Indian Journal of Chemistry, Section A — Inorganic, Physical, Theoretical and Analytical
Indian Journal of Chemistry, Section B — Organic Chemistry, Including Medicinal Chemistry
Inorganic and Nuclear Chemistry Letters
Inorganic Chemistry
Inorganic Materials
Inorganica Chimica Acta
International Journal of Chemical Kinetics
International Journal of Environmental Analytical Chemistry
International Journal of Quantum Chemistry
International Sugar Journal
Israel Journal of Chemistry
Journal de Chimie Physique
Journal de Microscopie et de Spéctroscopie Électronique
Journal für Praktische Chemie
Journal of Adhesion
Journal of Analytical and Applied Pyrolysis
Journal of Analytical Chemistry of the USSR
Journal of Applied Chemistry

Journal of Applied Electrochemistry
Journal of Applied Polymer Sciences
Journal of Carbohydrates — Nucleosides — Nucleotides
Journal of Catalysis
Journal of Chemical Education
Journal of Chemical Information and Computer Sciences
Journal of Chemical Research — S
Journal of Chemical Technology and Biotechnology
Journal of Chemical Thermodynamics
Journal of Chromatography
Journal of Colloid and Interface Science
Journal of Coordination Chemistry
Journal of Crystal and Molecular Structure
Journal of Dialysis
Journal of Electroanalytical Chemistry and Interfacial Electrochemistry
Journal of Fluorine Chemistry
Journal of Heterocyclic Chemistry
Journal of Inorganic and Nuclear Chemistry
Journal of Labelled Compounds and Radiopharmaceuticals
Journal of Liquid Chromatography
Journal of Macromolecular Science — Reviews in Macromolecular Chemistry
Journal of Membrane Science
Journal of Molecular Catalysis
Journal of Molecular Spectroscopy
Journal of Molecular Structure
Journal of Natural Products — Lloydia
Journal of Organic Chemistry
Journal of Organometallic Chemistry
Journal of Photochemistry
Journal of Physical Chemistry
Journal of Polymer Science — Polymer Chemistry Edition
Journal of Polymer Science — Polymer Letters Edition
Journal of Polymer Science, Part C — Polymer Symposium
Journal of Quantitative Spectroscopy and Radiative Transfer
Journal of Radioanalytical Chemistry
Journal of Raman Spectroscopy
Journal of Scientific and Industrial Research
Journal of Solid State Chemistry
Journal of Solution Chemistry
Journal of Structural Chemistry
Journal of Supramolecular Structure
Journal of Synthetic Organic Chemistry Japan
Journal of Thermal Analysis
Journal of the American Chemical Society
Journal of the American Leather Chemist Association
Journal of the American Oil Chemists' Society
Journal of the Association of Official Analytical Chemists
Journal of the Association of Public Analysts
Journal of the Chemical Society — Chemical Communications
Journal of the Chemical Society — Dalton Transactions
Journal of the Chemical Society — Faraday Transactions I
Journal of the Chemical Society — Faraday Transactions II

Journal of the Chemical Society — Perkin Transactions I
Journal of the Chemical Society — Perkin Transactions II
Journal of the Chinese Chemical Society
Journal of the Chromatographic Science
Journal of the Electrochemical Society
Journal of the Indian Chemical Society
Journal of the Indian Institute of Science, Section B — Physical and Chemical Sciences
Journal of the Japanese Society of Starch Science
Journal of the Oil and Colour Chemists' Association
Journal of the Society of Cosmetic Chemists
Justus Liebigs Annalen der Chemie
Kémiai Közlemények
Khimiya Geterotsiklicheskikh Soedinenii
Khimiya Prirodnykh Soedinenii
Kinetics and Catalysis
Kobunshi Ronbunshu
Kristall und Technik — Crystal Research and Technology
Kristallografiya
Macromolecular Reviews, Part D — Journal of Polymer Science
Macromolecules
Magyar Kémiai Folyóirat
Makromolekulare Chemie
Marine Chemistry
Materials Chemistry
Microchemical Journal
Mikrochimica Acta
Molecular Crystals and Liquid Crystals
Monatshefte für Chemie
Nippon Kagaku Kaishi
Nouveau Journal de Chimie
Nucleic Acids Research
Organic Magnetic Resonance
Organic Mass Spectrometry
Organic Preparations and Procedures International
Organic Reacticity
Petroleum Chemistry — USSR
Phosphor and Sulfur and the Related Elements
Physics and Chemistry of Liquids
Physics and Chemistry of Minerals
Polish Journal of Chemistry
Polymer
Polymer Bulletin
Polymer Engineering and Science
Polymer Journal
Proceedings of the Indian Academy of Sciences, Section A 1
Progress in Solid State Chemistry
Przemysl Chemiczny
Pure and Applied Chemistry
Radiation Physics and Chemistry
Radiochemical and Radioanalytical Letters
Reaction Kinetics and Catalysis Letters

Recueil des Travaux Chimiques des Pays-Bas — Journal of the Royal Netherlands Chemical Society
Revue de Chimie Mineral
Revue Roumaine de Chimie
Review of Physical Chemistry of Japan
Rubber Chemistry and Technology
Separation and Purification Methods
Silikaty
South African Journal of Chemistry
Soviet Electrochemistry
Spectrochimica Acta, Part A — Molecular Spectroscopy
Spectrochimica Acta, Part B — Atomic Spectroscopy
Spectroscopy Letters
Stärke
Synthetic Communications
Synthetic Metals
Synthesis (Cambridge)
Synthesis (Stuttgart)
Synthesis and Reactivity in Inorganic and Metal-Organic Chemistry
Talanta
Tetrahedron
Tetrahedron Letters
Theoretica Chimica Acta
Thermochimica Acta
Transition Metal Chemistry
Ukrainskii Khimicheskii Zhurnal
Uspekhi Khimii
Vestnik Moskovskogo Universiteta, Seriya Khimiya
Vysokomolekulyarnye Soedineniya, Seriya A
Vysokomolekulyarnye Soedineniya, Seriya B
X-Ray Spectrometry
Zeitschrift für Analytische Chemie
Zeitschrift für Anorganische und Allgemeine Chemie
Zeitschrift für Chemie
Zeitschrift für Kristallographie, Kristallgeometrie, Kristallphysik, Kristallchemie
Zeitschrift für Lebensmittel-Untersuchung und -Forschung
Zeitschrift für Physikalische Chemie (Leipzig)
Zeitschrift für Physikalische Chemie (Wiesbaden)
Zhurnal Fizicheskoi Khimii
Zhurnal Neorganicheskoi Khimii
Zhurnal Obshchei Khimii
Zhurnal Organicheskoi Khimii
Zhurnal Vsesoyuznogo Khimicheskogo Obshchestva Imeni D. I. Mendeleeva

REFERENCES

1. De Solla Price, D., *Little Science, Big Science,* Columbia University Press, New York, 1961.
2. Moravcsik, M. J., A progress report on the quantification of science, *J. Sci. Ind. Res. (India),* 36, 195, 1977.
3. Cole, F. J. and Eales, N. B., The history of comparative anatomy, *Sci. Prog.,* 11, 578, 1917.
4. Hjerppe, R., Ed., *An Outline of Bibliometrics and Citation Analysis,* Report TRITA-LIB-2013, Royal Institute of Technology Library, Stockholm, 1980.
5. Pritchard, A., Statistical bibliography or bibliometrics?, *J. Doc.,* 25, 348, 1969.
6. Nalimov, V. V. and Mulchenko, G. M., *Naukometriya,* (in Russian), Izd. Nauka, Moscow, 1969.
7. Crane, E. J., Sharp rise in chemical publication, *Chem. Eng. News.,* 24, 3353, 1946.
8. Crane, E. J., Chemical abstracts, *Chem. Eng. News,* 27, 529, 1949.
9. Strong, F. C., Trends in quantitative analysis: a survey of papers for the year 1946, *Anal. Chem.,* 19, 968, 1947.
10. Fischer, R. B., Babcock, R. F., Conley, R. F., Cross, S. B., Paudler, F. A., and Guthrie, W. W., Trends in analytical chemistry, *Anal. Chem.,* 28(12), 9A, 1956.
11. Fischer, R. B., Trends in analytical chemistry 1965, *Anal Chem.,* 37(13), 27A, 1965.
12. Brooks, R. R. and Smythe, I. E., The progress of analytical chemistry 1910—1970, *Talanta,* 22, 495, 1975.
13. Braun, T., Some comments on "The Progress of Analytical Chemistry 1910—1970", *Talanta,* 23, 743, 1976.
14. Dobrov, G. M., *Nauka o Nauke,* Naukova Dumka, Kiev, 1966.
15. Menard, H. W., *Science, Growth and Change,* Harvard University Press, Cambridge, 1971.
16. Baker, D. B., Growth of chemical literature, past, present and future, *Chem. Eng. News,* 39(29), 78, 1961.
17. Baker, D. B., Chemical literature expands, *Chem. Eng. News,* 44(23), 84, 1966.
18. Baker, D. B., Worlds chemical literature continues to expand, *Chem. Eng. News,* 49(28), 37, 1971.
19. Baker, D. B., Recent trends in growth of chemical literature, *Chem. Eng. News,* 54(20), 23, 1976.
20. Braun, T., Bujdosó, E., and Lyon, W. S., An analytical look at chemical publications, *Anal. Chem.,* 52(6), 617A, 1980.
21. Moravcsik, M. J., Phenomenology and models of growth of science, *Res. Policy,* 4, 80, 1975.
22. Orient, I. M., Scientometric research in analytical chemistry (in Russian), *Zavodsk. Lab.,* 41(9), 1071, 1975; (English transl.), *Ind. Lab. USSR,* 41, 1327, 1975.
23. May, K. O., Quantitative growth of mathematical literature, *Science,* 154, 1672, 1966.
24. Holt, Ch. C. and Schrank, W. E., Growth of the professional literature in economics and other fields and some implications, *Am. Doc.,* 19, 18, 1968.
25. Orient, I. M. and Markusova, V. A., *Electrochemical Methods of Analysis* (in Russian), Sbornik, Izd. Metallurgiya, Moscow, 1972.
26. Orient, I. M. and Pats, R. G., *Polarography, Problems and Trends* (in Russian), Strabynya, Ya. P. and Majranovskiy, S. G., Eds., Sbornik, Izd. Zinatne, Riga, 1977, 388.
27. Braun, T. and Bujdosó, E., Growth of a journal reflects trends in the literature of nuclear analytical methods, *J. Radioanal. Chem.,* 50, 9, 1979.
28. Braun, T., Lyon, W. S., and Bujdosó, E., Literature growth and decay: an activation analysis Resumé, *Anal. Chem.,* 49, 682A, 1977.
29. Braun, T. and Lyon, W. S., unpublished results, 1983.
30. Orient, I. M., Artemova, O. A., and Davidova, S. L., Investigation of development of atomic absorption analysis by science metrics (in Russian), *Zavodsk. Lab.,* 43, 419, 1977; (English transl.), *Ind. Lab. USSR,* 43(4), 498, 1977.
31. Braun, T., "Skyrocketting" in the growth of publication productivity of a new analytical subject field. The case of ion chromatography, *Anal. Proc.,* 19, 352, 1982.
32. Orient, I. M., Analysis of publications on organic reagents (in Russian), *Zh. Anal. Khim.,* 32, 502, 1977.
33. Melikhov, I. V. and Berdonosova, D. G., Sorption by inorganic sorbents in analytical chemistry. Analysis of publications (in Russian), *Zh. Anal. Khim.,* 31, 809, 1976.
34. Magyar, G., Bibliometric analysis of a new research subfield, *J. Doc.,* 30, 32, 1974.
35. Moravcsik, M. J., Measures of scientific growth, *Res. Policy,* 2, 266, 1973.
36. Rescher, N., *Scientific Progress,* University of Pittsburgh Press, Pittsburgh, 1978.
37. Braun, T., Analytical literature: quality vs. quantity. Letter, *Anal. Chem.,* 55, 132A, 1983.
38. Cole, P. F., Journal usage versus age of journal, *J. Doc.,* 19, 1, 1963.
39. Line, M. B., The half-life of periodical literature: apparent and real obsolescence, *J. Doc.,* 26, 46, 1970.
40. Brown, P., Half-life of the chemical literature, *J. Am. Soc. Inf. Sci.,* 31, 61, 1980.

41. Burton, R. E. and Kebler, R. W., The half-life of some scientific and technical literature, *Am. Doc.*, 11, 18, 1960.
42. Price, D. J. D., Networks of scientific papers, *Science*, 149, 510, 1965.
43. Meadows, A. J., *Communication in Science*, Butterworths, London, 1974.
44. Boig, F. S. and Howerton, P. W., History and development of chemical periodicals in the field of analytical chemistry: 1877—1950, *Science*, 115, 555, 1952.
45. Futekov, L., Specker, H., and Stojanov, S., Development of analytical methods, illustrated by analysis of elements selenium and tellurium — statistical review of years from 1970 to 1975 (in German), *Z. Anal. Chem.*, 285, 353, 1977.
46. Futekov, L., Paritschkova, R., and Specker, H., On the development of analytical methods illustrated by the analysis of elements selenium and tellurium. A statistical review of the years from 1970 to 1981, *Z. Anal. Chem.*, 315, 342, 1983.
47. Brooks, R. R. and Smythe, L. E., Trends in atomic absorption spectroscopy, *Anal. Chim. Acta*, 74, 35, 1975.
48. Earle, P. and Vickery, B. C., Subject relations in science/technology literature, *ASLIB Proc.*, 21, 237, 1969.
49. Braun, T. and Bujdosó, E., Some tendencies of radioanalytical literature. Statistical games for trend evaluation. I. Distribution of information sources, *Radiochem. Radioanal. Lett.*, 23, 195, 1973.
50. Lyon, W. S. and Ross, H. H., Nucleonics, *Anal. Chem.*, 50, 80R, 1978.
51. Orient, I. M., *Trends in the Logics of Development and Scientometrics in Chemistry* (in Russian), Kabanov, V. A., Ed., Moscow State University Press, Moscow, 1976, 49.
52. Berezkin, V. G. and Chernysheva, T. Yu., Some trends in development of analytical chemistry and chromatography as deduced from analysis of literature, *J. Chromatogr.*, 141, 241, 1977.
53. Berezkin, V. G., Chernysheva, T. Yu., and Bolotov, S. L., Evaluation of the role of chromatography in analytical chemistry based on the analysis of the subject-matter of publications, *J. Chromatogr.*, 251, 227, 1982.
54. Orient, I. M., *Trends in the Logics of Development and Scientometrics in Chemistry (in Russian)*, Kabanov, V. A., Ed., Moscow State University Press, Moscow, 1976, 90.
55. Beyermann, K., Trace analysis of organic compounds, *Pure Appl. Chem.*, 50, 87, 1978.
56. Petruzzi, J. M., Scientific journals, *Anal. Chem.*, 51(1), 86A, 1979.
57. Kabanova, O. L. and Kurilina, N. A., Publications on electrochemical methods of analyzing inorganic material during 1955 to 1973 (in Russian), *Zh. Anal. Khim.*, 30, 2432, 1975.
58. Thomas, J. D. R., Electroanalytical horizons, *Anal. Proc.*, 19, 60, 1982.
59. Volkova, G. A. and Kontsova, V. V., Tendencies in development of methods for analysis of mineral ores based on study of information flow (in Russian), *Zavodsk. Lab.*, 42, 395, 1976.
60. Subert, J. and Blesova, M., Methods of pharmaceutical analysis. Analysis of information flows, *Pharmazie*, 31, 624, 1976.
61. Kara-Murza, S. G., Research techniques as the object of scientific analysis (in Russian), *Vestn. Akad. Nauk. SSSR*, (1), 44, 1979.
62. Preobrazhenskaya, G. B., Pruktova, H. M., and Granovskii, Yu. V., Analysis of statistical terms in publications. Dealing with spectral analysis and analytical chemistry (in Russian), *Zavodsk. Lab.*, 40, 1240, 1974.
63. Bradford, S. C., *Documentation*, Crosby, London, 1948.
64. Zipf, G. K., *Human Behaviour and the Principle of Least Effort*, Addison-Wesley, Reading, Mass., 1949.
65. Naranan, S., Power law relations in science bibliography: a self-consistent interpretation, *J. Doc.*, 27, 83, 1971.
66. Brookes, B. C., The derivation and application of the Bradford-Zipf distribution, *J. Doc.*, 24, 247, 1968.
67. Lyon, W. S., Ricci, E., and Ross, H. H., Nucleonics, *Anal. Chem.*, 44, 438R, 1972.
68. Lyon, W. S., Ricci, E., and Ross, H. H., Nucleonics, *Anal. Chem.*, 46, 431R, 1974.
69. Lyon, W. S. and Ross, H. H., Nucleonics, *Anal. Chem.*, 48, 96R, 1976.
70. Garfield, E., *Journal Citation Reports. A Bibliometric Analysis of References*, Vol. 9, 1976 Annual, Institute for Scientific Information, Philadelphia,
71. Narin, F., Evaluative Bibliometrics. The Use of Publication and Citation Analysis in the Evaluation of Scientific Activity, National Science Foundation Report PB-252, Washington, D.C., 1976, 139.
72. Narin, F., Pinski, G., and Gee, H. H., Structure of the biomedical literature, *J. Am. Soc. Inf. Sci.*, 27, 25, 1976.
73. Pinski, G., Influence and interrelationship of chemical journals, *J. Chem. Inf. Comp. Sci.*, 17, 67, 1977.
74. Petruzzi, J., Statistics, *Proc. 7th East. Anal. Symp.*, Hirsh, R. F., Ed., Franklin Institute Press, Philadelphia, 1978, 277.

75. Bujdosó, E., Braun, T., and Lyon, W. S., Information flow in analytical chemistry journals, *Trends Anal. Chem.*, 1, 268, 1982.
76. De Solla Price, D., Some remarks on elitism in information and the invisible college phenomenon in science, *J. Am. Soc. Inf. Sci.*, 22, 74, 1971.
77. Juhász, S., Calvert, E., Jackson, T., Kronick, D. A., and Shipman, J., Acceptance and rejection of manuscripts, *IEEE Trans. Prof. Commun.*, PC-18(3), 177, 1975.
78. Zuckerman, H. and Merton, R. K., Patterns of evaluation in science: institutionalisation structure and functions of the referee system, *Minerva*, 9, 66, 1971.
79. Harnad, S., Ed., Peer commentary on peer review, *Behav. Brain Sci.*, 5, 185, 1982.
80. Petruzzi, J. M., Peer review in analytical chemistry, *Anal. Chem.*, 48, 875A, 1976.
81. Ziman, J. M., Information, communication, knowledge, *Nature*, 224, 318, 1969.
82. Gordon, M. D., Disciplinary differences, editorial practices and the patterning of rejection rates for UK research journals, *J. Res. Commun. Stud.*, 1, 139, 1978.
83. Gordon, M. D., Evaluating the evaluators, *New Sci.*, 73, 342, 1977.
84. Zsindely, S., Schubert, A., and Braun, T., Editorial gatekeeping patterns in international science journals. A new science indicator, *Scientometrics*, 4, 57, 1982.
85. Braun, T. and Bujdosó, E., Gatekeeping patterns in the publication of analytical chemistry research, *Talanta*, 30, 161, 1983.
86. Geller, W. L., DeCani, J. A., and Davies, R. E., Lifetime citation rates as basis for assessing the quality of scientific work, paper presented at the Belmont Conf. Use of Citation Data in the Study of Science, Elkridge, Md., April 1, 1975.
87. Garfield, E., Is citation analysis a legitimate tool?, *Scientometrics*, 1, 359, 1979.
88. Zsindely, S., Schubert, A., and Braun, T., Citation patterns of editorial gatekeepers in international chemistry journals, *Scientometics*, 4, 69, 1982.
89. Ziman, J., The paradoxical conventionality of traditional scientific papers, in *Coping with the Biomedical Literature Explosion*, Goffman, W. and Shaw, W. M., Jr., Eds., Working Papers of the Rockefeller Foundation, New York, 1978, 20.
90. Frame, J. D., Narin, F., and Carpenter, M. P., The distribution of world science, *Soc. Stud. Sci.*, 7, 501, 1977.
91. Merton, R. K., The Matthew effect in science, *Science*, 159, 56, 1968.
92. Garfield, E., *Journal Citation Reports, 1980*, Institute for Scientific Information, Philadelphia, 1980.
93. Braun, T. and Nagydiósi-Kocsis, Gy., The publication lapse of papers in Radiochemical and Radioanalytical Letters, *Radiochem. Radioanal. Lett.*, 52, 327, 1982.
94. Garwey, W. D., *Communication: The Essence of Science*, Pergamon Press, Oxford, 1979.
95. Petruzzi, J. M., Publication times, *Anal. Chem.*, 51, 277A, 1979.
96. Kuhn, T. S., *The Structure of Scientific Revolutions*, 2nd ed., Chicago Press, Chicago, 1970, 10.
97. Goffman, W. and Newill, V. A., Generalization of epidemic theory. An application to the transmission of ideas, *Nature*, 204, 225, 1964.
98. Snyder, L. R., Continuous-flow analysis, present and future, *Anal. Chim. Acta*, 114, 3, 1980.
99. Stewart, K. K., Flow injection analysis. A review of its early history, *Talanta*, 28, 789, 1981.
100. Mottola, H. A., Continuous flow analyses revisited, *Anal. Chem.*, 53, 1312A, 1981.
101. Margoshes, M., Flow injection analysis revisited, *Anal. Chem.*, 54, 678A, 1982.
102. Ruzicka, J. and Hansen, E. H., Flow injection analysis and its early history, *Talanta*, 29, 157, 1982.
103. Ruzicka, J., Hansen, E. H., and Mosbaek, H., Flow injection analysis. IX. New approach to continuous-flow titrations, *Anal. Chim. Acta*, 92, 235, 1977.
104. Skeggs, L. T., Jr., An automatic method for colorimetric analysis, *Am. J. Clin. Pathol.*, 28, 311, 1957.
105. Spackman, D. H., Stein, W. H., and Moore, S., Automatic recording apparatus for use in the chromatography of amino acids, *Anal. Chem.*, 30, 1190, 1958.
106. Nagy, G., Fehér, Z. S., and Pungor, E., Application of silicon rubber-based grafite electrodes for continuous-flow measurements. II. Voltammetric study of active substances injected into electrolyte streams, *Anal. Chim. Acta*, 52, 47, 1970.
107. Stewart, K. K., Beecher, G. R., and Hare, P. E., Automated high-speed analyses of discrete samples. Use of nonsegmented, continous flow systems, *Fed. Proc., Fed. Am. Soc. Exp. Biol.*, 33, 1439, 1974.
108. Eswara Dutt, V. V. S. and Mottola, H. A., Novel approach to reaction-rate determinations by use of transient redox effects, *Anal. Chem.*, 47, 357, 1975.
109. Ruzicka, J. and Hansen, E. H., Flow injection analysis. I. New concept of fast continuous-flow analysis, *Anal. Chim. Acta*, 78, 145, 1975.
110. Stewart, K. K., Beecher, G. R., and Hare, P. E., Rapid analysis of discrete samples. Use of nonsegmented continuous flow, *Anal. Biochem.*, 70, 167, 1976.
111. Garfield, E., *Essays of an Information Scientist*, Vol. 4, Institute for Scientific Information, Philadelphia, 1981, 255.

112. Crane, D., *Invisible Colleges: Diffusion of Knowledge in Scientific Communities,* University of Chicago Press, Chicago, 1972.
113. Ruzicka, J., Flow injection analysis. From test tube to integrated microconduits, *Anal. Chem.*, 55, 1040A, 1983.
114. Lotka, A. J., The frequency distribution of scientific productivity, *J. Wash. Acad. Sci.*, 16, 317, 1926.
115. Vlachy, J., Frequency distributions of scientific performance. A bibliography of Lotka's law and related phenomena, *Scientometrics*, 1, 107, 1978.
116. Vlachy, J., Time factor in Lotka's law, *Probl. Inf. Doc.*, 10, 44, 1976.
117. Lutz, G. J., Maddock, R. J., and Meinke, W. W., *Activation Analysis; a Bibliography,* NBS Tech. Note 467, National Bureau of Standards, Washington, D.C., 1971.
118. Coile, R. C., Lotka's frequency distribution of scientific productivity, *J. Am. Soc. Inf. Sci.*, 28, 366, 1977.
119. Kessler, M. M., Bibliographic coupling between scientific papers, *Am. Doc.*, 14, 10, 1963.
120. Garfield, E., *Citation Indexing, Its Theory and Application in Science, Technology and Humanities,* John Wiley & Sons, New York, 1979.
121. Hjerppe, R., Ed., *A Bibliography of Bibliometrics and Citation Indexing and Analysis,* Report TRITA-LIB-2013, Royal Institute of Technology Library, Stockholm, 1979.
122. Kaplan, N., The norms of citation behaviour: prolegomena to the footnote, *Am. Doc.*, 16, 176, 1965.
123. Cole, J. R. and Cole, S., *Social Stratification in Science,* University of Chicago Press, Chicago, 1973.
124. Orient, I. M., A statistical study of citation of papers on analytical chemistry (in Russian), *Zavodsk. Lab.*, 33, 1383, 1967.
125. Preobrazhenskaya, G. B., Semantic analysis of references in metallurgical articles (in Russian), *Nauchno Tekh. Inf. Ser. 2,* 10, 10, 1965.
126. Kara-Murza, S. G., Scientific research techniques (in Russian), *Nauchno Tekh. Inf. Ser. 1,* 1, 7, 1979.
127. Axén, R., Porath, J., and Ernback, S., Chemical coupling of peptides and proteins to polysaccharides by means of cyanogen halides, *Nature,* 214, 1302, 1967.
128. Porath, J., Axén, R., and Ernback, S., Chemical coupling of proteins to agarose, *Nature,* 215, 1491, 1967.
129. Cuatrecasas, P., Wilchek, M., and Anfinsen, C. B., Selective enzyme purification by affinity chromatography, *Proc. Natl. Acad. Sci. U.S.A.*, 61, 636, 1968.
130. Schön, D., *Technology and Change,* Oxford University Press, Oxford, 1968.
131. Laitinen, H. and Ewing, O. H., *A History of Analytical Chemistry,* American Chemical Society, Washington, D.C., 1977.
132. Orient, I. M., Scientometric research in analytical chemistry (in Russian), *Zavodsk. Lab.*, 41, 1071, 1975.
133. Orient, I. M., A scientometric study of A. K. Babko's legacy, *Ukr. Khim. Zh.*, 42(10), 1068, 1976.
134. Belcher, R., Resurgence of analytical chemistry, *Analyst,* 103, 29, 1978.
135. Garfield, E., *Essays of an Information Scientist,* Vol. 2, Institute of Scientific Information Press, Philadelphia, 1977, 7.
136. Garfield, E., *Essays of an Information Scientist,* Vol. 2, Institute for Scientific Information Press, Philadelphia, 1977, 23.
137. Bujdosó, E., Lyon, W. S., and Noszlopy, I., Prompt nuclear analysis: growth and trends, *J. Radioanal. Chem.*, 74, 197, 1982.
138. Bird, J. R., Campbell, B. L., and Cawley, R. J., *Prompt Nuclear Analysis. Bibliography,* 1976 Report, AAECIE 443, Australian Atomic Energy Commission Research Establishment, Lucas Heights, May 1978.
139. Bird, J. R. and Campbell, B. L., *Prompt Nuclear Analysis. Bibliography Supplement,* 1979 Report, NPA-9, Australian Atomic Energy Commission Research Establishment, Lucas Heights, June 1979.
140. Bird, J. R., Campbell, B. L., and Price, P. B., Prompt nuclear analysis, *At. Energ. Rev.*, 12(2), 275, 1974.
141. Lyon, W. S. and Roberts, P. P., Technical oral presentations: what happens to them, *Anal. Proc.*, 20, 341, 1983.
142. Mulkay, M. J., *Sociology of the Scientific Research Community, in Science, Technology and Society,* Spiegel-Rösing, I. and De Solla Price, D., Eds., SAGE Publications, Beverly Hills, 1978.
143. Bujdosó, E., Modern trends in activation analysis (in Hungarian), *Fiz. Szemle,* 25, 85, 1975.
144. Girardi, F., Radioactivation analysis. Past achievements, present trends and perspective for the future, *J. Radioanal. Chem.*, 69, 15, 1982.
145. Gillis, J., The mathematical theory of epidemics, *Interdisc. Sci. Rev.*, 4, 306, 1979.
146. Bailey, N. T., *The Mathematical Theory of Epidemics,* Griffin, London; Hafner, New York, 1957.
147. Goffman, W. and Harmon, G., Mathematical approach to prediction of scientific discovery, *Nature,* 229, 103, 1971.

148. Bujdosó, E., Lyon, W. S., and Braun, T., Scientometric study of "Health Physics", *Health Phys.*, 41, 233, 1981.
149. Sullivan, D., Koester, D., and White, D. H., Survival in particle physics: toward a simulation of the manpower and research productivity of experimentalists, paper presented at Conf. Evaluation in Science and Technology, Theory and Practice, Dubrovnik, Yugoslavia, June 30 to July 4, 1980.
150. Burger, M. and Bujdosó, E., Oscillating chemical reactions as an example of the development of a subfield of science, in *Oscillations and Traveling Waves in Chemical Systems*, Field, R. J. and Burger, M., Eds., John Wiley & Sons, New York, 1984.
151. Sullivan, D., White, D. H., and Barboni, E. J., The state of a science: indicators in the specialty of weak interactions, *Soc. Stud. Sci.*, 7, 167, 1977.
152. Shaw, W. M., Jr., Statistical disorder and the analysis of communication graph, *J. Am. Soc. Inf. Sci.*, 34, 146, 1983.
153. de Beaver, D. D. and Rosen, R., Studies in scientific collaboration. II. Scientific co-authorship, research productivity and visibility in the French scientific elite, 1799—1830, *Scientometrics*, 1, 133, 1979.
154. Amsel, G., Nadai, J. P., D'Artermare, E., David, D., Girard, E., and Moulin, J., Microanalysis by direct observation of nuclear reactions using 2 MeV van de Graaff, *Nucl. Instrum. Methods*, 92, 481, 1971.
155. Amsel, G., This week citation classic, *Current Contents (Phys. Chem. Earth Sci.)*, 33, 12, 1980.
156. Pinski, G., *Subject Classification and Influence Weight for 2,300 Journals*, Proj. No. 704R, Computer Horizons Inc., Cherry Hill, N.J., June 30, 1975.
157. Schubert, A. and Glänzel, W., A dynamic look at class of skew distributions. A model with scientometric applications, *Scientometrics*, 6, 1984.
158. Schubert, A. and Glänzel, W., Prompt nuclear analysis literature: a cumulative advantage approach, *J. Radioanal. Chem.*, 82, 215, 1984.
159. Hulme, E. W., *Statistical Bibliography in Relation to the Growth of Modern Civilization*, Grafton, London, 1923.
160. Science Indicators 1972, 1974, 1976, 1978, 1980, National Science Board, National Science Foundation, U.S. Government Printing Office, Washington, D.C., 1973, 1975, 1977, 1979, 1981.
161. Frame, D. J. and Sprague, A. N., *Indicators of Scientific and Technological Efforts in the Middle East and North Africa*, Computer Horizons Inc., Washington, D.C., 1978.
162. Schubert, A., Glänzel, W., and Braun, T., *Scientometric Indicators for Comprehensive Evaluation of Basic Research in Science of 32 Countries, 1976—1980* (in Hungarian), Library of the Hungarian Academy of Sciences, Budapest, 1983.
163. Freeman, C., *The Measurement of Scientific and Technological Activities*, Proposals for the Collection of Statistics on Science and Technology on an Internationally Uniform Basis, UNESCO, Paris, 1968.
164. Freeman, C., *Measurement of Output of Research and Experimental Development: A Review Paper*, UNESCO, Paris, 1969.
165. UNESCO, *Annotated Accessing List of Studies and Reports in the Field of Science Statistics*, UNESCO, Paris, 1968.
166. UNESCO Office of Statistics, *Division of Statistics on Science and Technology, Statistics on Science and Technology*, UNESCO, Paris, 1981.
167. Carpenter, M., *International Science Indicators. Development of Indicators of International Scientific Activities Using the Science Citation Index*, Computer Horizons Inc., Cherry Hill, N.J., 1979.
168. Carpenter, M. P. and Narin, F., The adequacy of the science citation index (SCI) as an indicator of international scientific activity, *J. Am. Soc. Inf. Sci.*, 32, 430, 1981.
169. van Heeringen, A., Some research on publication-counts as a measure for the output of research, in *Science and Technology*, Indicators Conf., Organization for Economic Co-operation and Development, Paris, 1980.
170. Schubert, A. and Glänzel, W., Statistical reliability of comparisons based on the citation impact of scientific publications, *Scientometrics*, 5, 59, 1983.
171. Bujdosó, G. and Braun, T., Publication indicators of relative research efforts in physics subfields, *J. Am. Soc. Inf. Sci.*, 34, 150, 1983.
172. Narin, F. and Carpenter, M. P., National publication and citation comparisons, *J. Am. Soc. Inf. Sci.*, 26, 80, 1975.
173. Dieks, D. and Chang, H., Differences in impact of scientific publications: some indices derived from a citation analysis, *Soc. Stud. Sci.*, 6, 247, 1976.

INDEX

A

B

C

G

H

I

J

K

L

M

N

O

P

Q

R

S

T

U

V

W

X

Y

Z

Milton Keynes UK
Ingram Content Group UK Ltd.
UKHW051950071024
449327UK00026B/2244